From Galileo to Boltzmann:

A History of the Fragility and Resilience of Science

By

Rodrigo Fernós

For more information, visit www.rodrigofernos.org

"From Galileo To Boltzmann: A History Of The Fragility And Resilience Of Science" by Rodrigo Fernós. ISBN 978-1-62137-822-8 (softcover).

Published 2016 by Virtualbookworm.com Publishing Inc., P.O.Box 9949, College Station, TX, 77842, US.

For E.L. and M.R., Rest in Peace

Table of Contents

Preface

WHEN ONE SURVEYS THE RISE AND DEVELOPMENT OF SCIENCE, one cannot help but notice how fragile its origins had been. It held a condition not unlike that found in the words of Marcus Aurelius in the film "Gladiator" (2000) when describing the birth of Rome: "There was once a dream that was Rome. You could only whisper it. Anything more than a whisper and it would vanish…it was so fragile. And I fear that it will not survive the winter." Its hopeful beginnings in the Classical Greek world did not assure its continued existence for posterity; just as men do not lean towards democracy in the historical record, they also tend to prefer power over truth.

Although 'another history of science survey,' this book will focus on precisely those crucial turning points in its history when either the birth of science itself or of some subdiscipline were dangerously fragile, possibly stillborn. It is a detailed history which has been easily forgotten in light of the success of modern science.

Some of these incidents are very well known, the most famous of which is perhaps the conflict between Galileo Galilei and the Catholic Church or between Arab natural philosophers and their surrounding Islamic cultural milieu. Others are less so: the conflicts between Isaac Newton and Robert Hooke, Charles Darwin's brushes with death or misfortune in Latin America, for example. (No, we will not provide any spoilers.)

This aspect of the story is an important one which hopefully provides the reader a greater appreciation of the historical fragility of science; with a notion of how, at any given moment in time, we might have ended up living in an intellectual landscape noxiously different from our contemporary one. It's outcome should not be taken for granted.

At the same time, it will also help impress upon the reader just how resilient science has been. Again somewhat akin to the character "Maximus" in the same movie, our historical figures also persisted in spite of all odds, showing a clarity of mind and determination of will that is in and of itself a testament of the human condition. (Again, no spoilers.)

Not all were successful. But even in the cases of those who 'failed,' they were able to endure long enough to have a remarkable impact on their era's intellectual

1

landscapes. This book might be considered a testament to the heroes in the world of ideas.

-Rodrigo Fernós

November 9, 2015

Acknowledgments

I WOULD LIKE TO THANK THE PHYSICS DEPARTMENT at the University of Puerto Rico in Rio Piedras for their generous support. Its director, Prof. José Nieves, as well as its administrative staff, Ileana Desiderio, Nilka Ellis, and the students Rafael and Maria made the difficult process of writing a book a much more satisfactory one.

I would also like to extend a warm word of appreciation to Carmen Pantojas, Mayra Lebron, Ana Heliva Quintero, Hector Rosario, Libia Gonzalez, Julia Mignucci, Fernando Gilbes, Maria Juncos, Ruben Davila, Socorro Cruz, Seppo Tervahartiala, Isabel Maldonado, Felix Ojeda Reyes, and Liliana Cotto Morales for their kind gestures or conference invitations.

As always, the librarians of the Universidad de Puerto Rico at Rio Piedras were of enormous assistance, as was Roger Stuewer at the University of Minnesota. Finally, I would like to recognize Angel Casiano, Jonathan Pérez, Ernesto Quiñones, Estefanía Santos, David Silva, and Shawn Zografos for their valuable comments and observations.

Part I: Seedlings

Introduction

What is Science?

IN HIS BOOK, *WHAT IS PHILOSOPHY?* José Ortega y Gasset noted that the best way to get to a topic was via a spiral: an indirect and circular route whose points are breached repetitively as one circles around a topic. This is a useful tactic that we will take as our starting point.

What is science? Where does science come from; the brain? Science is usually associated with geniuses, as Isaac Newton or Albert Einstein. There can be no doubt that they were intelligent. May we then conclude that science is due to human brain size? Is science restricted only to geniuses?

Human Evolution

The size of primate brain can be determined by the Dunbar number showing a close correlation between its respective group size and the neocortex. The larger the social group, the greater the cranial mass and "IQ" of the species. Tamarind monkeys show a 2.3 ratio with groups of 5 members. Macaques have a 3.8 ratio with a 40-member group. Humans, with the largest neocortex (90%), have an average group size of 148 members; so accurate is the relationship, the 'normal group size' of humans was predicted simply based on the average human brain size. David Attenborough in his documentary "The Life of Mammals" (2002) uses plasticine to effectively show the relationship.

This relationship can be easily explained: more interconnected points lead to a near exponential growth in the number of connections, defined by the simple formula $x=(n)(n-1)/2$. The same phenomenon can be observed in the world of telecommunications. Better computers are needed as the total number of telephones increase, which otherwise would overburden the system. This relationship is akin to the area/volume ratio of a sphere: as a sphere's volume

6

increases as a cube (v^3), its area increases as a square(a^2), $a^2 : v^3$, with endothermic benefits. As animals grow bigger, they can retain heat more easily, with reduced metabolism, generating an evolutionary pressure to increase weight. Similarly, the increase in the number of telephones acts as a spur towards technological development.

The large homo-sapiens brain was a necessary factor for the emergence and creation of science. But was it a sufficient one?

Obviously, humans are not born with scientific knowledge, unlike most animal species with instincts. Newton was not born with the gravity equation engraved on his forehead; Einstein did not suckle on a relativity bottle as an infant. Two counter examples also help demonstrate this point. There is the strange case of a 'wolf boy' in Europe: the discovery of an abandoned infant reared by a wolf/wolf pack, also called a 'feral child' which survives but is found deformed and suffering from malnutrition and poor teeth. It never acquired capacity to speak, and dies shortly thereafter. In a crude way, it illustrates the point that we are not born with scientific knowledge; cultural content is not passed down genetically.

The capacity for mathematics varies greatly across cultures, as shown in anthropological studies conducted by Jack Goody. Primitive societies with oral traditions lack the ability to count above low numbers, larger ones above five are routinely referred to as a 'cluster.' Mathematics, particularly advanced mathematics, is inherently graphic and requires a written language; one simply cannot do calculus in society where oral language is its only means of communication. Newton, the inventor of calculus and the law of gravity, simply could not have emerged in a tribal village setting—not for lack of intelligence but simply due to his immediate cultural context.

Humans (*homo sapiens*) emerged 200,000 years ago, meaning that "Western history," roughly dated at 2,000 years, accounts for only 0.1% of its totality. Our history is but the utmost thin crust atop an infinitely vast 'deep history.' Even if a shorter time span is considered, with the emergence of *homo sapiens sapiens* some 40,000 years ago and the dying out of the Neanderthals 10,000 years after that, the image of an apple's rind remains. A modern person placed in that era would not be biologically distinguishable from the rest, in spite of the marked difference in lifestyle and worldview. We are not as different from our ancestors as we would like to believe.

Technological Evolution

Today we can fly to Europe in less than 24 hours, have access to records thousands of years old, and even have a relatively good picture of the vast universe. The 'stone age' past had none of these things; they had no medicine and hence an average life expectancy 35 years. The most sophisticated tools were the spear, arrow, and flint. The key difference between us is science and technology, which is the principal source of wealth of the modern world—or at least of its potentiality.

7

We may draw three immediate conclusions. The first is that the existence of science is conditional on various factors which are not 'inherent' to human nature (i.e. genius). The second is that science is a cultural code acquired through education. Individuals are their society, figuratively and literally; they make up the society and in turn are made by the collective. Psychology cannot validly study a person isolated from their social environment. The third conclusion is that as science is cultural, it is hence external to man. This third observation requires elaboration because of its importance.

Science is of a cultural nature, and by definition external to us. It is retained in books, articles, but also in education and schooling. This means that there is an inherent material nature to culture/education in a deep sense which we all too often tend to take for granted, not appreciating its true significance—a lesson many high school teenagers should heed. Modernity cannot exist without science, and it without schools.

Yet the loss of science is often irreplaceable. That the enormous advances are today are seen as commonplace is as tragic as the foolish teenager, and has just as frightening consequences. We fail to appreciate the enormous effort and revolutionary developments underlying common everyday tools.

The common cellphone today has the computing power of a million-dollar computer from the 1960s, yet costs only a fraction of its predecessor. The IBM 370, housed in a large room and divided between various distinct cabinet units, had a memory of 2 'million bytes' (2 megabytes of RAM) and a storage space of 32 million bytes, costing around $2,500,000. Today's advanced smartphone on average costs some $400, has 2 gigabytes of memory and can easily hold more than a thousand million bytes of information. In the 1970s, a 5-megabyte hard drive was the size of a filing cabinet and required a crate to be lifted onto an airplane for shipment. Today's 128 gigabyte smartcard is smaller and lighter than a paperclip, easily fitting in a shirt pocket. As computers developed, there was even concern by United States military leaders during the 1990s that these could be used to guide intercontinental ballistic missiles. But, as "Marty" and "Doc" noted in a recent televised anniversary of "Back to the Future" (1985), today we only use our cell phones today to take pictures, chat, and see tragic videos of "Jackass" (2002) copycats jumping onto cacti.

At around 49 BC, the Alexandria Library was burned to the ground, losing perhaps the largest collection of recorded human knowledge available at the time, some 400,000 scrolls. The library had suffered a series of continuous attacks, which ultimately destroyed its vast wealth of learning. This is perhaps not unlike the Budas of Bamiyan, destroyed by the Taliban. The impressive sculptures, ancient cultural relics dating back thousands of years to the days of the Silk Road uniting the Orient and Europe, was destroyed by unruly Arabic teenagers. Similarly, the war in Iraq after 2001 ultimately led to civil chaos, whereby many cultural treasures thousands of years old were destroyed. Irreplaceable relics of humanity's cultural legacy were forever lost.

Could such a degree of cultural savagery reduce Western civilization back to the stone age? Given that culture is to some degree external, this is not an unrealistic scenario, with 200,000 years of learning wiped away from history. The external character of science does not preclude such an outcome, however improbable it might be.

But, What Is Science?

Science could be described by its sub disciplines: physics, astronomy, biology, chemistry, geology. It could also be described for its essential traits, symbolized by the anagram EEMM. It is empirical, based only on information provided to senses, directly or indirectly; spirits and mediums are not allowed. It is experimental; scientists have to trick nature into revealing her secrets. It is also mathematical, the key to deciphering nature's secrets. Finally, science is also mechanistic or model based; we can only know something until we have constructed a model of the phenomena.

But, is this a viable explanation? While suggestive, it does not tell us that much. It is only by tracing its origins and emergence that a genuine understanding of the nature of science can be obtained. Cliché as it might sound, only by looking at the past can we understand the present.

For example, the history of the discipline of chemistry is intertwined with many different practices that today are defined as wholly different from it. In the past, chemistry was part crafts and part religion. Processes that today we define as chemical were defined as the "affinities" between substances: anthropomorphic projection of human psyche onto the natural world, our all too innate propensity to detect faces on obscure innate objects.

We may also note that the 'scientific method' typically described in textbooks, of hypothesis, experiment, and theory is an oversimplified description. Very few historical examples actually exist that accurately fit it. The history of science is full of complicated twists and turns, in which the context and questions asked constantly change—thus making for a constantly shifting territory without clear boundaries or rules. As Stephen Toulmin and June Goodfield describe, protochemists first began looking for incorporeals and then, as if by magic, chemists were plotting elements on what would become the periodic table: a rabbit out of a hat magic trick.

The scientific discovery of the new is, by definition, unpredictable, and in a sense the use of the term 'pseudoscience' to describe intellectual efforts from the past are unjust anachronistic characterizations, reading back onto the past modern contemporary values and expectations. The very notion of 'pseudoscience' can only be conceived of post facto, after science has been formally created in a process which took centuries. In this sense, the development of science is not as rational as it is often portrayed. For example, diseases as 'malaria' (bad air) were nonspecific, and were identified by a whole host of changes in temperature and fevers of the diseased; as such, the long list of well-defined illnesses today were then grouped by broad categories as "quartians."

This oversimplification of the history and character of science is also an inherent problem of all science policies which hope to stimulate scientific advancement. Given that the search for truth cannot be determined beforehand, one obviously cannot predict scientific breakthroughs.

As every historian of science knows, the term 'science' emerges in the 19th century when it was first coined by William Whewell in recognition of its professionalization. Prior to that, science was 'natural philosophy'; for Newton, scientific practice was tantamount to a religious exercise, seeking God's rules in the universe. The early modern religious motivation for scientific activity makes sense in light of the fact that there were few financial and economic incentives for its practice.

This book will try to get at that deeper and richer muddle which is history, hoping to create a reevaluation of imperfect concepts, which will hopefully result in a better appreciation of its difficulties. While science strives for ever greater perfection and exactitude, its history is imperfect, discontinuous, and erratic. But that is the beauty of the story of the quest of man to understand the world and himself in it. This book deals with some of the most profound questions humans have ever asked themselves; the early history of science is often a history of philosophy. These questions might appear to be easily answered but in fact are very hard to do so. How big is the universe; how old is it? What is the nature of time and space? The understanding of time and space have drastically changed in history, shifting from the closed world to the infinite universe.

But, again, what is science?

Overall science is not one "thing," but is rather composed of multiple elements, each of which have their own unique and particular histories. Each 'science' underwent their revolutionary periods at different moments, and do not necessarily coincide at a single 'explosion' of activity. Biology, for example, had its 'revolution' during the 19th century with Darwin's notion of evolution. Medicine's 'revolution' emerged during the late 19th century when Robert Koch and Louis Pasteur discovered bacteria. Organic chemistry emerged during the second half of the 19th century with the science of polymers. Genetics did not truly emerge until the mid-20th century, finally solving the Darwinian puzzle of evolution.

In spite of this diversity, the Scientific Revolution is the grandfather of all, equally complex and strewn throughout various centuries. The term alludes to changes in astronomy and physics, taking some 200 years from 1453 (first printing press) to 1687 (Newton's *Principia*). The key traits of modern science (EEMM) were developed during this period. Newton becomes the crowning glory of prior developments, and hence truly 'standing on the shoulders of giants,' as he often personally noted. His work was the culmination of all those who preceded him: Copernicus *De revolutionibus* (1543), Brahe *New Astronomy* (1588), Kepler *Astronomia Nova* (1609), Galileo *Dialogue of Two World Systems* (1632), Descartes *Discourse on Method* (1637), and Bacon *Great Instauration* (1620).

This evolution is referred to as a "revolution" because the world view that existed prior to it had drastically changed by the time it was completed.

Civilization

Precursor to Science

ONE OF THE UNQUESTIONABLE FACTORS for the emergence of that cultural form we all 'science' is civilization. As we have noted before, the 'deep history' of humanity occurred outside of a 'civilized' context. You can put a modern human today into a community tens of thousands of year ago, and there would be no marked physical difference between the two—except for their vastly different cultural levels. Yet, how did humanity shift from small primitive clusters of nomadic tribes to the megacities with atomic bombs of today?

A 'civilization' can be defined as life in a large social grouping with enough resources to allow for specialized labor; it is the development of specialists whose activity is focused on a particular topic that allows for any type of scientific development. Early astronomers could not study the skies if they had to spend most of their time worrying about food, as most other animals do. Complex civilizations hence form the bedrock of science; without civilization, there is little chance whatsoever for modern science to arise.

As most readers likely already know, the first sciences—math and astronomy—belong to the first civilizations: Sumerian, Babylonian, Chinese and Indian. There were a few cases also in the Americas: Maya, Inca and Aztec. The rise and fall of civilizations is a fascinating issue with multiple interpretations, but one where each new view contributes to our understanding of this complex topic.

Factors in the rise of civilizations

We will look at the ideas of four key thinkers: 1) Arnold Toynbee (historico-social aspects), 2) Jared Diamond (biological aspect), 3) Joseph Tainter (mathematical analysis), and lastly 4) Joseph Goody (technological analysis).

Arnold Toynbee's (1889-1975) interpretation can be summed up as the "Challenge-Response" theory. He was historian at the London School of

11

Economics and the University of London (1934-61), whose magnum opus was a 12 volume masterpiece called *A Study of History*. (There is a more accessible one volume abridged edition prepared in 1947 by Somervell.)

In his work, Toynbee looks at 26 civilizations, attempting to find common factors for their emergence.

> Society ...is not a collection of persons but is a network of relations; it is the field of interactions of two or more agents....the medium of communication through which human beings interact with each other...

For him, society is a dynamic entity, whose inter-social relations and daily exchanges play a very important role in their development. It has a markedly anti-geographical character given that it was a response to geographical theories predominant at the time. It was obvious to many academics that the largest civilizations had been formed in river valleys with rich alluvial flooding, thereby providing a stable agricultural food source. The principal earliest civilizations are hence tied to their nearby rivers: Nile river: Egypt (3200 BC); Tigris and Euphrates River: Mesopotamia (3500 BC); Indus River: India (2200 BC); Huang Ho River: China (1500 BC).

One might suppose that challenging traditional geographic-centers theories would have been foolish. But by enlarging the total number of groups studied Toynbee was able to get a better and more accurate notion of the factors underlying formation of complex societies. He identified favorable and unfavorable conditions and is able to trace distinctive historical patterns. Societies with favorable conditions did not necessarily emerge to greatness as traditional geographical theory suggested, Boethia and Calchedon being two examples.

Instead, he found the opposite to actually be true: difficult conditions tended towards the growth of civilizations. Perhaps the ideal case is that of Greece. The region of Greece is relatively inhospitable, with a mountainous terrain and dry semi-arid climates that did not easily allow for agriculture. Hence the Greeks were forced to take to sea, creating new opportunities in the process. Other examples along this same train of thought were the Phoenicians, seafarers who also used water trade routes and to the exchange of ideas culminating in monotheism and the consonant alphabet.

He also noted that, contrary to expectation, the conquest of Mesoamerica was not as easy as one might suppose. Modern biologists have shown that much of its wildlife actually occurs on the forest canopy, under which lies a vast desert. The Mayas, Toynbee noted, had to be constantly fighting the forest; whenever they stopped, forest took over territory, as attested by the 'discovery' of Inca by Hiram Bingham. A similar example was that of Angkor in Cambodia.

Toynbee then generalizes these cases into a broader notion of 'challenge-response' in that civilizations emerge when social groups confront difficult circumstances, and produce adequate reactions—leading these into a positive cycle of continued growth. The "environment" was not just geographic but rather ncluded a whole host of secondary factors. Toynbee enlarges its definition to also include the social context, such as legal circumstances. Legal sanctions against Jews prohibiting these from entering into traditional professions in government and politics, their 'environment,' led these to successfully focus on others: banking, commerce, medicine, etc.

For Toynbee, societies were not static entities, but where rather characterized by a continual dynamism, either growing or decaying. Positive creative responses, which had led to the formation of such societies in the first place, had to be repeated on an ongoing basis given that threats and challenges of some kind always existed or emerged. It is in its continual response to these challenges in which 'civilization' grew and maintained itself. Civilization in this sense meant 'forward moving'; 'precivilizations' lacked this growth given their emphasis on the past. There could also be the case of "arrested growth" which was typically seen when creative responses had at one time emerged but were no longer being created. Emerging societies could thus stagnate and, ultimately, collapse.

One should also not be fooled into associating a great deal of external activity with the dynamic process of 'civilization; much activity can occur in the absence of appropriate changes in lifestyle or ways of living. Various examples of arrested growth were the cases of the Eskimo or Eurasian nomadic societies as the Sami in Lapland.

Dynamic 'civilizational' societies were characterized by particular features. The conquest of nature often led to the conquest of the psyche, creating a distinctive culture. A series of internal conquests as the creative formation of laws and social institutions to meet their respective needs was a key aspect of his theory. It was certainly a continual uphill battle; the creation of civilization did not have an inherent 'inertia' and autonomous growth, but rather was a process which continually needed intellectual responses adequate to the needs of their respective circumstances. Civilization was something that had to be continually fought for, and which simply could not be taken for granted. Leaders which rested on their past laurels would ultimately doom their respective civilizations.

By definition, expanding social groups had 'creative minorities' who found solutions; others in the community noting their success would hence follow their example. This was a process referred to as 'mimesis' by Toynbee. It was not blind imitation but rather imitation of character and spirit in the relation between leaders and the masses. By contrast, office holders who blindly repeated old responses were not suited for leadership, often typified by an inadequate passivity when action was required. In this, a ruinous and excessive use of force was also both a cause and symptom of the loss of leadership. In these circumstances, the role of mimesis no longer applied, given the lack of credibility and moral authority of its officials, with few genuine followers and little overall influence.

The popularity of Toynbee's ideas is likely due to affinities with its existing cultural setting; one cannot help but notice the very "Protestant" character of his interpretation, with its focus on individual will and merit. Certainly, his notions contributed to the popularity of Greek studies at mid 20thn century Europe and United States. There is an underlying social dynamic that is undeniably captured by his theory. It was a nuanced psycho-cultural interpretation which was very

13

good for its time, but not as 'scientific' when compared to the other theories that emerged afterwards.

Joseph Tainter (1949-) is an anthropologist of some 60 years who has taught at Utah State University and at University of New Mexico. In 1988 wrote *The Collapse of Complex Societies*. At its core is a mathematical modeling of rise and fall of societies, from a cost/benefit analysis point of view. Some of its ideas are curiously illustrated in Zeno's paradox such as the race between turtle and Achilles. In spite of his enormous speeds, Achilles can never catch up with the turtle, who always takes one half step for every one of his, suggesting a freezing of their motions in an eternal race. Zeno is, of course, tricking the reader by situating both figures in two different frames of reference, which could be classified as a 'description error.' The key lesson to the story, however, is the law of diminishing returns: the harder we try, the less we achieve.

Tainter applies the law of diminishing returns to a broader social scale. The early formation of societies implied inherent benefits against the typical difficulties of human life: protection against floods, fires, and drought with surrounding community. However, these benefits cannot grow indefinitely, and reach a peak of maximum efficiency as more institutions are added onto the body politic. These institutions withdraw greater benefits from society than they contribute to it. This detrimental pattern continues until the costs have increased to such an extent, that individuals within the collective find it more personally beneficial to remove themselves from the community rather than to remain within it. There is a delicate balance between the costs and benefits of social complexity.

This is an important contribution. It used to be believed that social collapses led to a decline of welfare of its members, whereas Tainter shows that in fact, the opposite is true; the arrow of causality had actually been reversed. Medical and biological studies have shown that the consequent 'atomism' following a broader social dissolution actually resulted in the increased health and well-being of its participants. Striking as it may sound, the nutrition of individual Maya increased after the collapse of empire.

There are many examples of this dynamic, and can be seen in contemporary society. Over time, institutions tend to show a marked loss of efficiency as the percentage of administrators increase relative to its productive personnel, as in the United States whose percentage increased from 6% to 22% over half a century (1900-1950). As is often said in Puerto Rico, the institutions had acquired many chiefs and few Indians (*"muchos caciques y pocos indios"*). The dynamic is also visible in higher education as well as the United States health care system, as shown by the vast increase in its total share of the gross domestic product (nearly 20% of GDP)—without visible net gains in the lifespan and life quality of North American citizens.

In the particular case of the Maya, Tainter makes the observation that the homogeneity of the territory also worsened the rate of decline of the Mayas. In the absence of domesticated animals which served at supplementary food sources, drought had a compounding effect throughout the region, increasing the likelihood of warfare and (inversely) decreasing social stability. This was a similar

14

dynamic observed by Jared Diamond in his book *Collapse* (2005), where he describes the clash between Hutu and Tootsie in Rwanda. Violent and horrific incidents could be directly linked to famine in the region.

Similar ideas can be found in Roland Wright's *A Short History of Progress* (2004). Wright's key notion is that of a "technological trap." Acts of creation can have negative impacts in that they introduce new problems as well as benefits, raising the severity of a collapse given their compounding effects. As noted by James Burke, a subway car which facilitates the transport within a city could become a coffin in the absence of electricity. An elevator in a skyscraper, which allowed for greater concentrations of individuals in a given geographical location, would become a 'technological trap' during a blackout, literally speaking.

Wright is highly critical of notion of "progress." As modern society encompasses the entire globe, its collapse implied the doom for all of mankind. By contrast, as ancient societies were limited in scope and size to relatively minor geographical spaces, collapse impacted only the immediate group, without any 'global repercussions' as in the contemporary world.

Jared Diamond, a 78-year-old physiologist (b. 1937) undertook a different approach in his *Guns, Germs and Steel* (1997), winner of the Pulitzer Prize. If, as Toynbee recognized, the vast differences in development between different societies were not due to 'race,' what other factors were there? "Why the West and not the Rest"? as Diamond was asked by a young New Zealander. Why did the West conquer world and not the other way around?

Diamond seeks to account for this outcome on a 'biological basis,' more 'scientific' and 'concrete' than prior models. He looks at the preconditions for civilizations, specifically at factor nergy in this system when crossing layers. The importa s that would encourage formation of large complex societies. The title of his work is actually misleading as 'guns, germs, and steel' were only proximate factors; it is certainly the case that a book with the name "Farming and the Rise of the West" would not have sold as quickly as his provocatively titled book did.

Diamond analyzes the underlying factors which led to food production, the ultimate precondition of all civilizations. He notes that domesticable animals were a key element; none of the animals in Africa are domesticable. We cannot ride zebras as horses because of their violent character; similarly, cute looking hippopotamus will all too readily kill men who dare to venture too close to their territories. Of the 72 large mammals in the world, only 18% are domesticable—the horse, cow, sheep, etc.—all of which were originally found near Europe. The Americas, by contrast, had only one species: the alpaca.

Particular social traits of species were also necessary, allowing these to be co-opted by humans. Domesticable species are characterized by the tendency of having an alpha leader who heads the group, thereby allowing another, man, to serve as its substitute. With the new 'alpha,' all others follow, akin to a substitute mother duck whom is followed by all her ducklings (Konrad Lorenz).

Key traits identified by Diamond include the presence of: a) herd communities, b) a dominant leader, and c) overlapping territories. Secondary traits included: d) the capacity to breed in captivity, e) a short breeding cycle, and f) being non carnivorous. The importance of this last trait was not to eliminate

15

aggression, but rather that it would have been too expensive, requiring the growth of a secondary species to feed the domesticable one. As shown by the application of thermodynamics to the study of nature, there is an enormous loss of energy in this system when crossing ecological layers. The importance of a short breeding cycle can be obtained by considering that elephants breed every 15 years—far too long to wait to use as a food supply.

Secondly, Diamond also emphasizes the particular genetic traits of domesticable plants. Very few plants can actually be consumed by humans; of the millions of plant species that exist, only 0.1% can be digested by humans. As the case of domesticable animals, nearly all of the consumable plant varieties, (90%) come from the Fertile Crescent: wheat, barley, etc. While humanity was fortunate in the initial genetic foundation of domesticable crops, these have been 'selectively bred' over millennia to obtain their particular contemporary features and properties. Wheat, for example, had to be gradually selected, as the grain of the first variants tended to easily fall off when slightly shaken.

Finally, Diamond identifies key geographical factors to account for the rise of the West. He notes that the Fertile Crescent is characterized principally by an east-west axis orientation. This might not appear to be as important, until one considerers that populations distributed along lines of latitude (horizontal) rather than longitude (vertical) have the advantage of crops being easily transferable throughout their territory. Similarities of climate and condition characterize such regions, thereby allowing for a faster expansion and population growth. The Americas and Africa, by contrast, have axes that are oriented principally along a north-south direction, making for the transplantation of food much more difficult, and consequently, human expansion.

The impact of print on intellectual development is another important factor. The Greeks criticized print, complaining that it led to a decayed memory. Daniel J. Boorstin in *The Discoverers* (1985) describes how Simonides was able to identify all of the bodies after the collapse of a building in which he had minutes ago recited a poem. Recent studies curiously demonstrate that chimpanzees shown better memory than humans; our over-reliance on 'external technologies' appears to have had an evolutive cognitive cost.

However, it is certainly the case that print has also contributed enormously to Western development. Words are the instruments of thought, and their nature has an undeniable impact throughout history; they 'shape' to some degree our thought as well as the pattern of intellectual development of a community. It is a dynamic that occurs on a macrosocial scale and is not obvious on a daily basis.

For example, Roman numerals (I, II, III, V, X, C, M) are impractical mathematical tools given that the absence of decimals—unlike the original Indian number system. Newtonian calculus under this setting will also not emerge. We might take the example of Chinese print versus Western alphabet. The former has a symbol for every word/syllable (logo syllabic), resulting in an enormous alphabet of some 4,000 characters. One cannot just learn key pieces (26 letters), and as a consequence, a large portion of an individual's education is taken up simply to learn the alphabet. Note as well that new unique words have to be created for every new objects and idea. As strange as it may sound, the 26 letter

16

alphabet allowed for a faster rate of cognitive change and growth in Western societies. Understandable compound words can be easily created to account for new notions and concepts; a greater part of the educational process can thus be spent on the acquisition of new knowledge and ideas, and on the exploration of this knowledge base—so much so that the Western alphabet is today used as a learning aid for their own native language in China.

Jack Goody (b. 1919) is an anthropologist who specifically studies role of print in socio-cognitive development. Goody was influenced by Robin Horton's work in Africa, *Patterns of Thought in Africa and the West*. He departs from Claude Levi-Straus's notion that society is *suis generis* whereby all cultural traits are accounted for at the at macro level with relatively minor analysis of the individuals that compose it and their interactions. Africa is often taken as a case study because of the readily available evidence, but it's not the only location in this field of study.

Print introduces a level of objectivity that did not previously exist in oral societies. Asymmetrical and emotional relations of power develop between student and teacher in the absence of print. Because the student is bound by what teacher says, the oral tradition reinforces need for personal interaction and emotional dependency, and its hierarchy sets the rate of learning as one defined by the teacher rather than the student.

With the presence of the printed work, however, the student can learn on their own, at a faster pace if desired given that they are no longer dependent on an external figure to obtain information. Print also allows the student to verify the information received from the teacher, rather than the blind obedience typically seen in oral traditions. Differing positions, ideas, and notions in the world of print also allow for comparisons. A student can compare two pieces of writing, calmly and objectively without the presence of the teacher—even having the ability to verify the validity of what was taught by the teacher.

More importantly, the written word allows all readers to trace changes or the historical development of concepts that have been accumulated in writings over years, decades, or centuries. Changes of ideas can be easily identified with the written word that are impossible to do with oral tradition. Critically, this allows for innovation to flourish in that originality can be traced and its creators rewarded. Intellectuals have existed in all societies; but in the context of oral societies, the creator is subsumed by it as his innovations are routinely usurped without credit to their original creator. A good example is Homer's *Odyssey* and *Iliad*, which were the accumulation of multiple contributors. They are both long poems whose rhyme allowed for easier memorization, but inversely made it harder to identify exactly whom contributed which verse. Since Homer is the last poet prior to its written form, he became its 'author.'

There are other benefits identified by Goody. The printed word introduces new categories of organization. Preliterate tribes in Uganda have no word for 'word,' only 'speech'; there was also no notion of grammar: noun, verb, adjective, etc. Print allows for the analysis of language itself, broken up into concrete units whose relationship can be analyzed.

Secondly, the printed word allows for the unique reordering of language as alphabetization, which is a purely graphical feature. While there are negative aspects of this trait, lists and tables are unique to print. They are graphical tools which allow users to sort information. However, it has to be noted that all lists tend to have an implicit 'hierarchy bias'; items do not 'coexist' on the same plane but are rather placed 'higher' or 'lower' in relation to one another. Inevitably, a 'hierarchical interpretation' naturally emerges. Any list of items implies some sort of assessment—up/down true/false—while also providing the appearance of neutrality.

Words in oral traditions are also unstable, widely changing meaning and value according to their context, as for example 'left-handedness.' A man with stick in left hand is a positive sign of diviner in Africa, whereas a man with stick in left hand in a group confrontation is perceived negatively, with the alleged intent of attacking. To some degree this helps account for some difficulties in the history of science: concepts vary in meaning over time. Words in a table suggest fixed relationships that may not actually be the case.

Tables and lists also tend to suggest simple linear relationships, while hiding the complexity of relations behind these. They are inevitably tied to power relations and the rise of empire. The first printed words were not poems but rather administrative tax lists, allowing for the register and long-term preservation of debts and obligations for all to see. Empires cannot be constructed without printed 'lists' akin to those of the Inca's 'quipu.'

It is important to emphasize the complexity of that phenomena known as 'civilization.' Intellectuals from the 1950s tended to grossly overgeneralize and simplify cultures and cultural dynamics. They were usually from the United States with ties to elite schools of modernization theory. This does not mean that the body of knowledge was false or uninteresting, but rather that it was typically limited in its scale to a few factors; the context of policy-making implied a corresponding requisite for simplicity that undermined its cognitive accuracy and cultural sophistication of its analysis.

It goes without saying that multiple factors play a role in the creation of civilizations—in complex and mutually interacting dynamics. Some are psychological (Toynbee), some are biological (Diamond), while others are cultural (Goody) or historical (Tainter). Any one single factor, in and of itself, will not lead to 'civilization.' There is no 'magical formula,' and any calls and promises are destined to fail. Some factors can often be identified only a posteriori, but are hard to see at their moment of impact.

Factors in this sense are cumulative or 'synergistic.' The existence of preexisting factors contributes to the emergence of new ones. The river valley led to the creation of print (written word), which in turn lead to the creation of empire. The imbalance of social tools between two competing groups favored one group over another. The 19th century was "Britain's Century," literally ruling the globe, because it was leading in science and technology. Yet its inherently complex character also implies unstable social dynamics. Empires do not last forever given the constant change in human affairs. Overinvestment in military, as noted by Robert Kennedy, routinely leads to the collapse of empire. China in the 1970s was grossly poor, but now appears to be on the road to becoming global behemoth. It has transformed from a backward rural nation to a dominant one, whose cyberespionage exceeds that of the United States.

Finally, it goes without saying that advancement is not related to race. As we have shown, intellectual progress and economic prosperity are driven by external factors. Inversely, the only thing that separates modern man from his stone age ancestor is education.

Mesopotamia

Babylonian society arises from Sumerian civilization in 3000 BC. While the Tigris and Euphrates Rivers had alluvial flooding, they lacked the natural protection that as that afforded by the Egyptian desert. The region was hence characterized by a constant state of warfare and an initial lack of complex social organization. Two key cities emerged, and other protocivilizations in the region were quickly absorbed into Mesopotamia, such as the Jericho (Jordan River) and Oxus-Jaxarten (Karun River). Typical battles between Ur and Lagash, show a distinct cyclical pattern whereby a charismatic leader and his descendant would last for 2-3 generations, only to shift to new another cycle of family dominion. Two principal societies emerge: Akkad and Sumer.

Sumer, the earlier of the two, was more 'stereotypical' given the rise of a priestly class whose symbolic power was based on its special relation to the gods; however, much of their actual political power originated from administrative duties. Akkad developed a different, if more chaotic, style. The tribal and non-hierarchical power relationships forced many leaders into continual series of invasions to maintain their own personal power and control, even if they eventually adopted the Sumer style. However, its prosperity lead to the development of an important merchant class which traveled widely through the region and stimulated the development of cuneiform.

Mesopotamia's political dynamics are aptly reflected in the important ruler Sargon. He arises with a strong 5,400-person army but, due to the lack of writing, the king is forced to continually live with his own army, making regular patrols in order to retain power—a feature which can also be seen when his son and grandson inherit his leadership. The first cuneiform tablets made were simple accounting records: lands owned, rented, debts, taxes paid, etc. Interestingly, cuneiform also took the form of a 'contract,' whereby the clay tablet could make a 'faithful copy' for both parties at the moment of 'singing' by pressing one clay tablet against the other. (Many modern practices are not as unique as supposed.) These forms were then gradually modified to record names and more sophisticated mathematics.

Formal "Babylonian" civilization emerges around 1762 BC; 1,500 year later, Hammurabi establishes his famous code, one of the earliest known collection of laws, based on the principle of 'an eye for an eye' concept.

The sources of knowledge about Babylonian science emerge from some 500,000 tablets. While the number might appear impressive, they are but a minute fraction of all that previously existed. These tablets are found in a few key archives: Yale Babylonian Collection, Yale Plimpton Collection, Columbia University, and the British Museum. The largest collection is located in London, with some 46,000 tablets at the British Museum. Ironically, these tablets have been preserved because of the destruction of their societies. Their cities were not repopulated, but rather were absorbed back into the desert from whence they came. This cultural collapse allowed for the preservation of the historical material; had they remained thriving population centers, these tables would have likely perished.

Their sciences can be divided into two basic periods, each separated by a thousand years: 1) the Hammurabi period, 1750s, and 2) the Seleucid period, 300 – 0 BC (150BC) following Alexander the Great's conquests. Each period has its own distinctive scientific character. The earlier Hammurabi period was characterized by a significant amount of mathematical activity, with little trace of astronomical interest. By the Seleucid period, its astronomical work was more fully developed.

There is however, an underlying pattern to both periods. The Babylonian exact sciences were mainly undertaken for practical purposes, showing much sophistication in its record keeping, but was markedly simple at a theoretical level. The division of inheritance, the calculation of land, the volume of water for damns, or what we would refer to as 'astronomy' were all mainly 'algebraic': calculations of numbers and figures. That deep and rich theoretical analysis so typical to later Greek civilization was absent.

Certain myths, however, exist, such that the Babylonians were excellent observers taking as its basis the detail of their astronomical predictions. This claim is hard to sustain when one considers the difficulty of observing horizon events; deserts and storms and dust in air made for poor sight of the landscape. These are required for observation of heliacal stars, which set just right before sunrise. The accuracy of their 'measurements' were mainly due to the long time span of their observations and agile mathematics; all ambiguities could be 'calculated away.'

Babylonian number system

The main problem with the Roman numeral system was that it had no place value, making complex calculations difficult, as mentioned earlier. Their number system had symbols which were cut in half: 5 (V) was half of 10 (X); 500 (D) = half of 1,000 (circle with line in the middle). By contrast, Babylonian numbers were written in columns on clay tablets typically the size of hand: 3 1/8 x 2 inches. The reed left fine minute markings, a triangle like shape, a single vertical line or a diagonal triangle. While they did not invent the zero, they did have something similar: an empty space used to help prevent numerical misreadings.

There was an interesting variation in their number system. Our modern number system is decimal, all based on factors of 10. The Babylonian number system, however, used a diverse base which varied along various multiples of 60:

10, 24, 40, 60. The predominant sexagesimal system was used in astronomy for it allowed the quick calculation of large numbers; the value of digit was dependent on location on an imaginary table, much like ours: "1, 20" = 60 + 20 = 80.

As one might imagine, Babylonian mathematics had a relatively high level of sophistication. Implicit in it is the discovery of Pythagorean formula right triangle ($a^2 + b^2 = c^2$), which can be observed in both 'old' and 'late' periods. That typical 'Pythagorean' number patterns regularly appeared in their solutions to quadratic equations was used to identify this formula. They had a diverse series of solutions for quadratic equations, from very simple to more complex forms. The remaining tablets suggest that student actually practiced exercises where they had to 'reduce down' quadratic equations. Record keeping over long periods of time allowed for accurate determinations of astronomical events.

In spite of these impressive results, however, there were many limitations to Babylonian mathematics and astronomy. As mentioned before, their exact sciences were theoretically simple, relative to that of the Greeks. They had no notion of an irregular number (square root of 2), reducing all such figures in their algebraic calculations. Theoretical geometry as that of Euclid or Archimedes did not exist. They did have the notion of 'pie' (π), but only an approximation (3) was used (rather than 3.14.....). It was 'good enough' math, but not 'exact' in a strict sense of the word.

Egyptian astronomy

The Egyptian civilization is one of the oldest ongoing civilizations in world history. The area contains around 138 world-renown pyramids, the most famous of which are situated at Gaza, which were built around 2500 BC. Its success was due to a combination of favorable factors: the Nile river, on one hand, provided food security through its annual floods, and the Saharan desert, on the other hand, provided military security by making a 'lateral invasion' prohibitively expensive. While the Nile could have served as an invasion highway, there are a series of water falls along the Nile River preventing such attacks. Under these favorable conditions, Egypt developed very quickly, but equally as quickly stagnated into formalized cultural forms. Menses was the first to unify 'upper' (south) and 'lower' (north) kingdoms, using a mixture of religious symbols from all groups for political expediency. By definition, however, this tactic resulted in an 'illogical character' to its ideology with many discordant elements from various places that were never reconciled.

The priestly class actually emerged into a quasi-merchant class which controlled the Nile River's flow of goods and people in region. The Nile was divided into 42 nomes, providing a degree of homogeneity of rule that was not possible in Mesopotamia with its sharp division of priestly class and merchant class. One of the most well-known and odd social traits was that of the inbreeding of its leaders. While the marriages of brother and sister might keep wealth and power within the family, these degraded into gross genetic deformations. King Tutankhamen, typically characterized as a young athletic male

21

on pyramid walls, was actually a feminized fat adolescent showing deformities as poor ankles and wide hips—all typical signs of inbreeding.

Feeling too secure and peaceful, the Egyptian civilization was unable effectively respond to external invasion, and hence its collapse as a collective.

Their mathematics was simple and nothing of note. They were made up of a series of 'additions' and subtractions, lacking true multiplication. Large numbers were broken up into sub units for calculation. Their contribution was more significant to Western astronomy, however.

Their metonic solution to the problem of the calendar was so useful, that it was quickly adopted by Hellenistic astronomers. Lunar cycles are the most easily detectable astronomical patterns, and were a key to the Mesopotamian civilization calendar. However, calendars based on it lose 11 days every year—a loss that accumulates over time; in the short time span of 3 years, an entire month is 'lost.' The metonic cycle was created as a way of solving this 'loss of time' simply by overlapping lunar and solar cycles every 19 years or 235 lunar cycles. (This solution was also independently created by the Mayans.)

Yet the importance for the Egyptians in 'solving' the calendar issue was not a result of 'astronomical advances per se' but rather due to the need in determining the annual flooding of Nile. Their unique solution was the use of 'heliacal stars,' which ultimately gave us our current 24-hour day/night cycle.

A 'heliacal star' refers to the last star seen at night prior to dawn, and was used to define the 'end of night.' The star was 'destroyed' right after its 'birth.' Sirius is a heliacal star, with a close association the regular annual flooding of the Nile, appearing one month prior to the event and hence used to predict it each year. Sirius is a bright star on the horizon that could be clearly and distinctly seen, thus made for a good reference point. Sirius was known as Sopdet to the Egyptians and marked beginning of the calendar.

But the particular 'heliacal star' shifts over the course of the year because of the tilt of the Earth's axis. As a winding clock, the position of the heliacal star gradually changes location, thus the star assigned the heliacal 'function' would change over the course of the year. Heliacal stars switch every 10 days, as in a game of tag. This period itself was referred to as a 'decan.' The entire 'heliacal cycle' actually took 36 decans over the course of a year, which were divided by two as only the night position is what counted. (Heliacal stars during the day cannot be seen, as their light is swamped out by the sun's.) However, the 18 decans were further reduced to 12 given that only 12 decans would appear at night in the summer given the longer amount of daylight. The customs were gradually adopted over the ages, hence giving us our 2 x 12-hour cycle.

Over longer periods of time, the calendar inevitably sank out of place, and reforms were regularly called during the Roman period. Copernicus himself was called forth to Rome by the Catholic Church for the same reason, all due to the difficulty of establishing accurate measures of time. Without a good clock, a calendrical method could be very close and "good enough," but over time, it would eventually lose sink with the solar cycle.

Customs created under particular circumstances often obtain a static character, even when the very circumstances which motivated those customs in the first place change. Culture in this sense acquires a conservative character, with beneficial and detrimental outcomes. While its static character renders stability to social institutions, it also makes the adaptation to changing circumstances and conditions difficult. Inversely, when we consider the brevity of man's lifespan, their static character is usually to the benefit of their respective civilizations.

Social institutions are a double edged sword.

Greek 'Science'

Universals and the Problem of Change

MODERN SCIENCE HAS ITS ORIGINS IN GREEK PHILOSOPHY; the importance of Greek philosophy cannot be underestimated as it established the key questions and underlying 'world view' of science. It might be pointed out that the loss of Greek corpus coincided with the Middle Ages, whereas the "Renaissance" was ushered due in part to its recovery. Nearly all of the important figures of the Scientific Revolution, from Copernicus to Newton, were influenced in one way or another by the Greeks.

Yet while Greek literature is closely linked to modern science, it is not the sole determining factor. Science is not born fully grown in Greece. There was no notion of rigorous experimentalism in Athens, for example; love of reason and debate were its principal tools. Hence, while Greek philosophy is a necessary cause of modern science, it is also an insufficient one.

It is hard to fathom that Aristotle was the dominant philosopher for 1,000 years in Western Civilization, roughly from 400 BC to 1700 AD. But, as one might suppose, Aristotle did not emerge in vacuum. There are three general periods of Greek philosophy. The more well-known is the Classical Greek (480-380 BC) or the epoch of the great thinkers such as Socrates, Plato and Aristotle. The three periods of the Greek golden age can be easily bookmarked around the figure of Socrates: before (Presocratic) during (Classical) and after Socrates (Hellenistic).

There was an enormous time span between all of these thinkers. They are usually portrayed as contemporaries, but in fact an average of 166 years separated Thales, Plato, and Archimedes, the main thinkers of their respective periods. The entire time span is roughly as long as the entire Spanish Colonial period of Puerto Rican history (1492-1898); the chronological distance between Thales and Archimedes is as that between Christopher Columbus (caravel ships) and Elihu Root (steel warships).

Contrary to common presumption, the Greeks were not talking directly to each other, as in a conversation. Rather, they were reacting to the writings of authors who had died many lifetimes prior their own births; to intellectuals whose originality of thought had been so striking and insightful so as to lead to their deep study and continued preservation. The 'Greeks' tend to be clumped together today only because of the historical distance from our postmodern era.

The Presocratics, spanning from the 6th and 5th cent BC, are referred to as the "Ionian philosophers," located mainly around Magna Grecia and surrounding regions. They are known for their search for 'universals.' Thales claimed that 'all is water' and Pythagoras that 'all is number,' but there were enormous differences amongst them. Plato at one point studied with the Pythagoreans and was greatly influenced by them.

Classical Greek philosophy occurs mainly during the 4th century BC. Socrates (469-399 BC) was a widely respected lower class Athenian. His intellectual emphasis focuses on moral philosophy rather than 'natural philosophy,' leaves no writing whatsoever, and becomes famous for his imprisonment and assassination by the Athenian state (399 BC). His student Plato (429-347 BC), who by contrast belonged to the upper class, placed Socrates as the central figure in all of his writings. Plato is most well-known for his theory of "forms": pierce the veil of reality to get at its underlying structure. He creates the Academy, center of learning on the margins of Athens.

While Aristotle (384-322 BC) appeared destined to be Plato's successor at the Academy as he had been the best student, he is overtaken by Plato's nephew Speusippus. This rejection marks Aristotle, who then leaves for Asia Minor, previously having served as tutor to Alexander the Great in 339 BC. Upon his return to Athens, he forms the Lyceum with an emphasis on the empirical study of nature. Aristotle's key focus was on the biological sciences.

Hellenistic science, dating roughly between the 2nd century BC and the 1st century AD, is demarcated by the concurrent death of both Aristotle and Alexander the Great in 323 BC. There is a marked shift in the location and character of Hellenistic science, whose center of activity moves from Athens to Alexandria in Egypt, and whose focus is the practical application of the sciences. Although not 'revolutionary,' it was no less significant. Its key figures include Eratosthenes, (275-194 BC) who calculated the size of Earth and Archimedes (287–212 BC), a Hellenistic 'Leonardo da Vinci.' Archimedes is killed at his desk by the solder of an invading army who is told by Archimedes to wait while he finished solving a problem. Recent investigations suggest that Archimedes might have invented the calculus before Newton.

Other important Hellenistic figures include Euclid- (300 BC) and Ptolemy (2nd century AD). Both men published compendiums in their fields (mathematics and astronomy), *Elements* and *Almagest* respectively, which made all prior works fall 'out of print.' Ptolemy establishes the predominant view of the cosmos until it is superseded by the work of Copernicus.

Why did such revolutionary philosophical innovations occur in Greece and not elsewhere?

The Greek puzzle

Greek society was unique for its era; even a comparison to our world reveals of its distinct historical character. Had the Mediterranean been occupied by a prior empire, it is certainly the case that the Greeks would not have emerged as a civilization; they had the fortunate historical circumstance of being able to feely explore the Mediterranean 'unchecked.' As is well known, they were dedicated to the 'polis' which is often incorrectly interpreted to mean 'city state,' but rather provided the principal cultural focus.

The Greeks were 'outwardly' driven. For example, the notion today that each individual should privately live their own lives would have been a strange notion to them. The small geographical size, the open nature of their politics, and the requisite participation in their 'legislature' meant that the focus of life was the collective (polis). All events could be observed by its citizenry and hence debated; the most ordinary citizen, a humble farmer, could as easily occupy an important political position as the wealthiest of Athenians. Whereas today we use entertainment to withdraw from reality—movies, television, and now the internet—the Greeks viewed as vile any amusement that did not have local issues as its principal concern. The notion of citizenry and public virtue were drawn from Homer's works, a literature whose role was similar to that of the Bible today. As Toynbee noted, a society is a web of relations, and the strong bonds of its citizens made for a unique cultural experience, something which helps explain why all non-Greeks were seen as 'bar-bars' (barbarians).

That being said, the emergence of natural philosophy in Greece is still a difficult issue to properly explain, and constitutes a key subset of the historiographical question "why did the Scientific Revolution occur?" It is, for example, difficult to answer why the small island of Miletus became a central location for Ionian philosophers as Thales. In his writings, Aristotle abundantly recognizes his debt to Thales as thinker. Yet Miletus was not all that different from other Greek city states; it was prosperous, but not significantly more so.

A comparison of Thales and Solon is perhaps suggestive. Solon was an important Greek 'politician,' a contemporary of Thales, whose key legal reforms included the making of laws publicly known, and in the process opened the door to an objective analysis of such laws. Another of Solon's legal contributions a was prohibition against a man being made a slave for his debts, going directly against a common social pattern of the Ancient world. One may observe similar traits in Thales's rigorous discussion of prior thinkers, making their ideas available for all to see.

The inherent process of rational debate in Greek democracy is a key to understanding the rise of classical Greek philosophy, principally in its culture of open debate before peers of alternative solutions. Pericles, for example, openly discusses the costs and benefits of Greek policies during the Peloponnesian War, at one point warning against unjust military actions as genocide.

This rigorous open debate was also typical of philosophers, who tended to mention precedents and critically analyze their predecessor's ideas. Aristotle is perhaps the best example in that he tended to make historical reviews of prior thinkers while exploring his own solutions. This procedure is in fact the principal mechanism by which the writings of centuries old authors were preserved for posterity; we know the ideas of the Presocratics simply because these would be cited by later philosophers; the original writings have often been lost to the sands of time. Aristotle would evaluate particular topic by seeing what had been done before not just for 'history's sake' but to clarify problems and identify mistakes done, and thus to help him trace a path to an answer. His nephew Callisthenes wrote constitutional histories to account for the rise of Sparta for his study of political systems. Aristotle even had histories of science written at the Lyceum and might be considered the first historian of science.

The Greek city states certainly did stand in sharp contrast to other societies of the time. As we have seen, Egypt was a very hierarchical society, with priestly control of trade, and where all decisions flowed top down from its rulers. "Greece" is a misnomer in that it was a loose alliance of cities, each as its own center of power. However, all Greek city states shared a common trait: their democratic systems where decisions were undertaken by equals; those affected by laws would have a say in the definition of such laws. It is suggested that the equality of its members was reinforced due to their participation in the phalanx, a military formation where the lack of uniformity became its weakest point. It was a 'democratic' city state in that it was composed by a community of men with common interests. By definition, debate promoted an emphasis on rational analysis and critical thought, as debate became the key arena by which legal changes were enacted. Debate forced an exposition of assumptions, a contextualization of ideas, and an analysis of logical argumentation. By contrast, the rule by power so common in autocratic societies tended to justify itself arbitrarily, leading to a typical rhetorical pattern that was 'illogical' and 'erratic' by nature given that no steadfast criteria could be established to reach consensus.

Democracy and science, broadly speaking, are closely interrelated.

Other factors contributing to its unique social organization was its sea-trade economy, based on the naval exportation of olive oil and wine throughout the Mediterranean. Olive oil served as the 'petroleum' of Classical world. Its wide ranging maritime focus exposed the Greeks to new ideas, such as Egyptian and Babylonian astronomical concepts. That its geography was characterized by series of separate valleys stimulated the formation of distinct and independent social units, all residing within 40 miles from the coastline.

Classical Greece begins in 480 BC with the defeat of the Persians; prior to that point, many Greeks had even served in Persian armies, referring to themselves as "Hellenes." Its earliest origins date to Crete. A phalanx can be identified in their pottery as early as 750 BC and the first Olympics occurred in 776 BC. The defeat of Leonidas at Thermopylae in 480BC did teach the Greeks an important lesson: seemingly impossible battles against much larger armies could have successful outcomes. The consequent expansion in the Mediterranean directly benefited Athens and its powerful naval forces by the collection of tributes. Its rivalry with Sparta is well known—two city-states which historically diverged.

Sparta, where all free men had to serve in the military, contained the most powerful army in all of Greece, and could have easily dominated the region had they chosen to do so. However, the fear of weakening its army due to an exponential growth inhibited such efforts; all foreigners tended to be rejected within Spartan communities. Their military training was very harsh and prohibitive of ordinary sensibilities, which make for some rather amusing stories and anecdotes. Whereas the Athenians knew how to think, it was often said that the Spartans knew how to act. Yet the Spartan's over-reliance on slavery, in contrast to Athenian's light hearted treatment, also generated important differences in world view and political concerns. In Athens the army was open to slaves, thereby opening a path of social mobility to the members of its lowest economic strata—a feature which led many Spartans to ironically comment that in Athens, a free man could not be distinguished from a slave. The vast majority of slaves in Athens were in fact domestic workers, and it is to be noted that Aesop had been slave. Slavery in this sense did not have a permanent, inter-generational character that it was to acquire in later world history.

The size of the Greek city state was relatively small, with a population of 20,000, about the size of Caguas, Puerto Rico (18,000). The largest city was Athens with 300,000, roughly the size of contemporary San Juan. Athenian victory in the Persian War led to network dynamics from which it enormously benefitted; Athens became the jurisdictional and legal hub of Greece. Citizens from the whole region would have to go there to settle disputes or serve in government, thereby contributed to a constant influx of income to the city. This populational flow also fed the diversity of ideas. Athenian hub had a direct benefit to Greek philosophy and to its well-being in general.

The Peloponnesian War (431 BC – 404 BC) became a decades long conflict between Sparta and Athens, and played a distinct role in the emergence of Greek Philosophy. Its principal thinkers would be affected by the war in one way or another.

Socrates (469-399 BC) fought in the Peloponnesian war as a hoplite, and was well known for his impressive physical stamina. He tended to live a life of relative asceticism, simple clothing, walking barefoot, rejecting shoes even in winter, and of relative poverty. The wide respect for his intellect amongst the upper classes limited the brunt of poverty's impact. It appears that his military experience, accustomed to enduring harsh conditions as a soldier, influenced his world view. He came to reject the external world and, in turn, to the diminution of empirical analysis. Unfortunately, Socrates came to be associated with Critias, an unjust tyrant during the Rule of 30 who had been responsible for some of the worst abuses during the war. Socrates actually opposed Critias when the latter tried to use him as a justification for corrupt acts.

Plato's mother was an upper class descendant of Solon, and Plato (429-347 BC) had been introduced to Socrates by Critias and Charmides. Upon Socrates' 'trial' and consequent death, Plato leaves Athens. In his travels, he passes by Syracuse (Sicily) in 387 BC, when he is first introduced to Pythagorean thought, leaving a distinct philosophical influence. At one brief point, even Plato is himself enslaved, but his friends are able to pay and obtain his liberty.

The role of the Peloponnesian war is much more indirect with Aristotle (384-322 BC). Upon Plato's death, Aristotle's inability to obtain the leadership of the Academy leads him to a similar self-exile as that of his former teacher. Aristotle goes to Assos in Asia minor, which was ruled by Hermias, who was in association with the Persian empire. Hermias is killed by the Persians and Aristotle flees again to the island Lesbos with his new bride Pythias, daughter of Hermias. These travels enormously contribute to his interest in natural history. At Lesbos he identifies a species of fish in one of its lagoons which exists nowhere else in the world. Aristotle's father, Nichomacus, had been a physician and was also of enormous influence in his scientific orientation towards natural history.

While it may go too far to argue that classical Greek philosophy emerged as a result of the Peloponnesian war, it is certainly the case that its historical context had a substantial impact on the particular scientific outlooks of its participants. Socrates' anti-empiricism was greatly influenced by his experiences as a soldier. Plato's abstract Pythagorean ideas can be linked to his implied expulsion from Athens upon Socrates' death. Aristotle's father in law's death at Persian hands encouraged his study of biology. Contingent accidental historical circumstances expanded the experiences of each thinker, leaving a distinct philosophical mark in their intellectual characters.

There is also a role played by the relationship to political power by each thinker. Direct ties to political power often had unforeseen consequences upon the downfall of leaders with whom they had been associated. In spite of Socrates' formal rejection of Critias, Socrates is still brought before trial likely as a result of it, and it is unclear whether Plato himself may have felt threatened by the Athenian state given his close association to Socrates.

That Aristotle tutored Alexander, however, gives a false suggestion of greater influence and intimacy than actually existed between the two men. Calliphanes, Aristotle's nephew, is assassinated by Alexander for treason. Yet paradoxically, Aristotle himself is also forced to flee Athens upon Alexander's death due to the animosity towards all things Macedonian. While Aristotle commented that he would not allow a 'second crime to philosophy' committed, he dies only a year after his self-imposed exile at the age of 62. Plato's relationships to powerful political figures in Syracuse also resulted in his brief enslavement. The intellectual's relationship to power is a double-edged sword, providing immediate benefits at long term costs.

There can be no question as to the validity of Toynbee's analysis—at least with regards to Greece. The highest forms of thought did not arise in regions of 'peace and tranquility' but rather in a society characterized by clash and conflict, sometimes appearing to Egyptians as barbarous hordes constantly fighting amongst each other.

The Presocratics

It is all too easy to overly simplify the past, given the long time spans involved; large chunks of history covering various centuries tend to be clumped together into small and discrete units when seen from 'afar.' This tendency, common in the study of prehistory, is particularly common in discussions of the Presocratics.

It is striking to consider that Miletus, starting point of the era, has a territory of only 220 acres, smaller than a regular university campus. Their philosophical claims of world often appear to be overly simplistic. For Thales, all is water; for Heraticlus, there was nothing but change. How could one ever pretend that modern science emerged from such primitive beginnings?

In order to fully appreciate the originality and intellectual contribution of the Presocratics, one has to get "into their minds" to truly understand why such claims served as the foundations of modern science. In spite of being separated by vast segments of time, there was an active debate going on between them, and it is only by looking at these debates that we can come to better appreciate their enormous contribution. Their profound influence on "Classical" Greek thinkers, as Socrates, Plato, merits attention; even if ideas of the Presocratics had been unoriginal, the strikingly innovative character of their disciples would inevitably force their study. Even Aristotle, a 'millennial philosopher, publicly recognizes his debt to Thales, as previously mentioned. So rather than focusing on individual ideas strictly speaking, it is important to turn attention to the persistent strands of their long running debates.

One of the most common feature of the Presocratics is that they tended to deal with exceptional natural phenomena—lightning or earthquakes—and consistently so. Thales, for example, accounted for earthquakes by arguing that the Earth rests on water; underlying wave caused tremors on the Earth sitting above it. Prior mythological explanations for earthquakes were based on the power of the deities, earthquakes typically associated with Poseidon and lightning with Zeus. As every child knows, Atlas held the Earth on his shoulders.

In this sense, the Presocratics specifically sought to reinterpret preceding myths by explaining such events upon different terms. Nature was to be understood on its own terms, avoiding religious explanations based on human emotion and arbitrary behavior. There was the operative presumption of an underlying order to nature, and it is important to point out the Presocratics dealt with classes of objects rather than specific events accounted for in traditional mythologized explanations. The latter can be understood as a natural human psychological reaction to suffering; specific instances of loss were attributed to Zeus and other deities, helping its victims to emotionally cope with their grief.

Even then, it is remarkable how intellectually rich the explanations of the Presocratics had been relative to their predecessors. If the Earth was held up by water, what in turn held water in its place? Any material would in turn lead to infinite regressions seen in Indian mythology, where the world sits on the endless column of animals standing upon another. Anaximander raises the quality of explanation up a notch and proposes that the Earth sits where it did because it is equidistant from all other points in the universe.

Regardless of whether we accept the validity of their ideas or not, it certainly points to a cross-generational debate with an increasingly higher order of abstraction. With each generational cycle, ideas were pushed a step upward, leading to conclusions which today we take for granted, but which could not have been predicted at the beginning of the deliberative process. The context is altered, new questions are proposed, leading to a recontextualization of the issues. The notion of a linear 'hypothesis-experiment-conclusion' cliché does not appropriately capture this intellectual dynamic, so pervasive across the history of science.

It is clear that Thales is trying to provide a unifying principle of experience, a bringing of order to the complexity of the world. It was natural for Thales to make this argument. Water, after all, is associated with all living things as they cannot live without it. Water is also pervasive throughout the geographical world: oceans, rivers, etc. More importantly, in the search for visible manifest properties rather than arbitrary deities, water has a unique quality far above most other substances: it is formless, colorless, odorless, and can take on the attributes of other substances. This inherent lack of 'intrinsic features' made it a natural candidate for a 'universal.'

This trait curiously would come to characterize many universals, and led to the distinct problem of their causality, shifting again the focus of the debate. Since universals lacked specific traits, it was hard to specify and detect what the exact nature of its influence in the world was. How could form emerge from the formless? How could a body lacking any particular traits imbue distinct properties to others? If all is water, how do fire and water cancel each other out? The problem of universals was hence tied to the problem of change in the universe from the very beginning of the debate. Accounting for change in the universe would become one of the most important issues for the Presocratics, a problem ingeniously sidestepped by Aristotle.

Anaximander's (c. 610– c. 545 BC) 'pendulum theory' was something akin to the idea of 'energy'; the world was made of extremes in constant recombination of energies but never differentiated between themselves. Its most obvious manifestation was the change in seasons, being a phenomena common in temperate climates but not in tropical ones. Anaximenes (c. 546 BC) expanded on this to argue that all was air, in that all resulted from different states of compression of air. The most compressed air became rock, and its gradations were associated with different forms of matter. Anaximander's interpretation is also based on a commonly observed natural phenomenon, using precipitation as its inspiration, where continuous change between multiple forms can be observed. Air turns to rain, drops to form lakes and oceans, evaporates, and reinitiates cycle.

Heraclitus (c. 500 BC) goes one step further to claim that "All is change." That we cannot step into the same river twice might sound overly simplistic, as there is obvious change in world, but the idea pointed to a much more important underlying notion in that the problem of change becomes directly associated with the problem of knowledge or epistemology. How do we know that our senses are not fooling us? Heraclitus observed that a bow and string do not materially change upon the shooting of an arrow, but will return back to their original form. For him, it was clear that there was underlying change in the material of the bow that was not perceptible to man, bending and contracting; the appearance of

31

stability was illusory. Heraclitus thus rejects sense experience as guide to knowing the world.

However, the Presocratic who would most forcefully deal with this issue, and establish the paradigm for the rest of the period, was Parmenides; everyone after Parmenides would be forced to account for change in some fashion or other. For him, change was completely illusory. In a simple poem of two parts, "Way of Seeing" and "Way of Truth," Parmenides made the apparently innocuous observation that something cannot come from nothing, and hence that there was no fundamental change in world. This apparently trivial observation became the principal metaphysical obstacle of era. The following Presocratics would have to address this issue in some way or other before going onto other problems, and would even plague Aristotle, who comes up with his own ingenious solution and is able to bypass the issue completely. Its solutions would lead to most radical and revolutionary ideas of period.

One of these was the notion of an 'element,' by Empedocles, a doctor-philosopher (493 - 433 BC). In contrast to the lost writings of other Presocratics, his poems were very popular, of 500 lines of verse, 350 have been preserved. For Empedocles, all change was the result of mixtures of the elements, of which there were four: fire, air, water, and Earth. The various mixtures resulted from two factors, 'love' and 'strife,' an anthropomorphic explanation which obviously contributed to its popularity. Change in this sense was reducible to fundamental 'universals,' in this case elements.

The notion of proportion was very important in his intellectual construct, in that amount of each element in a given unit of matter is what gave the resultant matter its distinct properties. While the idea of the four elements had long preceded Empedocles, his rationally comprehensive explanation is what made him original. In contrast to Heracles, Empedocles accepts sense experience as valid evidence. While our perceptions were valid, there also existed an underlying structure beyond our senses. Nonetheless, we could use logic to deduce its internal dynamics. One should point out that that his notion of elements was not the same as ours, alluding more to states of matter. For example, water alluded to all liquid forms, including molten metals. Empedocles also established what we today would call a research paradigm: systematic investigation into the mixture of various elements—to some degree.

As Empedocles, Anaxagoras (c. 500 BC) provides an explanation to the problem of change, and was actually a historically important figure in that he was the teacher of Pericles—for which Anaxagoras, as Socrates, is attacked in order to discredit Pericles. Curiously, Anaxagoras accepted the basic premise of Parmenides and expanded it with a very original logical analysis.

It was obvious, notes Anaxagoras, that hair cannot come from non-hair, flesh cannot come from non-flesh, and blood cannot come from non-blood. However, instead of deducing the absence of change, he concludes that at the beginning of time all was in each other, in the "nous." The nous was the original undifferentiated mass of the universe, in which some part of all substances and objects existed. Because all substances exist in each other to some degree (the 'nous'), hence a mechanism was provided accounting for change in the world.

While it sounds like a very complicated theory of infinities within infinities, in fact, its simplicity rested in that its entire complexity could be accounted for by a single predominating principle (the "Nous").

Pythagoras (c. 580–500 BC) does not fit to well into this historical progression in that he preceded many of these thinkers and was also rather unusual. Pythagoras established a religious sect with strict religious beliefs, as the prohibition the eating of beans. Pythagoras is born on island of Samos, but is forced to flee because of a tyrant. In contrast to the "universalist" notions of other Presocratics, his idea that 'all is number' alluded to a formalist cause rather than a material one. However, the belief that numbers were 'real' in some sense led to many difficulties akin to those of other universals, such as the exact nature of their interaction with the world of matter. The Pythagoreans were widely criticized by Aristotle, which helps account for his departure from Platonism; changing a number of something will not change its fundamental character, Aristotle observed.

Pythagoras came up with his ideas from a study of music, specifically noting that chords (fifth, fourth, octave, etc.) form exact numerical ratios between themselves, 1:2. 2:3. 3:4. To him, numbers revealed an underlying order in the universe. This intellectual orientation did push the Pythagoreans to explore and develop mathematics. There was an obvious emphasis on number theory, the definition odd and even numbers, the identification of 'square' and 'oblong' numbers, and generally the relation between numbers and geometric shapes onto which these were described. Irrational numbers, as the square root of 2, was referred to as a diagonal that could be presented as a ratio (instead of our modern 'number'). Proofs of various sorts were developed, as every high school student learns: presume the validity of a claim, and then show the consequent contradictions that emerge therefrom.

The group also had some interesting cosmological ideas that would play a role later during the Scientific Revolution. At the center of the universe there existed an 'central fire,' about which a counter Earth revolved, as a counter to the Earth's own revolution. This particular feature was criticized by Aristotle in that it violated known evidence. However, it had been postulated purely as a 'logical' argument to account for the higher number of lunar eclipses to those of solar eclipses, given that lunar occur at twice the frequency of the latter. The Pythagorean world view might be referred to as an 'anti-empirical' world whereby the sensorial information was illusory, and hence required the use rigorous logic (mathematics) to decipher its truths. These notions would have a strong influence on atomists and Plato.

The rise of atomism as a philosophy is perhaps the culmination of the solution of Parmenides's 'dilemma.' Whereas others had accepted Parmenides' presumption of the impossibility of the void, Leucippus and Democritus turns it around on its head. They not only accept the existence of the void, but incorporate it into the very center of their philosophy. The void was the 'space' in which atoms moved and interacted with each other, giving the world and its objects their particular features. Atoms by definition were 'indivisible' objects in which a substance could be divided into.

The range of interaction of atoms were reducible to three aspects. When they clash, atoms could either repel one another or latch onto each other by 'hooks.' They could also be affected by arrangement (AT / TA) as well as orientation (H vertical, versus H horizontal). Atomism was too rich a theory in that it provided a limitless number of combinations of atoms, and avoided the problem of infinite regression in Anaxogoras's 'nous.' Yet so infinitely wide-ranging were the arrangements of atoms, that they could not be systematically studied. Leucippus did not seek to account for specific instances of theory; no 'research paradigm' was ever developed because of the futility implicit in the exploration of infinite combinations. His student Democritus did this to some degree, but in a very limited manner. According to Democritus, 'sharp' objects obtained their property due to the 'sharpness' of their atoms—a philosophically inconsistent feature that would plague atomism during the 17th century.

Atomism was one of the least popular of the Presocratic theories, made known in Leucritius's *On the Nature of Things* (50 BC) five centuries after the original thinkers had lived. Its poor public reception is relatively easy to understand: it had an abstract nature and alluded to unseen entities. By contrast, most popular natural theories tended to be those which directly alluded to the senses and played upon anthropomorphized human themes. Aristotle's philosophy, discussed in the section below, is a good point of contrast to the neglected atomic theory. Empedocles, the most popular, used senses as the starting point of his ideas. Atomism would remain relatively dormant until the 19th century.

We may thus point out a couple of things about the Presocratics. They took the fundamental step of using naturalistic explanations to account for classes of objects, by presuming there existed a coherent order to the universe which could be found by reason and rigorous logic. This attitude was actually unusual in world history. Secondly, their exploration of universals and the problem of change gradually evolved into groundbreaking ideas and epistemologies. The notion that mathematics was the key to understanding the universe, or the idea that underlying all visible change is a coherent order accounted for by atoms would become fundamental tenets of modern science.

Socrates / Plato

We only know of Socrates through the writings of Plato, yet Plato is such a good writer that it is hard to distinguish where "Socrates ends" and "Plato begins." Socrates's ideas are likely represented mainly in *The Republic* and *The Trial of Socrates*. Generally speaking, Socrates does not make a direct contribution to our story, as he was mainly focused with the moral life of man. His influence on science was indirect, through the now known "Socratic method" and his personal impact on his disciples.

By contrast, Plato's influence is more direct and pronounced. Plato's philosophy of science left a lasting legacy, specifically his Theory of Forms and his epistemology. Institutionally, by instructing his students at the Academy to design a model using at its core perfect circles to account for stellar and planetary motion in the Universe, Plato helped create the fundamental cosmological model until the Scientific Revolution. The problem of 'saving of the phenomenon' was solved by his disciple, Eudoxus; Ptolemy's improvements are based upon Eudoxus's cosmology.

Socrates was a stone mason, but his fame was of such a degree in his day, that he is routinely mentioned in a number of Greek plays. As we have mentioned, he fought in Peloponnesian Wars as hoplite, or spear holder; spears were not used for throwing but mainly for piercing at a distance. He is known to have gone into deep trances while thinking about a problem, allegedly visited by Delphic oracles for hours and days on end.

In the *"Trial of Socrates,"* Socrates is forced to drink hemlock after he is accused of corrupting Athen's youths. The common mythologized view, however, that Socrates had been without a defense is an incorrect one. He actually had an opportunity to 'escape' its harsh outcome, as was often in these types of cases, but his proposal that he be fined 30 minas was such a small sum that it ended up offending his jurors. His trial in 399 BC comes shortly after the finalization of the Peloponnesian War in 404 BC. His faith in reason and subservience to the state thus ends up in his demise, the death of one of most original thinkers in world history. Whatever the facts of the case, Socrates becomes a martyr of individual rationality to public and political excesses.

Although *The Republic* is not directly concerned with science, it was not without impact. The principal aim of Plato's work (Socrates) was that of determining the moral conditions necessary for the creation virtue of its future leaders. Socrates (Plato) was not undertaking a structural analysis of the best forms of government, as that which can be found in Charles de Secondat Montesquieu during the Enlightenment or the Federalists in the United States, but rather a cultural/psychological one. The study of 'science' as astronomy for example, was as good insofar as it helped establish the virtue of leaders. This was perhaps no different from the suggestion that youth should spend time in exercise, a practice seen to elevate an individual's moral fortitude aside from his physical conditioning.

The key contribution of natural philosophy to a person's virtue was through its development of abstract thought. For Socrates, science gave its practitioners the ability to peer at reality and look at its true causes, allowing these to detect abstract patterns in the chaos that is the reality most men live; it gave men the ability to isolate the wheat from the chaff. Most individuals are full of false ideas that are never rigorously questioned, living as if in a cave where shadows were mistaken for the truth. The self-examined life, Socrates tells us, allowed men to be free of false notions. The sciences of astronomy and acoustics he alluded to were all forms of abstract geometry, rather than the empirical sciences viewed as today. As in geometry, its truths could not be established by the empirical measurement of drawings, but rather could only be logically deduced—thereby fortifying man's capacity for abstract reasoning.

It has been suggested that Socrates/Plato at times did have a negative effect on scientific development by discouraging empiricism. Certainly the dynamic is more complex than that as Plato's student Aristotle would be the foremost empiricist of the era.

Plato's key contributions to science was through the "Dialogues" (*Timeaus*), which are famous for revealing the 'Socratic method.' Here, their respective roles begin to change.

This method is likely what made Socrates such a respected and beloved by his students. Its aim was to get an individual to verbalize all their presumptions and notions, and to expose the full extent of what they knew (or believed they knew). Once verbalized, each presumption would then be analyzed in order to detect contradictions, logical incongruences, and implications. (Plato's fascinating dialogues can be obtained online for free at the *Project Gutenberg* or at the *Internet Archive*, archive.org.) After each one of the preconceptions had been put to the test during the rigorous process, the individual would then have arrived at a higher level of understanding from that which he previously held.

It is unfortunate how often the Socratic method is misunderstood and improperly practiced. Certainly, anyone being subject to it would experience an unpleasant experience, and likely contributed to some of the hatred for Socrates. Regardless of social status, political position, or role in society, deeply cherished notions would be attacked. An individual would be made to look the fool before his peers, and hence the inherent criticism of the process was likely perceived as a personal attack, rather than the rigorous logical analysis of ideas for which it was defined. Yet it is to be emphasized that the procedure was in spirit an exploratory one, used to discover new points of view that were not present at the beginning of its inquiry.

In this sense, Socrates' approach greatly contrasted to that of the sophists, whereby prior conclusions were already held and the sole aim of the argument was not an honest encounter between two minds, but rather force another mind to predetermined conclusions and beliefs. It was for this intellectual strait jacket that Socrates and Plato detested the sophists, whom in fact sold their services in order to train politicians. These tactics could be used not only to sway individuals in a court room, but to sway public opinion as well. The sophists betrayed their subjects by giving the appearance of an open exploration and discussion, when in fact the conclusions had been predetermined long before the discussion was initiated; the 'sophist method' was not a true exchange of ideas, but an unequal encounter used to impose beliefs on unwitting victims.

Plato had been deeply influenced by the myth of Sparta, or the notion of self-sacrificing individuals living frugally without private property, in which women received the same training as men. It is curious to note that Aristotle's analysis of Spartan constitution revealed a very different society from what had been presumed. There was a great deal of corruption in Sparta, for example. Although its king was supposed to be overlooked by a 'senatorial' group, in fact these were routinely bribed. There was also a pattern of land concentration with huge disparities of wealth whereby women lived in the lap of luxury while the poor starved around them. *The Republic* had tragically been based on incorrect information about Spartan society and history.

36

Upon Plato's return from exile to Athens, he forms the Academy (385 to 370 BC est.), which became a formative institution where leading thinkers of the 4th century would meet on the outskirts of Athens. As its focus was mainly theoretical, the level of funding required to administer it was rather small, and therefore allowed its economic sustenance to be based on donations and student work. Plato compliments Anaxagoras for teaching without pay, while the sophists of his time charged a great deal and made fortunes. It is likely that the Academy did not have patronage support, as this form of financial support first formally emerges in Alexandria.

The Academy became a recognized meeting ground of minds and a space for intellectual discussion and analysis that could not be found within confines of city, thus leaving an important legacy for all future educational institutions.

Plato also creates a philosophy of science that would ripple through the ages. The Theory of Forms is his own particular response to the metaphysical and epistemological problems implied by the problem of change that afflicted prior thinkers. What is change, and what is constant all around us? Plato takes Socrates and fully expanded on his work.

The world had been created by a magnificent craftsman (the demiurge), but Plato's god is not like the God of Christianity. The traditional Christian God created the matter of the world and had absolute power over it. He can make any change at will, without reason or purpose, thus making God unknowable—an issue dealt with in Book of Job and his attempt to come to terms to the difference between the promise of God and his own personal suffering.

By contrast, Plato's craftsman did not create the material of the world, and hence does not have absolute and arbitrary control over it. His role is to impose order onto this material but is constrained by the limitations inherent it; as a result, only the 'best of all possible worlds' that can exist with the available matter emerges from his work. The Craftsman does the best that he can with what is available to him—but no more.

It was therefore the purpose of the philosopher to determine the underlying order which the Craftsman had imposed on world, as well as the limitations under which he operated. A good example is Plato's analysis of the human head.

Flesh and bone, according to Plato, are insensitive; they provide enormous protection but reduce the intelligence of man. As a result, the final form of the human head is thus the balancing of the two contrary dynamics at hand. Men could live for 200 years with thick and powerful heads, but would have been brutes. However, as the amount of tissue and bone is reduced, man obtains his intelligence, at the cost of his longevity. Thus the final design of the human head endowed men with intelligence, but at the cost of a relatively short span of life.

The good philosopher could not be limited by the immediate particulars of reality but had to look deeper at the balance of forces within nature to understand the world. Critical to the philosopher was the analysis of function: to what ends and to what purposes was a thing constructed? This line of thought would be incredibly influential for Aristotle, even while disagreeing was Plato's principal research agenda.

Plato also pushes atomism to a more rigorous form than that found in Leucippus. In contrast to Socrates, Plato makes a direct contribution to the theory of matter in classical Greece. While characterizing his response as a tentative solution, it was certainly of a much more rigorous character than the original atomic doctrine. Mathematicians as the Pythagorean school had proven the total number of perfect solids, which Plato obviously took for inspiration. He argued that atoms were actually geometrical shapes, from which objects obtained their particular traits. Fundamental to all atoms was the triangle, and goes at some lengths to describe the many shapes a triangle could take to form more complex arrangements. Instead of describing simple shapes, Plato takes the most complex scenarios possible

In hindsight, it easy to tell that Plato was seeking for the underlying universal geometrical forms behind reality, akin to his instruction to Eudoxus at the Academy to use the perfect circle to account for planetary paths. Geometric forms were simple 'perfect forms' and the influence of the Pythagoreans is notable.

More importantly, by assuming geometrical shapes, Plato imposes a theoretical limit to the total number of interactions between atoms, with the benefit that it would again allow one to establish some sort of coherent research agenda. Plato was no longer swimming in an infinite ocean of forms and combinations of atoms, but was instead walking along well defined boundaries established by the small number of perfect solids. Remember that for Leucippus, as the total number of atomic interactions were in theory infinite, there was no logical place to start an inquiry and no clear delimitation of its end.

In this sense, Plato's contribution to atomism via Pythagorism is concrete, and helped provide legitimacy to a body of unpopular ideas. Yet curiously, Plato is not a 'strict' atomist, in that he also incorporated Empedoclean notions of elements into his work. How the two interacted was a topic that was not explored by him. It is curious to point out its similarity to modern proteinomics; proteins obtain their unique physiological properties because of their distinct shapes, some playing roles as hormones, some as catalysts, etc. The study of proteins has its own order of complexity, that is not reducible to the chemical components of which they are made. The shapes, although not infinite, are so numerous and interactions so complex, that supercomputers are needed only to begin to model their interactions.

Aristotle

It is incredible to think that the writings of one of history's most influential philosophers was preserved mostly in the form of student lecture notes. Had the original writings not been lost, they would have more accurately transmitted Aristotle's original expressions in lucid, articulate, and artful construction of ideas. Lecture notes by their very nature tend to be dry, of a staccato nature. However, in spite of this impediment, the originality of Aristotle's thought flows through and is readily recognizable.

"The Philosopher," as medieval Arabs used to call him, made a number of innovations, the first of which was his study of 'dynamics,' or the science of motion. Motion, according to Aristotle, was the state in which nature existed; to deny this as some did was to show intellectual weakness. He was also first to undertake a systematic analysis of what we could call today the 'chemical' behavior of matter in *Meteorologica*. Aristotle introduces logical tool of syllogism a=b, b=c, therefore a=c, and it was his writings in logic were influential during the Medieval period.

Most important of all, he was a comprehensive biologist with roughly one fifth of his corpus dedicated to natural history. His father Nichomacus had been a physician who plays an important role in his life—a father-son pattern repeatedly occurring in the history of science. Such was the accuracy of his anatomical descriptions, that some were denied as scientifically valid until the 18th century, as viviparous fish with placental-like chords. Aristotle recognized the inherent difficulty in the study of biology. Living entities were difficult to analyze as one could not directly observe the interior processes or functioning organs. When cut up, internal structures as arteries collapsed, thus inevitably leading the researcher into errors. Aristotle was not immune to these errors, himself adopting mistaken popular folk beliefs. For example, he did not believe that the brain held blood, and hence rejected the claim that it was the 'seat of the soul.'

His creation of the Lyceum was also of enormous importance in the history of science as it was the first institution dedicated to systematic research. A large corpus of men contributed to Aristotle's findings, and was thus the first example of 'collaborative research.' Theophrastus eventually became its leader after Aristotle's death; both had met in the island of Lesbos while Aristotle was traveling through Asia Minor.

One of the most beautiful appeals ever written to promote the study of science appears in Aristotle's writings. He is well aware that most individuals found the study of biology disgusting, and does recognize that the contemplation of the perfect heavens was eminently more satisfying. However, he points out that we do not have direct access to stars or planets, contrary to the case of biology. In spite of its unpleasantness, the study of biology revealed that each creature was unique and special. For him, the study of all science in the end was pleasurable because it allowed men to understand the causes of things, one of the greatest sources of joy. The experience of reason, after all, was man's highest state of happiness.

Taxonomy had been the key biological problem in his day, and would remain so for various centuries: how do you classify living species? There were no clear methods, and much of Aristotle's studies in biology and data collection were focused precisely on trying to elucidate this issue. He recognized that the demarcation between life and non-life was not clear cut, but unfortunately his biological studies were themselves cut short with his self-imposed exile and consequent death.

Oddly enough, his most prominent legacy was not in biology, but rather in cosmology. One could argue that chemistry became the foundation of his cosmology in that his theory of the four elements were used to account for the dynamics of motion in the universe.

As previously mentioned, Aristotle analyzes what has been written down before, preparing histories of science not for history sake, but rather to clarify issues, and detect problems. There were a wide variety of theories of matter at time, thus Aristotle is able to 'build from the ground up' using binary-like contrasts to arrive and establish a topic's key bases. Once these key aspects are identified, he then uses these to build up a complex conceptual structure. One might suggest that Aristotle formalizes the Socratic method with regard to natural world. His emphasis on categories might appear absurd today but it had distinct benefits, as it allowed Aristotle allows to completely bypass Parmenides 'paradox.' His emphasis on sense experience allows him to also discard the abstract notion of the Pythagoreans. For him, the philosopher has to 'save the phenomena' by providing a logical explanation for what at first sight mistakenly appears contradictory.

There was thus an intimate relationship between chemistry and astronomy in his worldview. The world was composed of 4 elements: fire, air, water, Earth, combinations which of created world see today. While the suggestion that Earth might arises from liquid sounds peculiar, consider the formation of salt crystals when water evaporates. While the notion of four elements in and of itself was not original, his use of it is. His physics was defined by the "Doctrine of Natural Place," whereby each entity sought to move in a certain direction according to the elements it was made up of. The two extremes of elements, Earth and fire, thus had two extremes of motion, downward and upward, with water and air falling in between. By contrast, 'violent motion' was that motion which prevented an entity from fulfilling its 'natural place.'

Rectilinear violent motion was not natural, and required an external agent in constant contact with the moving object, via 'efficient cause'—a process in which the mover was affected as well. Given that the medium in which an object moved became a limiting factor, he defines speed as inversely proportional to the density of this medium—a suggestion which is 'reasonable.' An object falls more quickly through the air than in water. Whereas objects undergoing natural motion increased speed as they neared their destination, objects undergoing violent motion naturally slowed down. These observations lead to some particular conclusions.

Thinking on these dynamics, Aristotle concludes that 1) the universe is not infinite, 2) there could be no void, and 3) stellar motion required a 5th element referred to as the 'aether.' The universe could not be an infinite void because, as objects moved in this void, they would pick up speed over time and ultimately gain 'infinite speed' and 'infinite weight,' which Aristotle believed to be an impossibility. His solution to the problem was the creation of a fifth element (aether), which had a natural circular motion, and operated in the absence of what we today would call friction. Aristotle's 'craftsman' was an 'unmoved mover' who turned the aether, but was himself not influenced by it as he did not operate by 'efficient cause' but rather by thought ('final cause'). Upon moving the aether, the entire chain of consequent movements in the universe were themselves set in motion via crystalline spheres which held all of the planets in their respective locations.

Aristotle, as one might imagine, is a contemporary of Eudoxus, and improves on his model by taking the required 26 circles and creating a total of 55 nests spheres to account for the particular motion of the lower planets. The moon divided the Earthly and heavenly terrains, with natural/violent motion ruling the sublunar region, and the eternal circular motion of aether in the superior region.

There were inevitably problems with his all-too coherent model. How exactly did the two regions interact at the sublunar / supralunar boundary? If the aether's natural motion was circular, and rectilinear was the main characteristic in the sublunar region, how did objects interact at the boundary between the two? How did light from the sun reach the Earth, at the very center of the system, if it had to cross through the diverse crystalline spheres?

There were other problems with his concepts of motion, which were later criticized by Renaissance authors who began issuing challenges to his cosmological model. For Aristotle, arrows moved in triangular motion. As there could not exist a void, when an arrow was launched, it pushed air to the side, which in turn pushed it from the rear in its path through the air. When its influence stopped, the arrow fell to the ground in a direct line. The problem of falling bodies would be one of the key weaknesses of Aristotelian 'physics' and worldview.

In spite of these problems, it is easy to see why 'Aristotleism' became such a predominant philosophy for more than 1,000 years. The vast breadth of explanation and its comprehensive scope made it an effective intellectual 'purchase.' In effect, Aristotle solves the problem of maximization of intellectual goods for a 'low price'; his worldview provided an explanation for nearly everything, thus did not force others to seek elsewhere to account for phenomena that was not present therein.

It would be impossible today for a single thinker to obtain as much influence as Aristotle had during millennia in the history of Western Civilization—and with good reason.

Part II: Intermediaries

Hellenistic Science:

Is There Progress in Science?

ALEXANDER THE GREAT's conquests set the context for all of Hellenistic science. The Hellenistic period begins with the death of Alexander in 323 BC, which also coincided with that of Aristotle in 322 BC, and ends three centuries later. He extended Greece society and culture far beyond Athens, and into Persians lands where he went so far as to even defeat old rivals. The Persian King Darius III runs from battle in spite of having superior forces, and is consequently killed by his generals. Upon this death, Alexander is named the new king of Persia. The story is full of irony as the attack by Persia is what first led to formation of Greece. Alexander moved far beyond Persia, to reach India. Yet Alexander was not "Greek" properly speaking, but rather a "Macedonian" to the north.

The Macedonians had adopted Greek knowledge and policies for advancement, so as to initially limit colonization from Athens herself. Alexander's father Phillip II had been taken captive in Thrace, where he learns to appreciate Greek culture; later as king he becomes the source of this cultural transformation. As a father, he does everything possible to provide the highest education for his son, accounting for how Aristotle ended up as Alexander's tutor. Phillip also took his son into battles, giving Alexander the opportunity to lead the forces on one occasion, decisively winning. Alexander idealized Achilles in his desire to be remembered for posterity.

However, when Philip II is eventually assassinated, Athenians begin distinguishing Greek from non-Greek, showing prejudice against the Macedonians, the worst example of which occurred in Thebes that had never accepted Philip II's rule. In this power vacuum Alexander is able to rise to the top, and ultimately trashes Thebes. His emphasis on strategy helps account for his decisive expansion, demonstrated by his attack on Thyre, a supply point for the Persians. Alexander literally burns it to the ground during a period of two years, which prevents direct Persian attacks on Greece. This action, in turn, allows him to expand east without worrying about the security of his home base.

He establishes "Alexandria," a thoroughly Hellenized city in Egypt, which is used as the base power in lower Mediterranean. After the Egyptian conquest, he establishes a series of supply-cities throughout Asia which allow for the provisioning of an army continually pushing eastward, and which is also used as a channel of communications throughout the expeditions. Curiously, he tended to marry conquered king's daughters so as to strengthen social bonds bond, applying this policy as well to his soldiers. At one point, a wedding between 10,000 soldiers and Persian women is held.

The vast region he conquers is ultimately divided into three kingdoms upon his death by his three principal generals. Ptolemy Soter, who control Egypt, also led what was at the time the most powerful of the three. Alexander's body is taken back to Alexandria. A vast and meticulous tax system is established, serving as Soter's power base.

As one would expect, the existence of multiple ethnic groups undermines the stability of the Hellenistic empire, and multiple "Alexandrias" are established for Greeks to migrate into and Hellenize these regions. Political power is thus distributed between Greeks and the local Hellenized nobility.

With Hellenism, the character of Greek civilization changes in a significant manner.

The large urban size of Athens, which had been the exception, becomes the rule. The 'amateur' character of Greek politics is transformed into a career. The notion that farmers would routinely vote on an issue, only to return to their properties—a practice which provided for a certain equality and homogeneity—is replaced by enormous bureaucratic machines.

The philosophies which emerge during this period reflect their new social contexts; they are 'negative' philosophies. Both Stoicism and Epicureanism seek the avoidance of disharmony, which has a limiting effect on scientific development (natural philosophy). Philosophy was good only in so far as it helped to establish peace of mind, given that ignorance of the causes of violent events as earthquakes or thunderstorms led to panic and unease. Thus while these philosophies stimulated a certain level of science, they did not encourage a persistently active inquiry into nature to find its fundamental causes.

Alexander's territorial expansion leads to a consequent shift of scientific activity from Athens to Egypt (Alexandria), part of the general expansion of Greek culture throughout western Asia. This vast reach helps secure the preservation of culture Greek. Incidentally, while Alexander's march east ends with his death in what is today India, mathematics is carried over into the region, resulting in a gold age of 'Indian mathematics.' The resulting innovations serve as the basis for the creation of 'Arabic numerals,' originally called by the Muslims of the medieval period 'Indian numerals.' Natural philosophy (or 'science'), so previously closely tied to the Academy and the Lyceum, becomes a non-Athenian activity.

While there is no doubt about the Hellenistic recognition of Greek achievement, there is an overall decline of philosophical activity along with an increase in 'technical' work during this period. The focus of Hellenistic science is not so much on the "why," as on the "how." We do not see an attempt at a

45

comprehensive systematized view of the universe, but rather its efforts are delimited and less ambitious--which results beneficial for astronomy.

Hellenistic Astronomy

The notion that Christopher Columbus believed the Earth to have been flat as a pancake greatly undermines the sophistication of Greek science. Columbus's error actually rested in using a lower value for size of the Earth, a figure taken from the work of Eratosthenes of Alexandria. A less radical difference existed between the ancient world and the early modern period than is generally appreciated, undermining the significant achievements of premodern science.

Eratosthenes calculated the size of the Earth in a relatively simple manner. Finding that the sun at noon in Syrene and Alexandria cast respective shadows in their wells led him to realize that estimating the size of the Earth became a simple geometric exercise. Find the angle, determine the distance between the two cities, calculate the percentage of 360 degrees, and multiply to get the total circumference. It was a brilliant and powerful 'tour de force,' if albeit brief. Eratosthenes's error, however is that he presumed the two cities to have been on same line of longitude, which they are not, and hence the inaccuracy of his calculation. It is important to note as well that the unit of measurement, the 'stade,' was not consistently used. That being said, it was still a surprisingly accurate estimation at 39,000 km versus 40,008 km today.

Eratosthenes's case is important because it demonstrates a key feature of astronomy (and science) during this period: it was a subset of geometry, and would remain so for many centuries. Mathematics was not just a tool of science, but was science itself.

Perhaps more impressively, Hipparchus calculates the precession of stars. The Earth's axis tilts and slightly wobbles, and thereby the position of stars change very gradually over long spans of time. Hipparchus is impressively able to calculate that a complete cycle occurred every 26,000 years. In order to reach this feat of astronomy, Hipparchus used Babylonian star charts which had preceded him by 160 years, allowing for a calculation with fair degree of exactitude. The world becomes an enormous problem in geometry for Alexandrians: establish givens and determine for missing variables.

"Theory" in a platonic sense is no longer the hard part of science, but rather obtaining the data became its principal challenge. While their measurements do not have the exactness of today, given the limited technologies available, their methods and approaches in principle were the correct ones. Using simple right angle geometry, abstract proofs of properties of right triangles, Hellenistic natural philosophers were able to calculate the size of the universe, the respective sizes of the Earth and moon, and the relative distances between them. These were all impressive feats whose inaccuracy should not be mistaken for the rigor and soundness underlying their calculations.

Today we take it for granted that the Earth revolves around the sun. To us this is an 'obvious fact,' but we do not personally know that it is actually true. We have obviously not traveled to edge of solar system to look at its structure nor did we personally establish its proof. Rather this belief is to some degree a cultural construction that is passed down to from generation to generation via our educational system. We accept it as a truth of nature because of we are taught it is so and it appears to be in concordance with our daily experience. This is a dynamic for many accepted common beliefs, but this modern day "fact" was not self-evident during the Hellenistic period. Many convincing arguments were made for either side in an important series of debates.

Aristarchus of Samos (310 – 230 BC) claimed that the sun was at the universe's center. He calculated the respective size and distances of Earth, moon, and sun using right hand triangles. The moon was half the size of Earth by his study of lunar eclipses. The sun to him was 19 times away from Earth as the Moon-Earth distance, with a similar ratio for its size. We do not know if his heliocentrism emerged out of these calculations, but the suggestion is unlikely as the distances do not necessarily determine orientation in space. He also appears not to have proposed it as the universe's actual structure, but rather as a hypothetical model. We know of Aristarchus only because he is mentioned by Archimedes in detail, who opposed the hypothesis in his own study.

One of the positive aspects of the heliocentric model however was that it helped account for the unevenness of the seasons. Fall, winter, spring, autumn, do not have an even number of days; Calipanus at the Lyceum calculated their exact days to be 94, 92, 89, 90 respectively. By contrast, the rotation of the stars about the Earth in a geocentric system would have led to even days per season.

Yet the heliocentric model had its own difficulties, particularly the problem of motion. If the Earth rotated and revolved about the sun, the Earths' axial rotation would create such speeds, that nothing would ever be able to move eastwards. Clouds certainly would never be able to freely move as they do in the sky. Some weakly suggested the problem of motion could be accounted for by airs, making a distinction between lower and higher airs. The air immediately above the Earth moved at the same speed, thus keeping everything on the Earth's surface moving at the same pace.

The arguments against geocentrism tended to be dismissed principally because of the coherence of the Aristotelian system. Aristotle's 'chemical physics' held geocentrism firmly in its place. However, work done in the Lyceum after his death was already beginning to suggest important and irreconcilable anomalies in his model.

After Aristotle's death, the directorship of the Lyceum was passed onto two important figures, the first of which was Theophrastus who heads the Lyceum for forty years (322 – 286 BC). Aristotle had met and befriended Theophrastus during his exile at Lesbos. Although not the most original of the two, Theophrastus did extend Aristotelian 'chemistry' in his study "*On Stones*," a systematic analysis of the reaction of diverse materials to fire. While some rocks cracked, others melted (alloys), and some had no reaction at all. Perhaps his most significant contribution was in the field of biology, where he looked at problem of spontaneous generation. He noted that it could be accounted for by the dispersal of seed in air or water but did, but ultimately did not question the notion. In contrast to his friend, Theophrastus does not propose a comprehensive interpretation of the universe.

Strato, who later heads the Lyceum for half the time (286-268 BC), however was the more original of the two. He makes detailed observations of motion and notices two things. The first is that falling water from a roof showed a continuous flow at the top when first leaving the roof, but became discontinuous at bottom, in that droplets formed prior to hitting the ground. He also noticed the enormous variation of impact in free fall, according to height. For example, a stone dropped from the distance of one finger had a negligible effect; however, the same stone dropped from 100 feet would have a much larger impact, as if it were a boulder. From these observations, Strato concludes that objects sped up as they fell. Droplets formed in falling water because water could not retain its cohesion as it picked up speed while falling through the air. A falling stone obviously did not change weight before and after the fall, and hence the enormous difference in impact could only be accounted for by an increase in speed. He also noted that a vacuum does exist. One could blow air into a perfectly sealed tube, and create a partial vacuum by sucking air out of it.

Strato's work seemed to suggest a revaluation of Aristotle's doctrines of motion, which in turn would have inevitably led to a reexamination of Aristotelian cosmology on which it was based. However, Strato never proposes a new global doctrine of motion, and in this sense the tone of his work is very similar to that of Theophrastus given the limited scope of their explanations.

Apollonius, a brilliant mathematician active between 220 and 190 BC, argued against the heliocentric system, an irony when one considers that his analysis of conic sections would ultimately contribute to the unraveling of Aristotelian cosmology. One problem with Aristotle's nested spheres is that it did not account for planetary retrograde motion, as planets or 'wandering stars' moved irregularly against the stellar background. By placing the planet on a secondary ring about the primary orbit (ellipse), Apollonius made the Eudoxian system more feasible, accounting for their backward variation in path. His mathematical device thus helps to maintain the coherence of Aristotelian model by accounting for some of its most noxious anomalies.

There were other arguments against heliocentrism as well. Ptolemy provided very strong arguments supporting geocentrism. He noted that if the doctrine of airs was correct—a system whereby the lower air carried all objects on Earth—then nothing would be able to move relative to everything else as all would be carried along by the air. Taking a page from Aristotle, given that speed was directly proportional to weight, Earth's massive size would also imply that the Earth would have left all of the objects on it far behind long ago.

John Philoponus during the 2nd cent AD expands on Strato's work by making two important observations. The first was that the variation in the time of fall between two stones of drastically different weights will not be proportional to their weight. A stone 20 times heavier than another will not hit the ground in 1/20th the time of the other. Philoponus also observed that if you held an object with a stick, but whose airs were moved behind it with a fan, the object showed no manifest motion whatsoever. The notion that an 'arrow is carried by air' was a false one in that the medium only served as an obstacle to motion rather than its cause. Philoponus therefore introduces the idea 'impetus' or a motive force' that was transferred from hand/bow to arrow, as if it were a substance.

In essence, Philoponus begins to undermine the physics of the Aristotelian system in 2nd century AD.

However, in spite of his innovations, he had little impact philosophically. By his time, one can begin to see a decline in 'science.' Again, because of the comprehensiveness of the Aristotelian world view, minor 'attacks' would do little to unseat it from the common imagination. Even when the underlying physics was discredited, the inability to provide a coherent physics led to continued support of heliocentric model, a tightly packed intellectual system whose validity would not be undermined by attacking only one portion of it.

Curiously, the geocentric system became much more cohesive during the Hellenistic period, specifically in the work of Ptolemy. Given Archimedes's mathematical genius, it is somewhat odd he did not contribute to its history.

Ptolemy

Ptolemy's *Almagest* is the culmination of Greek astronomy, being a comprehensive model of the universe that accounted for its movements with far greater accuracy and precision than all prior work. The books' title originates from its Arabic translation 'the greatest work,' indicating that it did supersede all others that came before it. Its actual name was *Mathematical Composition*, not a book to be read on a couch during an afternoon. Its detailed mathematical exposition and its comprehensiveness helped guarantee the perpetuity of the heliocentric theory.

Its core structure of planetary arrangements is based on Eudoxus. Ptolemy also admired and drew from Hipparchus, erroneously attributing the epicycle to him. Hipparchus does offset the center about which the primary orbit of a planet moved or the "eccentric," helping to further account for discrepancies of planetary motion.

Ptolemy's own contribution emerged from dealing with the problems associated to the calculations of Moon and Mercury that appeared in error; Hypparchus had been unable to account for irregularities of calculation and observation of the Moon during the first and third quarters. Ptolemy ultimately introduces the "equant", whereby the rate of motion was fixed relative to an offset point, a stratagem similar to the eccentric, but which only regarded time rather than distance. The problem of the moon's motion is thus 'fixed' with a deferent on Earth that is moving in the opposite direction as the moon's own epicycle. Similarly, the problem of Mercurian motion is resolved by having an epicycle move on an eccentric, which itself is rotating around a third point. The complexity of the model is somewhat baffling.

Ptolemy did recognize that there were inherent problems with system. His own innovation of the equant violated the Aristotelian principle of 'save the phenomena' but curiously justified system on the basis of simplicity. While the constructed mathematical models were complex, to him the heavens were ultimately perfect. For him, the observation of the heavens made men nobler, leading the mind to higher aims and truth than those of terrestrial concerns.

The idea of progress

Is there Progress in human history? Is humanity destined to continually improve or is it destined for collapse? The idea of progress is also a cultural construct, akin to the notion of Earth rotating about sun. However, rather than being untouchable phenomenas that can only be imagined, progress is known because it is experienced. We believe in progress today because we have lived through a series of improvements in the present when compared to those of the past. Scientific and technological innovations during the last five years have been many. As we have previously mentioned, they include the discovery of the Higgs boson at CERN, the 'god particle' which creates mass. Cellular phones have been constantly improving. In the 1970s they cost $4,000 and their battery only lasted one hour. Computers have shown ever more miraculous improvement. There is more computing power in laptops today than were held in warehouses during the 1950s. The list is vast as it is extensive.

Yet the notion of progress is ultimately only a belief, a faith, which can never be proven or disproven. It is a projection of the past onto the future. With some 6,000 years of civilized history, we can project only 6,000 years into future. But we can rationally recognize that the future is relatively boundless and infinite. What about 6,000 millennia or 6 million years? Do we have the ability to continually improve science during that period? Is there some intrinsic limitation imposed by the human brain to scientific activity?

The human brain, after all, is not boundless, and there are actually signs of historically-relative deterioration. Recent comparative studies have shown chimpanzees to have better memories than humans, as we have become dependent on external devices for their cognition. Also, can the morality of man improve? Will new technologies only mean new abuses of power? Recent events involving the NSA and Edward Snowden have shown that certainly to be the case. The rise of the internet allows for an invisible intrusion into privacy which few citizens have the capacity to detect.

Technology has changed quickly, but the mind of man has not.

The idea of progress is thus based on a series of unprovable presumptions. Will the future exist if conditions of life are no longer viable: no humans and hence no progress? Climate change should perhaps be key issue of concern for the current generation. Yet, where did the idea of progress come from?

In spite of their impressive philosophical, astronomical and mathematical achievements, the Greeks actually did not believe in progress. Some approached it, but do not arrive at it. Plato is rather typical in this sense. For him, the Demiurge created a world whose original form was perfect. Plato thus alludes to a perfect classic age prior to his, and in this sense the general tendency was to perceive change as negative. The notion of a degradation from an ideal point of origin led to a natural resistance to change, and was common across many Greek thinkers. The Greek cultural notion of "moira" is suggestive of, but cannot be translated to "fate." This pervasive Greek notion held that the world has an innate order and structure, of which humanity was part but which it could not transcend. This notion is akin to the great chain of being, how each animal occupies a particular niche in the world: birds in the air, fish in the sea, etc. For the Greeks, to merely suggest the notion that men could supersede their place in the universe was to imply folly, to give men godly faculties they lacked. Man was ultimately destined to operate within his 'humanity,' and could not transcend his human form and place in the universe.

The first inkling of the idea of progress emerges with the atomists. Epicurus suggests the notion of progress, discarding the notion of a prior "Golden era." He believed that men had ascended from brutes, which through a gradual series of innovations improved their condition: fire, social institutions as government and marriage. However, it should be noted that the atomists also believed that all would end up in chaos, and in this sense there was no linear progression to history in their philosophy. Rather history was defined as an eternal series of cycles of chaos and order. Curiously, there was nothing to suggest that the future could not be a mere repetition of the past; the same history could be repeated over and over again, as in the movie trilogy "The Matrix" (1999).

Seneca (4bc – 65 AD) comes closest to the notion of progress, recognizing that genuine scientific discoveries had been made in the past. For him, humanity was merely at the door of nature, and believed that the future would yield riches far beyond what he could even imagine. Posterity would look back to past and not understand their ignorance.

> There are many peoples today who are ignorant of the cause of eclipses of the moon, and it has only recently been demonstrated among ourselves. The day will come when time and human diligence will clear up problems which are now obscure. We divide the few years of our lives unequally between study and vice, and it will therefore be the work of many generations to explain such phenomena as comets. One day our posterity will marvel at our ignorance of causes so clear to them.

In spite of this hope, Seneca could not escape the negative pessimism typical of the classical world view. For him, humanity would ultimately be doomed by vice; the wickedness of man would increase with the consequent development of science and technology as these would provide new avenues for its realization.

Yet the philosophy of stoicism perhaps best captured the Greek worldview. The Roman emperor Marcus Aurelius (121-80), who was also a stoic philosopher, expressed in *Meditations* that the patterns of life were so periodic and cyclical, that a 40-year-old man would not learn anything new as he would have witness all that was already likely to happen during the rest of his lifetime.

Were the Greeks right; is there no progress? As it is a question of belief, one cannot answer it any more than one can prove the existence of God. It is interesting to observe the relative absence of history of Greek civilization, given that Greeks had been a relatively 'new society' lacking a long record of history, relative to that of the Egyptians or Babylonians. For the notion of progress to emerge required a long span of history and detailed knowledge of changes in society. The Greeks in this sense are actually rather typical of most societies in that the notion of the inherent cyclical nature of history has been the predominant view throughout most of human history. Societies lacking a written tradition by definition lack the possibility of an awareness of progress in their inability to detect historical change.

The notion of progress only truly begins in the 16th century, during the Renaissance and a distinct awareness of positive change and intellectual progress.

Dogmatists and empiricists

The history of Greek medicine raises other issues about the question of 'progress.' Is there a hidden cost to the acquisition of knowledge that is not recognized? What ethics should a scientist hold? Do we have the right to destroy nature in order to obtain her secrets? Should there be moral boundaries with regard to what scientist are willing to do to acquire knowledge?

Hellenistic physicians provide important clues to these issues, as Hetophilus or Erasistratus. The most well-known 'Greek' physician is Galen who, akin to Ptolemy, compiled all of the existing medical knowledge and hence becomes the compendium of his field and its major point of reference. Their medical research raises the issue of progress. How do we resolve ethical dilemmas? Do we justify immoral acts because of knowledge thereby gained or do we deny this rationalization and abjure any immorality in the acquisition of scientific knowledge? These questions are still pertinent as ethical issues still exist today in medical research, so well documented in the case of Henrietta Lacks and other minority victims of medical abuse.

There were two broad Greek medical philosophies on these issues: the 'dogmatists' and the 'empiricists.' Note that these terms do not reflect their modern meaning; 'empiricism' is not to suggest a focus on sense data and 'dogmatist' does not allude to 'stubborn' physicians, although perhaps implicit in their philosophy. Dogmatists believed that vivisection was necessary to understand the inner working of the human body, whereas the empiricists believed that medical knowledge had to be qualified by dignity. For the empiricists, the physician could not breach moral lines, and as a consequence, knowledge acquisition was thus characterized by a slow pace as information could only be gathered through the opportunistic treatment of disease. Only accidental circumstances could be taken advantage of to learn about human anatomy and biological processes. Those who practiced dissection have historically refrained from vivisection. Galen is a good example of the empiricist medical philosophy.

Galen only obtains full human skeletons by sheer chance, as when he happened to be in a town when such an accidents had occurred. Two particular cases are mentioned by him. The first was that of a thief who was killed by his intended victim, and the thief's body was left to rot by the afflicted community. The bones were consequently picked clean by birds, leaving behind a perfect skeletal specimen. The second case is that of a tragedy when a father missteps while crossing a turbulent river, falling into it and drowning. The body was not recuperated for weeks, which is also picked clean by fish, providing another well preserved specimen for Galen. In particular, his ligaments even preserved. Galen takes advantage of such ill-fated opportunities. He did not kill either thief or father, but rather makes use of evidence provided by (mis)fortune to yield better data on human anatomy. The same might be said of Louis Pasteur.

Given the prevalence of disease in human history, naturally creating hundreds or thousands of case studies, this was a reasonable stance to take, and in fact the dogmatist attitude is seen rarely in history. Its most ardent Hellenistic exponent was Erasistratus, whom is routinely attacked by Galen.

It is important, however, to note that Galen also conducted vivisections—but in a very different style and character to those of a dogmatist. Vivisection for him was only to be conducted but under delimited conditions in order to possibly yield medical knowledge. Galen believed that the medical researcher needed to undertake many dissections before even beginning to think about performing a vivisection; otherwise would miss important details in procedures that by nature could only be undertaken under enormous time constraints. Dissection on apes, specifically Barbary apes who were purposefully drowned for these purposes, was stressed by Galen as a critical method to all anatomical studies. In other words, vivisection did not necessarily provide a good biological methodology, resulting in many medical errors such as those by Erasistratus.

The historical context is also important to understand the pervasiveness of vivisection during the Hellenistic period. Criminals and slaves were very poorly treated, lacking our modern notions of human rights. These, for example, were often publicly tortured in horrific court procedures so as to induce a desired statement. Slaves are also often used to test the effectiveness of poisons, suffering horrific deaths. This immoral social backdrop thus became a moral reference, reducing the instinctive reluctance towards vivisection that is naturally felt.

We should note that the practice was openly criticized by the early Christian theologian Tertulian. One can observe a similar dynamic in modern 20th century when Germany experiment on Jews and Japan experiment on war captives during World War II. That the human brain is a 'relativistic engine' implies that values are not absolutely set, but are established by a comparison to the surrounding context. The evaluation of facts and actions rarely have 'intrinsic' value, and only in context do they acquire their respective importance and meaning. Similarly, it is the theoretical background which gives significance and meaning to facts that would not otherwise have them.

Hippocrates

Hippocrates (460-377 BC) can be regarded as the 'Presocratic' father of medicine, best known for the 'Hippocratic oath' that all modern physicians have to take before entering their practice. The oath's principal point is that the physician is not to use his privileged position to purposefully harm or take advantage of the patient in his care; the physician will always seek to improve health, even when he cannot. Some of the statements in the oath abjure assassination. "I will not give a lethal drug to anyone if I am asked," whereas others forego the promise of a full life upon its breaching. "Should I transgress this oath and violate it, may the opposite be my fate." As in matrimony, the ritualized public statement serves to reinforce all too easily breached rules.

The "Hippocratic Corpus" is used to describe a group of some 50 texts, that were not all written by Hippocrates, and do have an uncertain authorship. Nonetheless, they embody a general and consistent creed in the calling for the empirical study of disease and the identification of its material causes. We find 'cold' but objective descriptions of observable phenomenon, the best example is book *Epidemics*, whereby day by day analyses of 42 cases can be found which are striking for their similarity to modern medical procedures. The information gathered included estimates of body temperature, body color, with a minimal amount of interpretation. Sixty percent of the patients listed in *Epidemics* actually die, but the authors still candidly describe their treatment and its efficacy. The benefits of this approach are rather clear. The physician is leaving a record for posterity, akin to those of Babylonian star charts. By placing their observations in writing, the Hippocratic authors expand the range of information and experience for future practitioners, whom are hence not be require to perform everything firsthand.

Disease for the Hippocratic authors was not a random occurrence. Illness did not happen in an arbitrary and unpredictable fashion, a will of the gods, but rather it was the sole task of physician to uncover this pattern and identify its cause. "Phlegmatic constitutions" tended to suffer illness, and were intimately tied to the constitution of the body. The human body was composed of four humors: blood, yellow bile, black bile, and phlegm. Curiously, they rejected the four elements in their analysis by noting that the answers tended to be too vague to be of any use, and in this sense their approach was more 'pragmatic.' They tended to measure fevers according to the period of emergence, "quartian" alluding to a fever that appeared every four days and a "tertian" to one that spiked every three.

It goes without saying that the early history of medicine did not have 'medical profession' per se, and hence could not tell who could cure. There were no clear criteria of whom to trust, and faithful prognoses tended to occur only if the 'doctor' could accurately describe what had already happened to the patient, thereby gaining the patient's trust. In other words, upon demonstrating prior knowledge of disease, the early physicians suggested their concrete knowledge of their capacity of prediction and cure.

Given that the sophists were a body of men who charged to prove any point in a lecture, writers began making the distinction between 'idiotes' and physicians. 'Idiotes' as one may well suppose, were laymen without any experience in 'art of medicine.'

54

Galen's battles

Very little is known about Herophilus and the facts known about him are only known as a result of Galen, a common Greek pattern. Nonetheless, Herophilus did make important contributions whose terminology still remain in much of today's medical terms for organs: retina, ovaries, and the duodenum. He was the first to systematically analyze the pulse, using it as a key indicator of health. He finds that extremes of pulse did not positively correlate with for health, and varied greatly over a person's lifespan.

While Herophilus is a dogmatist, perhaps the worst case was Erasistratus, who practiced vivisection with too much zeal. This is not to say that Erasistratus made no medical discoveries at all.

Whereas Aristotle claimed that an inner heat 'cooked' food in the process of digestion, Erasistratus identified mechanical causes of digestion in the intestines. His opening of living torsal cavities revealed intestines that wrapped themselves around food, pulling it along the way with its inner muscles. Whereas Aristotle believed arteries and veins to have been different and separate systems, Erasistratus showed they were connected somehow, and suggested the existence of capillaries uniting veins and arteries. It is important to note, however, that his was not a circulatory system. However, rejecting Erasistratus's notions, perhaps out of moral good will, led Galen to commit one of the worst critical errors he ever made, an enormous medical blunder which effectively paralyzed medical research for centuries.

What was the source of life? For, Erasistratus, the arteries were full of pneuma (air) which to him was the cause of life in all animals. The arteries in his view were literally full of air, which raised the question why the body bled so intensely when its arteries were cut. Erasistratus accounted for this outcome by arguing for the role of the vacuum. When an artery is cut, pneuma was quickly released from it, thereby creating an instant vacuum which pulled blood from the veins into the artery, thus accounting for the effect of bleeding. Erasistratus argued that the liver cleaned blood by removing bile via a series of gradient veins, each smaller would allow blood to pass but would prevent impurities as bile. He also argues the same procedure occurred in the kidneys. Finally, the heart was made up of one-way chambers, which allow blood to move only in one direction.

Erasistratus's intensive use of vivisection led many to question whether there were alternative routes to knowledge that humanely complied with basic morals, particularly Galen.

Galen (129-199 AD) was born in Pergamon. A great deal is known about Galen, much more so than others, due to fact that had been physician to the emperor Marcus Aurelius and his son Commodus. Galen's father had placed a lot of effort in his son's education, initially training him as a philosopher. It is not generally well known that Galen wrote multivolume critiques of Aristotle, which were unfortunately burnt in a fire at the Temple of Peace in Rome where they were kept. Galen had the misfortune of losing many original texts during his own lifetime. At the age of 16, his father has a 'vision' and changes his son to the study medicine. Upon his graduation, Galen returns to Pergamon and becomes physician to the gladiators. This would be an important experience, as he develops a firm appreciation for dissection. In order to put body back in place, one need to know how it was organized originally.

Social unrest in his home town forces his return to Rome, which he detested calling it a city full of greedy backstabbing charlatans. After three years Galen had enough, and was about to leave when he is called to become physician to Marcus Aurelius, who had been initiating a campaign upon Germania. Strangely, when an epidemic of plague strikes, Marcus Aurelius leaves the battlefront, while Galen stays with the troops. However, when the emperor reinitiates his Germania campaign, Galen is allowed to return to Rome where he establishes a successful medical practice.

As previously mentioned, Galen becomes the 'Ptolemy' of medicine by preparing a synthesis of all that had come before him. In this sense he was not truly 'revolutionary' but rather provides a comprehensive and systematic exposition of prior discoveries; reference books are valuable even if do not present new knowledge per se. The chaotic character of information leads to value in its organization and preservation. As with Ptolemy, his compendium 'erases' many prior medical texts in that his medical encyclopedia becomes a more efficient (and transportable) substitute for the originals. But to suggest that Galen was a mere compiler was to provide a false notion of him, as in the case of Ptolemy.

Galen is as critical of the Greek medical legacy as he had been of Aristotle. Aristotle had claimed that all objects had a purpose, but in Galen's 16 volume study, analyzing the purposes of animals and their structures, he discovers many features that appear to have no role whatsoever in nature. He notes that the contrary was often the case; some organs were actually injurious to animal's life, as excessively big horns in mammals would lead to death in challenge or get tangled up in branches, preventing escape. He is also critical of Aristotelian categories and elements, noting that categories were often hard to differentiate. Minute gradations of variations are often difficult to separate; it was hard to tell if an object was 'wet or 'dry.' Counter examples could be provided, as in the process of heat. An object might be physically cold, but have great potential for heat, in what today we would call a chemical reaction. A great deal of heat not present originally in two substances might suddenly appeared in their mutual interaction.

In contrast to Aristotle, Galen did recognize the presence of blood in the heart, and postulated it as essential ingredient of life. In his evaluation of Erasistratus's interpretation of digestion, he finds that food was not just mechanically processed, but had to first turn into bile before undergoing mechanical change. Food thus also underwent a change of substance that could not be reduced to mere mechanical action; food was not just being broken up into smaller pieces, but undergoing a more substantive change.

In spite of his many important contributions, Galen's biggest medical mistake pertained to the heart, claiming that blood flowed through the thick inner wall of heart. There were valid and logical reasons for this argument. He measured the artery to be wider, and thus presumed that outflow did not match inflow. He also noticed bumps on the surface of the septum (the heart's inner wall), believing these to be the place where capillary action was occurring. Knowing the outcome of the process, Galen speculated about the existence of an internal mechanism he could not directly observe.

Is there progress in medicine? It is clear that medical knowledge is not 'cost free'; physicians do not merely go to nature and observe the birds and the bees. The study of human anatomy was believed to require an active intervention, specifically the obstruction of normal process to reveal its inner workings. This in turn implied two costly procedures: 1) dissection, which for Galen meant that thousands of apes would have to be killed, and 2) vivisection which for Erasistratus meant sacrificing the lives of hundreds of prisoners in torturous procedures.

The legitimate and ethical grounds for the acquisition of knowledge is an issue which extends itself to contemporary period. The United States medical practitioner Cecil Rhodes in Puerto Rico during the 1930s was accused of injecting victims with cancer, apparently caused by a romantic dispute. Whatever the reason, the scandal forced to Rhodes to flee Puerto Rico, never to return. Although a formal inquiry revealed no conclusive evidence, three letters which had been found after the incident, written in Rhodes's own handwriting, voluntarily confessed the crime. His case is akin to the syphilis studies of Afro-American prisoners in Tuskegee, Alabama or United States medical practitioners in Guatemala whose victims were purposefully also infected with syphilis to study the disease: a crass violation of the Hippocratic oath. In all cases, physicians purposefully harmed their patients to advance medical knowledge. It is often the case that the weakest of society suffer in the hands of the privileged: minorities, criminals, and women.

The issue also raises the question of category. Should we extend human rights to animals? Do they feel any less suffering and pain than we do? How many dogs were needlessly tortured and sacrificed to discover the functions of the pituitary gland? Does humanity have a right to act as the gatekeeper of the natural world, with the omnipotent godlike powers these activities imply? What if we were the subject of these experiments ourselves?

We might ask Seneca's question: is humanity discovering new knowledge or is it simply discovering new ways of doing evil?

Science in the Medieval Period:

Is There a 'Clash of Civilizations'?

THE MILLENNIA AFTER 250 AD (250-1450) was marked by a decline of scientific activity, roughly known as the 'Medieval' period. Its distinct intellectual decline can be best seen by a contrast of two respective thinkers, John Philoponus and Isodore of Seville. Philoponus in 2 AD does work similar to that of Galileo. His original insights into motion, that the speed of a falling object was not proportional to its weight, led him to the theory of 'impetus.' In spite of its originality, the notion fell on deaf ears. By contrast, Isodore of Seville in 6AD, a well-regarded scholar in premodern Spain, undertook scientific studies by reading the dictionary. To him stars shone because they reflected the sun's rays. It is no wonder that Muslims in Spain would speak so poorly of European culture, classifying it on a similar level to that of the Sudan in Africa. This is not to say that scientific activity disappeared completely, but there can be no doubt that it had been relegated to the shadows during the medieval era. As late as the seventeenth century, it was not uncommon to be whispered at monasteries: "one preaches Aristotle by day, but reads Newton at night."

The period can also be characterized by a distinct shift in the foci of scientific activity from the West to the East. The medieval period saw the flourishing of Arab culture, which preserved the Greek legacy, without which it otherwise would have certainly been lost. Unfortunately, the Islamic Arabs did not come to what we regard as the Scientific Revolution. There were many commentators, good insights, and innovations, but these did not step over preexisting boundaries; their science was eternally 'evolutive' rather than 'revolutionary.' It is a curious historical puzzle: Why did they not? The issue is by definition a negative

58

topic, in that we are trying to account of the absence of something that, in theory, should have existed.

The period is thus marked by two associated key questions. Why did science decline in the West, and why was there no Scientific Revolution in the Arab world during its Golden Era? Implicit is a hypothetical 'why not?,' underlain by a broader query as to the relative absence of the Scientific Revolution in the Non-Western world.

Yet the period also raises another interesting issue, valid in our time. Can we legitimately speak of science as an inherent feature of Western culture? Is 'westernization' is the same thing as 'science'? In turn, what political implications do these definitions have? Science is sometimes depicted as an instrument of imperialism, a claim more strongly made for medicine. Is Samuel P. Huntington then correct in suggesting that the 21st century will inevitably be characterized by the "clash of civilizations?"

We may immediately observe that, since the non-Western world acquired science (Arab) and the western world 'lost' science (medieval), this history and its themes are not as clear cut as we would like them to be. Science, as a cultural form, is neither fixed nor eternal. Rather than seeing the period in a black & white or binary yes/no terms, we need to presume a more nuanced gray distribution curve occupied by a wider range of positions and combinations.

Unfortunately, this is a stance that is often ignored in the broader academic community.

The dynamism of culture and history

Contemporary scholarship regarding science and power reflect to some degree the animosity with which science is treated. In some academic quarters, such stances are surprising. We find passive aggressive attitudes; rather than criticizing science openly, criticism is implicit and subtle. Phrases such as "el afan de la modernidad" contain an implicit attack on modern science and technology, whereby modernity is characterized as an ephemeral trend no different from fashion in clothing. Worst of all, such critiques are often made by scholars who enjoy its fruits: driving the latest cars, using the latest computers and mobile devices, enjoying air conditioners at home and at work, etc. Scholars that obtain private benefit from that which they attack in public are patently dishonest and disingenuous.

A strong animosity to science can also be seen in contemporary Arab Muslim scholarship. In it, we find claims that reverse traditional definitions, amounting to a distorted play on words: science as 'fundamentalist' or the scientist as 'Taliban.' Science is characterized as a 'religion,' religious catechism or as an ideology justifying imperialism.

As a historian of science, it is shocking to read the gross distortions of its history. Some of these are based on poor historical research, often using August Comte (1798-1856). While Comte, a nineteenth century scholar who created the notion of 'positivism,' did invent a religion out of science, his story does not

59

represent the entirety of science, and the philosophical history of 'positivism' is more complex than that used to portray it. Comte's work can actually be divided into two periods. During the early period, we find a rigorous and serious analysis of the stages human cultural evolution. In a later phase, the 'religion' of positivism emerges. Comte suggests that only select few could study science, selects 100 books and burns all others—including all Greek works. Only the study of 'useful science' would be allowed, which meant that much of astronomy would also be thrown out of his bizarre utopia.

How can we account for the vast differences in these two periods of his work? During the first, Comte had been an outstanding student at Polytechnique, regularly reading newspapers and being an active participant in the world. During the second period, Comte suffers through the psychological crises of a love lost, Clotilde de Vaux, after which he decides to stop reading and loses contact with the public domain. Hence his second period is characterized by a morose social isolation. Extremist scholars in this sense do not take Comte as a whole but rather nitpick and choose those aspects of his life which are most convenient for their argument, doing injustice to their subject. Comte's well regarded contemporary, John Stuart Mill criticized the change, recognizing the enormous flaws inherent in the second phase. Comte is oddly similar to Alan Kardec founder of 'Spiritism.' Kardec had also been a student at a Polytechnique, and also forms a religion, but of a much greater impact. Kardec and Spiritism become predominant in Brazil and widespread throughout Latin America.

Our observations, however, do not preclude that science has been used for ideological purposes, as shown by Edward Said. His work *Orientalism* (1978) is a rigorous analysis of Western studies of the Middle East and India, typically in France and England. Peaking in 19th period, there emerged "Orientalist studies," which still exist in academic but are rather hidden away. Overly broad cultural simplifications existed in the field. Said notes that Napoleon's invasion of Egypt in 1798 took an army of scholars exhaustively analyzing the region, describing its pyramids, Egyptian culture, and so forth. These studies culminated in a 17 volume study, *Description de l'Egypte*, 1809-1826. Said points out that, in spite of their comprehensiveness, the French scholars all ignored Abd al-Rahman al-Jabarti, specifically his three volume interpretation of the invasion from Arab eyes.

Said's piece is one of the most widely translated books available. It gives voice to the Nonwestern subaltern, contextualizes the power relations of colonial ideology, and critically analyzes the rhetoric of power. Said, who died in 2003, however was surprised by the way his work was taken. Right after the book was published, the Khomeini revolution in Iran used cassette tapes to distribute message in a pre Napster, Limewire – PirateBay – P2P, postmodern information diffusion. Said had never intended his book to be a legitimization of Islamic fundamentalism: all West was 'bad' and all Islamic was 'good.' More importantly, Said criticizes the presumption of cultures as a static phenomena.

Another important analysis of the topic is that undertaken by Michael Adas. In *Machines as the Measure of Men* (1989), he evaluates how the technological sophistication of Westerners profoundly shaped the historical appreciations of the Nonwestern other. When William Smith was on a geological survey of Gambia River, local tribesmen were both aggressive and fearful. Upon seeing a cartwheel for first time, they were fascinated but would not run in front of it. *Machines* offers a detailed analysis of Western interaction with Africa, India, and China, and provides a picture one would expect. Africa received a low evaluation, given the simple housing construction of palm, clay, which contrasted concrete European fortresses. By contrast both India and China were held initially in high esteem. These were complex societies, social structures, and majestic palaces as the Taj Mahal (India). The Jesuit entry into China during the sixteenth century praise Chinese innovations, as gunpowder and paper. If one were to guess in 1500 which nation would have been ruling the world five hundred years later, one would have likely picked China.

What is most important about Adas's masterpiece however, is his evaluation of the dynamic character of these relations. The social appreciation and interrelationship between cultural groups are not fixed because their respective technological capacities change over time. The emergence of the Scientific and Industrial Revolutions led to an enormous change in the evaluation of Chinese and Indian societies. Although still valued above African tribal communities, the two declined after the seventeenth century. British colonial administrators as McCauley noted that a single shelf of a British library contained more knowledge than entire Indian libraries. The Chinese came to be defined as 'despots,' and were now criticized for inability to innovate or adapt.

The Chinese were hard working, but stubborn.

Rome and the Decline of Science

The Roman period (190 BC-640 AD) of roughly 830 years had relatively few scientific innovations, and were generally more focused on engineering and technical marvels. There can be no question about the impressive size of the Roman empire. During the late period, 235-641 AD, the population peaked at some 75 to 100 million persons, spread throughout enormous swath of territory 9 million km^2. To place this in perspective, both the United States and China have territories of approximately 3.6 million km^2.

Needless to say, the Romans also had impressive infrastructures. The Roman Coliseum held 50,000 spectators, and could be flooded for simulations of naval conflicts, with an equally quick capacity to remove large amounts of water in a short amount of time. Rome's vast aqueduct system, maintained at a gradual slope, guaranteed a constant supply of water. Its road infrastructure consisted of 60,000 miles, principally for the transportation of troops, which ended up becoming trade routes. The Pantheon is still today an impressive quasi-religious structure, designed as an enormous sphere encased in a cube.

Most of these important engineering innovations were based on a key invention: the arch. Its design distributes weight evenly rather than laterally, allowing for much larger and ample structures. Applied on a three dimensional framework, the dome leads to the construction of buildings with enormous

61

empty spaces, offering the feeling of awe and sublimity that would be incorporated into Roman Catholic religious buildings as St. Peter's Basilica. Roman scientific activity, by contrast, came nowhere in comparison to these engineering marvels.

As we have seen, the decline of empire is a complex topic. Joseph Tainter noted that Rome, as so many other empires, grew too big for its own good. Input/output flows were unsustainable given the burden of government and the costs of inefficiency. Its leaders, Toynbee notes, did not rise to confront problems, merely repeating old solutions. Julius Cesar realized the state was unsustainable. He tried to eliminate thousands of pounds of bread freely distributed, but was also accused of sacking public funds to buy private favor. Julius Cesar, incidentally, had been a historian.

Generally, there was a recognizable decline of elite culture, as power became ever more concentrated in a few hands. The myth of Rome was that it had been founded on heroic acts of sacrifice, endowing its citizenry a sense of virtue lacking in later periods. Coming to rise as a republic, based on rational public debate and consensus, the formation of a large nonproductive bureaucracy ultimately set its demise. An emperor is endowed with absolute power to solve Rome's ills, initially pledging to maintain a delicate allegiance with senators within a culture of democracy, but soon abandon their commitment. Power is completely usurped by the emperor.

The decline of state coincided with 'barbarian invasions' in a series of mutual positive feedback dynamics which were self-reinforcing. A weak state always invites attack, which further weakens it, promoting the 'barbarian horde' to move into a power vacuum, which continues to further reduce its power until none remains.

By 408 AD Alaric invades Rome, sacking it for 3 days. Although defeated and killed, the incident shows how greatly Rome had declined. Germanic kingdoms eventually form in Roman territory. Theoderic takes over 'Italy' in 493 AD, Cloveo takes 'France' in 510. With the barbarian conquest, a quasi-market society is transformed into a feudal one, which would last for roughly another millennium until the 13th century where we again see the formation and expansion of thriving urban markets. The end of the medieval period is formally marked by the Columbus venture to America in 1492.

The evidence of medieval intellectual decay is so abundant, that exceptions used to justify other interpretations are only that, exception to the rule. In 500 AD, the Roman senator Boethius tries to salvage Greek knowledge, but is imprisoned by Theodoric the Great and executed in 524 for allegedly complotting with Byzantine rulers. The senator had embarked on a lifelong project of preserving Ancient Greek Classical knowledge by attempting to translate all the works of Plato and Aristotle into Latin. While in prison, he is able to write *Consolation of Philosophy*. Boethius believed that, in spite of the inequalities in world, there was a higher form, merging platonic and Christian concepts

St. Augustine of Hippo wrote *City of God* (427 AD) shortly after the sacking of Rome, in order to account for Rome's problems. Augustine depicts global history as a constant war between the Devil and God. The Earthly 'City of Man' with its perishable pleasures contrasted to the heavenly 'City of God' where men were focused on eternal religious truths. Truth could not be sought in the world as its most significant achievements, empires, coliseums, were all perishable goods that would ultimately decay over the long run. Men had to thus turn away from the Earthly world to focus on the eternal afterlife. St. Agustin had created anti-empirical attitude, formally adopted by the Catholic Church as its official doctrine. This emphasis is repeated again in 1962 during the Second Vatican Council, declaring itself to be in a 'war' with the Devil, whatever than meant.

For Augustine, all objects in the world were religious symbols and hence to be interpreted as such. Everything had a double meaning and the purpose of intellectual effort to get at an object's double meaning, marking a distinct shift of view away from the objects themselves into a 'literary' interpretation of the world. Galileo's book of nature, mathematics, in this period was in effect the Bible.

The anti-scientific attitude was also greatly influenced by the predominant social structure of the Medieval period, feudalism, a social structure inimical to scientific activity. Its principal features were established by Charlemagne during the Carolingian period (742-814 AD) in what we now call Germany. Charlemagne became the First Holy Roman Emperor in 800 AD so as to save Rome from foreign attack. In order to increase his military strength, he created vassals, giving these land while moving them around to keep their own power in check. This tactic proved to be very successful, and Charlemagne leads 44 successful campaigns with 8,000 men. Naturally, the early success of feudalism led to its perpetuation throughout region, particularly so under the continually growing environment of chaos, violence, and disorder. Yet feudal institutions had actually preceded the fall of Rome, but were not as widely used prior to it. Europe during the Middle Ages becomes a military society.

Although lord and vassal made up only 1% of population, it held tight control over all resources. The relationship between lord and vassal was, in theory, a relationship of equals. In its formal ritual, the vassal pledged his life to the lord, and the lord in turn provided a benefit to the vassal, typically a usufruct or *'feudo'* (Spanish). The vassal did not own the land, but rather had a right to enjoy all of the benefits derived therefrom—benefits which became hereditary. Its impact on the European economy and social system were immense, leading to a gross overall reduction of agricultural production. The protagonist's main interest was that of social status rather than financial gain or productivity. While in theory a vassal could pledge allegiance to only one lord, he tended to have many—which could lead to dual conflicts. The stagnation of its economic system was somewhat alleviated by the Crusades.

The basis of it the entire system was, of course, the lowly commoner, whose rights and responsibilities were more similar to that of slaves than might be imagined. The commoner was locked into the territory he was born in. Initially the system had not been that bad, as the commoner was usually judged by his peers in public court. However, the gradual encroachment of lords in the courts meant that most decisions were only taken with due regard to the lord's own particular economic interests. A sense of justice and fairness is lost. The commoner bore the brunt of medieval society across too many a generation.

63

Medieval rebellions were often attacks on monasteries; both castle and monasteries became Janus-faced symbols of protection and repression.

The rise of the castle and the monastery coincided with the drastic decline of cities and urban areas in general. The population of Rome and Paris fell to 4,000; cities are essentially abandoned. We do see the invention of the steel plough, very valuable specifically in northern sections of Europe with higher humidity and rich undergrowth, leading to the creation of medievalism's distinctive long agricultural plots. Another innovation is the use of three crop rotation, allowing for soil to replenish itself with legumes. The abundant nitrogen in the air, 80%, could also be 'fixed' into the soil by lightning, but was obviously not a reliable source. The 'fixation of nitrogen' via synthetic fertilizers would not emerge until the 20th century in Germany.

Why was the social structure of feudalism so inimical to scientific development?

Broadly speaking, there were a lack of incentives to the practice of science that we find in modern society. Specifically, 1) there were a limited set of social roles an individual could play in society: commoner, monk or warrior (vassal), 2) there existed no market for technological goods, and 3) currency generally did not exist as most exchanges occurred in the former of barter.

A similar dynamic repeats itself during the Brazilian slave society of the colonial period with egregious levels of inequality. Slaves lacked education or discretionary income, while owners who had wealth and education were not interested in labor saving devices. Even if one assumes an abundance of any creative geniuses, no market existed for their innovations. As long as feudalism dominated the social landscape, there was little probability for science to emerge as a self-sustaining professional activity.

Fortunately, there were other underlying social determinants as urban capitalism, which emerged as a favorable precondition for the fifteenth century Renaissance. As market relations increased and urban regions emerged, there was a consequent reduction in feudal relations, indicating an inverse correlation between the two, feudalism and capitalism. The emergence of the marketplace in Europe is described in Ferdinand Braudel's masterpiece. The new sources of income freed the commoner from his dependency on the lord, allowing for a much greater movement to urban areas, increasing market demand, stimulating the creation of new goods and services, and so forth.

It is important to note, however, that this form of early capitalism is markedly different from twentieth century capitalism. They are two entirely different beasts, and the former cannot be used as a justification for the later, so dominated by huge corporate entities, vastly unequal power dynamics, markets, and capacities to enter into productive functions.

Early capitalism was characterized by exchanges between individuals, rather than their exploitation by mega corporations.

Catholicism and Science

What is the relationship between religion and science? Were certain religious sects inevitably tied to the Scientific Revolution? Was Christianity as a whole 'destined' to have a Scientific Revolution, in contrast to the 'idolatrous' Islam ?

Science has at times been used in attempts to give greater legitimacy to a particular religion or sect. Such propagandistic purposes imply that a given creed has a closer link to truth (and God), and hence somehow more 'legitimate' than other creeds. Given the cultural instability of science and religion, it is in fact difficult to prove a particular relationship.

Christianity is not a 'single religion,' but rather is a broad umbrella composed of various groups and denominations, each with its own sectarian subdivisions. Protestantism itself is a huge umbrella divided into further groupings: Baptist, Methodist, Episcopal, etc. Although Catholicism contains a simpler division at an organizational level, being more socially 'cohesive,' there does exist a range of variation within it: Jesuits, Franciscans, Dominican, Carmelites, etc. Similarly, there has been a substantial amount of doctrinal change and ideological complexity throughout its history. What is nature of soul? At one point, the Catholic Church's notions were more similar to Indian reincarnation than one would presume, but eventually evolves into its current view. One should not presume homogeneity within Christianity in its organizational or theological aspects.

It has been typically argued that Protestantism was more favorable to science, a key proponent being Robert K. Merton, who in turn derived inspiration from Max Weber's famous tract, *The Protestant Ethic and Spirit of Capitalism* (1904). Allegedly, one's inability to know if one would go to heaven encouraged continual good works on Earth. This psychological orientation, in turn, stimulated frugality and savings, serving as the foundation for capital accumulation and the formation of what today we call capitalism. As credit card indebtedness rates in the modern world indicate, the general behavioral tendency for individuals is a negative feedback loop: any income obtained is usually spent on current wants, with little or nothing to account for future needs. Protestantism's ideology led to capital accumulation and served as the primary engine of modern capitalism.

Similarly, Merton noted that Protestants had higher participation rates in early modern science, as revealed in the statistical data. 62% of the members of Royal Society were Protestants as were 80% of correspondents to the French-Catholic *Academie des Sciences*. Between 1543 and 1660, only 3 who adopted Copernicanism were Catholic, while during the period 1550 to 1750, the percent of Catholic scientists declined from 23% to 9% of the total, while the Catholic population as whole increased to 30% in Europe. Catholics are disproportionably underrepresented in science.

It is important to point out that Merton was not talking about Protestantism as a whole, but rather alluding to a subgroup within it: the Puritan sect. Certainly an argument for broader Protestantism is somewhat problematic. The founder of Protestantism, Martin Luther, openly rejected Copernicanism. Another important early leader, John Calvin, did so as well. United States Baptists as a whole have also shown very low science participation rates, while Protestant Sweden records a marked decline as a scientific nation during the nineteenth century.

Catholicism also reveals a complex relationship, as the case of the Jesuits demonstrates, formed as a part of the Counter-Reformation to defend the Catholic faith against Protestant encroachment. The Protestant revolution stressed that a relationship to God was direct, and all members had the right and obligation to personally read the Bible. By contrast, under Catholicism the relationship could only occur through intermediaries of the Church, and its members could not access the Bible personally.

The process now known as the Inquisition gave the Catholic Church enormous social power. Key dates to this history include the meeting of the Council of Trent in 1540, the formal formation of the Inquisition in 1542, and the *Index Librorum Prohibitorum* (Index of prohibited books) in 1543. Paradoxically, the founder of the Jesuits, San Ignacio de Loyola, had a philosophy which promoted education, proposing that knowledge and education were a means to God. As with Protestants, good works had to be known on Earth. However, the Jesuits were beset by a problematic epistemology: the doctrine of probabilism stated that if something was true if it had been previously theologically proven. This led to a bizarre eclecticism that although a rich in its diversity of ideas and notions, prohibited the resolution of contradictory views. A vast gamut of diverse ideas tended to coexist alongside each other in blatant contradiction.

By contrast, the procedure of modern science is a harsh one, where ideas 'compete' amongst each other and are rejected.; it does not consist of a relativistic 'holding of hands' *cumbaya* association. When an eclectic philosophy assists a heterogeneous community in maintaining coherence by reducing the probability of social conflict, it will have horrible consequences for its intellectual growth and development. Science requires that some ideas be rejected and others be accepted, denoted by an intellectual rigor of utmost honesty and fair debate/exchange of ideas.

Typically, an Inquisitorial accusation meant the ruin of individual, as these did not have the legal rights we commonly take for granted today. The accused could not know the nature of the charges brought against him. They did not have the right to counsel (lawyer), and did not even have the right to see all the evidence presented against them. Worst of all, they could be held endlessly in jail without any reasonable expectation of release.

An Inquisitorial body would confiscate entire libraries on the basis of a single 'unholy work'—tragic when we consider the high cost of books during the period. It might take decades before ever obtaining one's books back. Since the Inquisition as an institutional body actually lacked the manpower to evaluate each book efficiently, the author could basically assume his entire collections to have been lost. Without any clear idea of its finalization, the inordinate length of trial and uncertainty of its proceedings more often than not forced its victims to sell all belongings. Being accused meant that one's family would inevitably end up in bankruptcy, destroying families, and creating an environment hostile to open debate.

Of all Catholic countries, Spain was perhaps the worst case. The arbitrary nature of the proceedings and their extreme cost, led individuals to snitch on each other; it was better to preemptively accuse than to be accused. Rivals could be 'neutralized' in the process, never knowing whom the actual accuser had been. The Inquisition created an irrational 'climate of fear' with its predictable impact on intellectual and economic progress.

The Inquisition had an enormous amount of social power, the power over life and death by determining whether or not an accused would be burnt at the stake. It could be pointed out that the total number of persons killed in this manner was low. During the seventeenth century 'only' 1,402 persons were thus condemned, amounting to 2% of the population, in contrast to the 33% who died as a result of the Bubonic Plague. Nonetheless, there can be no doubt that the potentially high costs of arbitrarily being accused—the loss of one's life—would certainly place a chill on any form of conversation, be it scientific or not, and therefore on any sort of intellectual development.

The exact nature of the impact of the Inquisition, however, becomes hazy with regard to specifics of science. There certainly were scientific authors in the list of banned books, Galileo and Copernicus being the most well-known. However, it was not a strict 'scientific' denomination per se. That is to say, not all natural philosophers were placed in the list of prohibited books. Throughout the period, these tended to be defined principally according to religious denomination. Protestants were thus much more likely to end up in list. The Index also did not distinguish "magic" from "natural philosophy," "astrology" from "astronomy," etc. Of all the 5,420 banned works (1559-759), only 759 could be classified as being scientific.

Tragically, the Inquisition's most active period, 1616-1640, coincided with apex of the Scientific Revolution.

Samuel P. Huntington And George Sarton

Is there a clash of civilizations? The prior sections have hopefully shown that the topic is more complicated than is usually assumed. Societies and institutions are not homogeneous entities, but rather a mixture of interests in constant competition, perhaps not unlike Heraclitus's image of a bow under pressure. Thomas Hobbes is right to some degree, and naiveté is often blinding to internal dynamics.

While we might presume that science is inherently 'Western,' in that the Scientific Revolution did not emerge elsewhere, it is clear that there are many exceptions to the rule. The Romans were not 'scientific' in strict sense of the word, and actually developed an 'anti-scientific' view: the existence of Platonic mysticism or its orientation towards engineering and practical applications were ascientific in character. Both Protestantism and Catholicism show a rather ambiguous relationship to science. While the Jesuit sect was empirical and valued education, it was unwilling to identify and discard conflicting claims. The Protestant founders, contrary to what one might expect, were initially more antagonistic to Copernicanism than the Catholic Church itself. Fortunately, the underlying spirit of free inquiry found in the former stimulated its adoption over the long run.

However, the broader notion that modernization, as defined by science, is 'Western' is a problematic contention. History is far too complex for easy generalizations. But, what did Huntington actually mean by 'clash of civilizations'?

The notion first appeared in a *Foreign Affairs* article of his in 1993. The title was not a statement, but a question primarily directed to an interpretation of global politics during the post-Cold War era. The Berlin Wall had fallen two years earlier, marking a decisive shift in the world of international relations. For some fifty years (1945-1991), Cold War politics had dominated the air waves, defined as an ideological conflict between communism and capitalism. Huntington's main argument was that the lines of rupture and conflict would shift during the period, increasing in occurrence along cultural boundaries rather than ideological ones. The Soviet Union eventually broke up into more ethnically traditional lines. Global events thereafter, as the Hutu-Tutsi conflict in Rwanda, have demonstrated the validity of Huntington's notions.

But Huntington did emphasize in a later work that 'modernization' was not the same thing as 'Westernization.' Western nations were 'Western' before becoming modern. By this, Huntington alluded to a set of particular cultural traits: the tradition of political democracy where decisions were usually based on general consensus, and individualism, by which he meant the existence free choice in a community, and so forth. He notes that all nations pick and choose cultural elements which are deemed to be the most suitable. Humans in this sense are no different from the capuchin moneys of David Attenborough's "Planet Earth" (2002) documentary. One member discovered a plant with insecticide properties, and all others followed in imitation due to the plant's obvious benefits.

Yet Huntington pointed out the irony that modernization often results in 'anti-westernization,' suggesting the pattern of a typical bell curve. Modernization is adopted because of perceived benefits, but tended to create negative feedback reactions. At the state level, leaders often sought to distinguish themselves from former colonial powers by reasserting 'national values.' At the at individual level, the postindustrial Durkheimian anomie incurred by modernity or the anonymity of large urban spaces negatively impacted interpersonal relations, described by Georg Simmel, led to a rejection of the industrial society. As individuals 'modernized' and began to live more isolated lives, they reacted to this sociological vacuum by wanting to return to a former sense of community. The use of 'Orientalism' implied a mythic image of a volk pre-industrial tribal community ruled by social harmony.

Aside from its similarity to Nazi ideology, this view incorrectly portrayed the dynamics of tribal life, ignoring the resultant problems from having all spheres of activity—family, work, friendship—conjoined into one another. A rupture in one would and the loss of the support of the community would lead not only to social castration and possibly murder. One of the benefits of modernity, at least, was

68

the existence of discrete psychological spaces whose members do not interact with one another.

There were certainly transitory cases of 'torn countries,' whose leaders recognized the importance of modernization, but whose visions conflicted with that of the rest of the population. Trained in western elite schools in the United States or Europe, their leaders' 'competitive' presumptions often clashed with the collective values of traditional society, the best case being that of Turkey. Its Western-educated elites clashed with its Muslim majority, but at the same time were not recognized by Europe as truly 'Westernized' thus were often not allowed into the club.

Huntington also makes other important observation of these changing cultural dynamics. Consumerism was often misunderstood for Westernization, as the purchase of modern industrial goods do not necessarily entail adoption of the values of the societies which created them. A justification for a global trading system also was distorting its history. Specifically, greater trade did not necessarily lead to peace, and worst still was often used as a rationalization for the excesses of capitalism. Huntington points out that in fact, greater number of encounters tended to exacerbate unknown differences between societies. The high level of United States consumption of Japanese products during the 1980s, for example, led to a consequent rise in the animus felt towards Japan without making North Americans necessarily more "Japanese" during this period. In fact, the identity of an individual varies enormously within the context they find themselves in. Two women engineers holding a conversation might identify as female in room full of male engineers, but adopt their respective British and French cultures in an isolated exchange.

The notion that science is 'Western' has its own curious history. Schools in China and Egypt during the nineteenth defined science as 'universal,' and hence tended to take a broader 'natural philosophy' approach. This allowed for a closer linkage and association between modern science and the native culture (Confucian thought). Definitions which tended to be more technical, according to this view, tended to prevent this diffusion of science into a local culture. In the twentieth century, however, a marked change could be detected, specifically during the 1920s and 1930s as science was increasingly defined as 'Western.'

The formation of anthropology and the greater interaction with primitive groups also encouraged this differentiation. For Lucien Levy Bruhl, the 'savage mind' was a derogatory 'prelogical mind,' a view which was sharply rebuked by Bronislaw Manilowski. Living with Toriobran islanders for years, Manilowsky noticed that Melanesians were astute observers of nature, and would not have been able to survive without it—such as their impressive skills of stellar-guided ocean navigation. Although containing a comprehensive body of knowledge, the fact that they had not codified onto paper it in a Western manner could not be taken as evidence of its absence. The increasing recognition of 'Nonwestern science' actually led to an exploration of this knowledge base, specifically of Latin American Indians knowledge of surrounding plants. The twentieth century was thus marked by a notion of the divergent cultural variants of 'science,' and a greater interaction than that which had existed during the colonial period.

However, do the continuing cultural conflicts between groups imply that there is there no hope at all? George Sarton, founder of the field of 'history of science, believed there was hope for the future.

A Belgian doctorate in mathematics, Sarton worked at Carnegie Foundation for International Peace for a number of years, extensively writing in the field. While his *Introduction to the History of Science* is now rather obsolete in its historical methodology, parts make for an interesting read. Oddly enough, Sarton considered himself mainly an 'arabist,' entering into strong debated with Henri Pirenne. Pirenne argued that the Islam invasion led to fall of the West; by contrast, Sarton emphasized the intellectual contribution of the Arab world in the creation of algebra, their role in the diffusion of Hindu numeral system, and so forth.

Sarton believed science and its history to be the 'new humanism': a new 'universalism' that would unite humanity in places where men were inevitably divided by religion. While humanity would inevitably disagree on particular religious points, all could agree on fundamental scientific truths. In essence, science could provide backbone to international peace in our tumultuous world.

Islam And Science:

Al-Hazen and the Tragedy of Arabic Science

IT MIGHT SEEM STRANGE TO A STUDENT the notion that the rise of the Roman civilization does not necessarily constitute proof of a scientific culture. In fact, the history of science as a study of ideas cannot use urban growth as an indication of intellectual change. Humanity emerged out of Africa and spread throughout the world without an inkling of scientific knowledge. Similarly, populations have grown and cities spread without their sciences or technologies necessarily increasing in sophistication. Eighteenth century urban regions in the United States lacked the steel reinforced concrete that so characterizes their 20th century counterparts, but grew nonetheless.

Facts are not "out there," waiting only to be discovered and seen in order to be understood; they have to be interpreted, analyzed, digested and debated before a claim can be validly established. Arab scholars, for example, often criticized Ptolemy, validly noting that the equant's role could be interpreted using only old tools as the eccentric and the epicycle. While this might appear bizarre, it is no different to the dispute between Brahe and Copernicus's cosmological models: two different models accounting for the same facts. "Facts" do not speak for themselves; they are not necessarily clear cut, unquestionably demarcating their interpretation. The eyes, as Al-Hazen noted, have to interpret in order to be able to see.

Arab scholars paved the way to the Scientific Revolution in more ways than can be imagined—and hence the ultimately tragic nature of their history.

Arab civilization

Islam emerged from the Arab peninsula during the seventh century. Mohammed is a complicated religious figure whose religious expressions in the Koran appeared in a chaotic and disorderly manner. His death in 632 AD led to

71

an intercene battle, forming what are now its two principal variants, Sunni and Shiite. The following century is marked by a rapid expansion out of Saudi Arabia and into the Middle East under the Umayyad Caliphates, stabilizing by the middle of the eight century.

This vast expansion inevitably brought the Muslims into contact with Hellenistic culture, as that which still existed in relatively small pockets. There was some amount of translation of Greek material by Persian officials at first but lacked any sort of systematic character. Islamic culture peaks perhaps in southern Spain; Seville, Cordova, Toledo were advanced and sophisticated urban center with conveniences similar to those of modern life. These included illuminated street lighting, a vast system of aqueducts, and gravity fed water fountains.

Islam's early rapid and tumultuous expansion, however, meant that any type of rigorous scholastic, academic, or intellectual activity did not truly emerge until the very end of the Umayyad period, when a minor caliph requests alchemical translations in order to gain an 'edge' after failing in a bid for power. The Abbasid Caliphates thus initiate an "Arabic renaissance" after the 'empire' is established and relatively settled, creating a series of social dynamics not unlike those which occurred at a later date during the European Renaissance.

The expansion has unexpected cultural benefits, the first being that the Arabic language becomes the lingau franca between many different ethnic groups: Berbers, Persians, Hindu, etc. This common language drastically increases the pool of information available by allowing for the easier diffusion of ideas and concepts between each ethnic group. The expansion in and of itself also requires a general attitude of open-mindedness by Muslim caliphs in order to be successful. The Muslim caliphs were open to ideas and notions in other societies, given that the approach eased the unification of such diverse groups within its domain. That the expansion implied a broader social base, led to what might be referred to as the, 'Home Depot effect.' The unified character of a vast region implied a rich synthesis of intellectual diverse notions. Muslims meet new ideas from both 'West' (Greek philosophy from Alexandrian) and 'East" (mathematics from India). This dynamic is further stimulated by yearly migrations to Mecca, whereby diverse person from the distant corners of the Muslin empire were brought together into a single geographic space and, by definition, interaction.

Two distinct groups form at the extremes of empire. Whereas the 'eastern' camp at Baghdad acquires a more mathematical character, the western camp in Spain has a more philosophical orientation with strong Aristotelian influences. The Spanish camp is far more conservative than its eastern counterpart. Rather than rejecting Ptolemy's equant as in Baghdad, they abandon the use of epicycles and eccentric altogether, making a backward turn to Aristotelian nested spheres. However, by this time, the formal abandonment of Ptolemy was impossible given the accuracy of his astronomical predictions—calculations which simply could not be done under the old Aristotelian system, and why they had been abandoned in the first place.

It is illustrative to compare the Islamic expansion to that of the Mongol under Genghis Khan half a millennium later. In spite of its severity and extension, the Mongol expansion came and went without leaving a single cultural trace behind.

While the Mongols did in fact put an end to Islamic hegemony in the region, it is Arabic culture, through its written literature, that not only outlives the Mongolian invaders but is even adopted by them. In the middle of the thirteenth century, Muslims lose their grip over the region in what might be loosely characterized as a prolonged 'two front conflict': The Spanish Reconquista at one end, and the Mongol attack at the other.

Islamic interaction with Greek Hellenistic culture begins in Alexandria, being one of the most important centers but was not exclusively limited to it; Antioch (now Turkey), Harran in Mesopotamia and Bactria in Asia Minor had also been important point of cross cultural diffusion. As we have seen, the principal traits of Hellenistic science were its technical and practical outlook, which in turn facilitated its entry into Muslim culture. The relative absence of general philosophical traits in this particular case eased its 'adoption' into the Arab world by showing no apparent ideological conflicts, although conditionally as we will see later. Hellenism is also used as means for religious propaganda; it is turned into a mechanism to explain Islamic faith in non-Islamic terms, in a defense of faith through 'reason.'

The important translations from Greek to Arab begins under the Caliph al-Ma'mun of the Abbasid dynasty in the capital Baghdad, which had been founded in 762. The story is told that the caliph had a dream about holding a conversation with Aristotle. He asks Aristotle what was the true nature of 'the good,' to which Aristotle replies that the source of all good was reason, which superseded religion and public opinion. Al-Ma'mun then begins an organized and systematic translation of all known Greek works.

Key traits of Islamic science

The three centuries long Abbasid Caliphate after the final geographic expansion of Islam thus marks a 'golden scientific age' in Arabic society. Its leading thinkers write and study Hellenistic thought, typically replying in the form of marginal 'commentaries' in the translated texts. The House of Wisdom in Baghdad, where this activity occurred, is created in 814 AD by Harun al-Rashid and his son al-Ma'mun. The activity of translation had been preceded by Mohammed al-Fazari, who translated astronomical works form Sanskrit. The House of Wisdom becomes a leading research center, with Hunayn ibn Ishaq (808-873) as its director. The translations were often from Syrian into Arabic, the Nestorian Christians serving as first translators in this enterprise. To understand 'scientific texts,' it was often necessary to translate other philosophical texts beforehand; in the process, a vast number of commentaries were produced. A total of approximately 65 scholars worked in the House of Wisdom, 47 of whom exclusively dedicated themselves to the direct transition of Greek works The earliest translations were often undertaken by literary scholars, resulting in texts of a poor quality as they could not understand the scientific contents of the original versions. Ptolemy's works, for example, were undertaken over the years, with each new version fixing errors of previous editions: 827 AD, 885 AD, 901 AD, etc.

Original Muslim thinkers began recognizing the enormous value and influence of Greek philosophy. Al-Jahiz (d 869) in the *Book of Animals* gives credit to Aristotle, while al Al-Kindi (d 870) *On First Philosophy* notes that truth was a universal good that must be sought out regardless of where it would be found. The appreciation of Greek thought is aptly verbalized by al-Wathiq in 847 AD.

> Our share of wisdom would have been much reduced, and our means of acquiring knowledge weakened, had the ancients not preserved for us their wonderful wisdom, and their various ways of life, in writings which have revealed what was hidden from us and opened what was closed to us, thereby allowing us to add their plenty to the little we have, and to attain what we could not reach without them.

Inevitably perhaps, this interaction with Greek culture eventually produces original research and discoveries. One of its most important scholars is Al-Kwarizmi (850 AD), inventor of algebra. His book *On the Calculation with Hindu numerals* is also responsible for the spread of Hindu numerals throughout the Middle East, and ultimately Europe. It was loosely translated into Latin as *Algoritmi de numero Indorum*, giving us our term 'algorithm.' Ibn Rushd, also known as Averroes (1150 AD), is another important Muslim scholar. The large body of comments written by him on Aristotle leads to his nickname "The Commentator." Al-Hazen (1010) writes on the optics of Ptolemy, using both empirical and mathematical analysis.

The largest output was in the form of commentaries, which were observations written on the margins of books, which in turn would be followed by further comments. Following the Greek tradition, these were important examples of active criticisms. Al-Hazen, for example, produced a long critique of Ptolemy in his *Doubts on Ptolemy* as well as *Amending the Almagest*. Unfortunately, he does not generate an alternative cosmological vision. Other examples in this style include Al-Razi's *Book of Doubts on Galen.*

The relationship between the religion of Islam and Arabic scientific practice during the medieval period is complex, showing both a positive and negative influences.

Islam plays a role, for example, in shaping the questions asked by its natural philosophers. Given that all believers had to pray five times a day in the direction of Mecca, the determination of the exact geographical location of Mecca became an important topic during the Arab Renaissance. While traditional determinations made associations to particular astronomical phenomena, it was a 'more or less' system whose accuracy declined as the location's distance from Mecca increased. Similarly, the hour of prayer and the measurement of time acquired importance, given the detailed prescriptions of its practice. Finally, calendrical problems emerged as the end of Ramadan was marked by the crescent moon, whereupon a month's long fasting would end. The formation of a portable astrolabe along with other elaborate instruments helped to practically solve many of these questions.

However, it might be asked, was Arab science original? Did Arab scholars make important contributions or were they just another footnote in history? To answer this question, we will look at three scholars: Al-Biruni (cosmology), Al-Kwarizmi (mathematics) and Al-Hazen (optics).

Optics

Al-Biruni (d. 1048), from the region of Afghanistan, wrote 146 works, most of which were in the field of astronomy and mathematics. Some titles included *Exhaustive treatise on the shadow* and *Derivation of the Chords in a Circle*. In a series of letters with Ibn-Sina (Avicenna) discussing the Aristotelian cosmological vision, Al-Biruni questions its claims. In contrast to Aristotle, he accepts the existence of void pointing that the available evidence did not rule out its possibility. He also comes to reject Aristotle's notion of stellar 'natural motion,' specifically the idea that the circle was the only possible path of a planet or a star due to the aether. He notes that rectilinear heavenly motion was also possible. Most interestingly of all, he claims that the ellipse was most accurately description of planetary paths—foreshadowing the work of Johannes Kepler some five centuries later.

Al-Kwarizmi (d. 850), a scholar at the House of Wisdom, invents algebra, one the most original creations of the era. He was equally well aware of the originality of his contribution. His solutions to quadratic equations were not tied to specific problems, finding for x under all cases with same formula. He showed that once general formulas were derived, these could be applied in a large number of unrelated cases. Its creation became widely used to solve specific problems of inheritance.

Perhaps the most original Muslim natural philosopher, however, was Al-Hazen (d. 1040). Working in Andalusia (Spain), Al-Hazen created modern optics and perhaps had the greatest direct influence on Western thought. He was the first to portray the means by which humans actually see.

The Euclidian - Ptolemaic view proposed that that light emanated from eye in a 'cone of vision,' what is referred to as the extromission theory of light. In his study of vision, Al-Hazen combines multiple sciences, form what we would today regard as 'physics,' 'biology,' and even 'psychology' when he incorporates the role of human brain in the visual evaluation of information. He noted, for example, that damage to the brain could affect perception, in spite of having undamaged eyes. His intromission theory of light overturned prior interpretations, and his work became the most comprehensive on the topic. Translated into Latin during the twelfth century, it was used as a principal textbook throughout Middle Ages. The philosopher Roger Bacon becomes a follower of his in England.

Al-Hazen was the first to propose that light bounces off objects and travels in straight paths to the eye, specifically the retina which inverted the image, thus producing the phenomenon of sight. He also identifies the conditions under which sight occurs. Given that light travels in straight line, vision required a light source of some sort, from whose rays bounced off the object seen. Observed objects thus required a certain magnitude and solidity in order to be seen, and obviously light rays needed a transparent medium in which to travel. Al-Hazen undertakes a comprehensive and complex analysis of light reflection.

There were key arguments used against extromission theory. It was impossible for light to originate from the eyes, as these burned when 'luminous objects' as the sun were observed. Only by considering the notion that light originated outside of the body of the eye could we account for this injury. His most original argument, however, used a type of *reductio ad absurdum* logic. According to the extromission theory, when we look at the heavens, a body from the eye traveled through the immense magnitude of space, which by definition would have to fill the entire space between it and heaven. By prior definition, the eye had to be a body because only bodies with magnitude could detect each other. Since it was impossible for bodies to travel such distances in such short amount of time, the subject of vision therefore had to originate from an external source.

His study of refraction is 'mechanical' in that he uses an ingenious description to account for the angle of incidence in refraction. Today it is well known that the path of light is altered between two media of varying densities. The denser the media, the greater the projection toward the perpendicular of the interface; with an inverse effect with less dense media, whose path is projected outward from the plane of interface.

While Ptolemy and others and only looked at angles in their analysis, they did not seek material causes to this phenomenon. By contrast, Al-Hazen notices that the type of media had a direct impact on the angle of incidence. He defines reflection as a variant of refraction. Coarse media did not allow light to pass and thus reflected back; he referred to such media as having high 'resistance.' By contrast, the refraction of light occurred through a less 'resistant media,' which led to the modification of its path. He used various mechanical analogies to account for the phenomena. If you take a thin board and throw an iron ball directly onto it, the board will break; if you throw the same ball at an angle, the ball will change course without breaking the board. Similarly, if you take a sword and hit a rod directly, this could be cut easily. If hit obliquely, the greater the angle of strike, the less the sword's capacity to cut.

For Al-Hazen, light tries to follow path of least resistance: that which is perpendicular to the line of interface.

Four case studies

If there clearly were original Islamic contributions to natural philosophy, why then was there no Scientific Revolution in Arab world?

The short answer is that Arab science was an elite science, entirely dependent on caliph patronage within a social environment which perceived it as a threat. Contrary to our era, science was not a mode of education taught at schools, whose rote religious learning depended on pure memorization and any type of

76

questioning was not allowed. Towards the end of a lecture, the teacher would proclaim "Allah is great," and all students would bow to him as he left the room. Since natural philosophy was not generally accepted, the death of a ruler could mean either assassination of the scholar, the burning of an entire life's work, or at best flogging.

In short, scientific activity could aptly be described as an oasis in desert, operating in small pockets of privilege and wealth that stimulated such activity, yet within the confines of a hostile environment. To suggest that the relations between 'scientist' and 'citizen' were not positive would be a gross understatement. At any time, the dangerous sands from the desert could blow in and destroy nascent scientific activity.

Four cases aptly illustrate this environment.

Al-Kindi (d 873) was a brilliant philosopher whose 270 works were in the fields of math and physics. He argued that the Koran should not be taken literally, as it contained too many errors, but rather that it should be seen as an allegory for the masses. One particular passage in the Koran, for example, claimed that that world 'bent in prayer' to God. To al-Kindi, rather than imagine a twisted world, he suggested one should take the passage to allude to a universal law to be followed, a subtle piece of argumentation favoring the sciences that was defined as heretical. When the enlightened despot Caliph al-Mutassim dies, Caliph Al-Mutawwakil rises to power. He then confiscates al-Kindi's personal library and orders 50 lashes in public for Al-Kindi, then 72 years old. This radical change in environment naturally led al-Kindi into a prolonged depression—a reaction that would be fairly common to other Islamic natural philosophers.

Al-Razi (d 925), known as the 'Arabic Galen,' studied in Baghdad and returned to Tehran where he writes *Book of Doubts on Galen*, as previously noted. One may observe the originality of his thinking in *On Smallpox and Measles*, where he identifies these as distinct diseases. All should stay away from the patient upon the formation of a blister, as otherwise it would turn into an epidemic, suggestive of the notion of quarantine. A more radical stance of his was that God endowed man with power of reason, enabling him to understand the universe. For Al-Razi, reason was superior to revelation in order to arrive at truth. These claims were defined as blasphemous by the emir al-Mansour, who then ordered Al-Razi to be hit with his own manuscript until either manuscript or head broke. All of his works were consequently burnt and banned. As a result of this punishment, Al-Razi was blinded, and became depressed. When an oculist suggested a possible remedial by surgery, Al-Razi turns it down, noting that he did not want to see more of world.

Ibn Sina (d 1037) or Avicenna, had been a child prodigy, who enters medicine, publishing *The Canon of Medicine*, which became a principal medical textbook until the Renaissance. Avicenna also called for primacy of reason over revelation in the quest for truth, leading the emir of Hamadan to call for his execution by beheading. Ibn Sina hides in the house of a friend, later fleeing from Hamadan. His biography reads like a novel full of intrigues and dangerous traps. While this might sound for an exciting spy novel, there certainly could be no pleasure in being constantly hounded in real life. All of Avicenna's books were banned because he refused to repent, claiming that he would rather lead a "wide and short life" than "long and narrow one." Oddly, he is attacked even by al-Biruni. One should never underestimate power of the masses and their manipulation by religious leaders. Five centuries later, Avicenna is still labeled a heretic in the world of Islam

Ibn Rushd (d 1198) or Averroes is one of the most well-known Arabic philosophers, referred to as "The Commentator" for his abundant works on Aristotle, as we have seen. Averroes argued that the notion of a routine intervention by God was absurd, in that it would imply that man could never have any knowledge of world and effectively denied the efficient cause of things. Averroes was directly contradicting al-Ghazzali, known as "the single most influential prophet after Mohammed." One may well imagine that the cards were stacked against Averroes. When his patron, the enlightened despot Caliph Abu Yaqud died, his son took over, and ibn Rush fell into all sorts of political intrigues. He was ordered to exile from Cordova, whereupon all his writings were also burnt, being preserved only though Latin and Hebrew translations that had previously been made.

In Islam, the ultimate source of truth was God as written in the Koran. For the theologian Al-Ghazzali, cotton burnt because God willed it so. God played an active intervention by angels at all times. So powerful was God that if one were to shoot an arrow, one would have no guarantee that it would hit anything as God could destroy and remake the world in time that it took the arrow to land. The distraction from God, as mathematics, was akin to devil worship. This deeply religious view of world meant that nature was ultimately irrational, and hence could not be understood except by acquiescing to God's will—a view which still exists today. It is a belief system that has yet to be reconciled with obvious scientific facts.

Oddly, some Islamic writers try to attribute discoveries to Islam, alleging that all discoveries have already been previously mentioned in the Koran. The historian of science A. I. Sabra is rather critical of the nonsensical scholarship of Islamic apologists as Seyyed Hossein Nasr. If science has already proven what was already in Islam, then why repress it? Floris Cohen points to the internalization of religious view by Islamic scientists, tending to reject what logic and experiment demonstrate.

Neil deGrass Tyson notes that when Al-Ghazzali made mathematics blasphemous, he cut all and any scientific activity from Islamic world. Whereas 25% of all Nobel Prizes have been awarded to Jewish scientist, less than 1% have been given to their Arab counterparts—in spite of the enormous disparity in the sizes of their respective populations groups: 13 million versus 1.5 billion.

Yet the distorted role of religion can be also seen in the law, which was not made but rather 'found' or sought for within the Koran. All 'legislation' is established in accordance with sharia canon law, or what could be termed as religious constitutional laws. Its interpretation is ultimately dictated by theologians who serve as the ultimate voice of the social order, and hence with power of oversight over all social aspects. The inherent arbitrary and irrational nature of the legislative-political order is perhaps the complete opposite to that established by Thomas Jefferson who sought to embed rationality into the very heart of the political process and social institutions.

Non-religious factors

There were also non-religious factors at work in the relationship between Islam and science. There tended to be an excessive 'utilitarian' emphasis in the Arabic culture of the period; if a discovery was of no apparent practical application, it tended to be discarded, suggestive of the mentality of the modern urban poor. When the diplomat Mustafa Hatti Efedi was shown a thick French glass that would not break if hit, but would with a piece of flint in it, Hatti Efedi made the accusation of 'French trickery.' Even a man as gifted as Ibn Kaldun, the founder of 'Arabic sociology,' dismissed claims when informed in the fourteenth century that important philosophical activity was occurring in France.

There appear to have existed some amount of 'political jealousy' on behalf of religious leaders. Arabic 'scientists' had direct contact with the caliphs, being consulted on a regular basis on a number of issues. Such consultations were viewed as a very dangerous practice by Islamic theologians given the alleged nature of their heretical beliefs. One may thus account for the ultimate fate of Arabic scientist by an 'internal group conflict model.' As previously mentioned, all social groups are heterogeneously composed of competing interests in constant conflict with each other, vying for a greater share of power, wealth and influence. There were two principal groups in the Muslim world: the falsafa and the kalam.

The falsafa were a Hellenistic school that emerged in Islam, ideologically characterized as a mix of Aristotleism and Neo-Platonism. These pursued rational inquiry independent of religious doctrines, and everything, including religious doctrine, was open to inquiry. Members of the falsafa group were not theologians, but tended to be leading intellectual figures: Al-Kindi, Avicenna and others. As one might suppose, they were typically supported by patronage. While they were the leading innovators under model, their position was institutionally uncertain and unstable

The kalam, on the other hand, had a more theological orientation, and develop a series of public inquiries on a vast number of topics; these public debates were more sophisticated than one would presume. The kalam adopt some notions of atomism. Most importantly, they were supported in the madras, initially law schools which were later turned into charity schools. The enormous expansion of madras guaranteed the success of the kalams upon establishing 'state backing,' that was independent of the uncertainties of political change to which the falsafa were constantly subject. By means of the madras, the kalam also routinely had a much larger number of students, and hence a much more wide-ranging and longer lasting social influence that the falsafa.

The kalam's greater institutional influence meant they ended up ultimately defining the relationship of science to religion in the Muslim world.

The Renaissance:

The Role of Magic in Scientific Advancement

WHY DO CERTAIN IDEAS FLOURISH at particular moments in history? Why do some take hold and spread quickly throughout society, whereas others linger like ghosts, ever present but without having much impact? On the other hand, why do we personally adopt certain ideas and reject others? Do we see the world with 'our own eyes' or with the eyes of preconceived notions?

The case of Copernicus is illustrative in this regard. The heliocentric model of *De Revolucionibus* (1543) initiated a revolutionary transformation in Western thought, and is often taken as the formal starting point of the Scientific Revolution. It had a certain degree of popularity, and was prevalent in 16th century in the sense that was widely discussed. Martin Luther would hear of it in 1539, and even laymen knew about the ideas. The poet John Donne would come to lament world lost in *Anatomie of the World* (1621). Copernicus is actually cited in other literary pieces during the seventeenth century.

Regardless of whether one agreed with him or not, Copernicus was generally well respected as a scholar. But he had not been the first person to advocate this view. Nicolas of Cusa (1401-64) had also suggested in *On Learned Ignorance* (1440) that the human mind could not fathom the unbound cosmos. Every celestial object moved relative to each other, without a fixed formal center, offsetting the Earth from the center of universe. In fact, notes Cusa, the rotation of Earth could account for the movement of stars at night. Note, however, that as a metaphysician, he was proposing a theological argument which was not seen as a direct threat to Catholicism, in contrast to Copernicus's coherent cosmological model. We might portray it as a minor internal religious debate within the Church. Merely suggested an idea was not the same thing as proving it.

Marsilio Ficcino (1433-1499) was a wide ranging hermeticist who is influenced by Causanus. For him the sun had to be at the center of universe.

80

Nothing reveals the nature of the Good [which is God] more fully than the [light of the sun]... the heat which accompanies it fosters and nourishes all things and is the universal generator and mover... there is nothing which spread out so easily, broadly, or rapidly as light... like a caress, it penetrates all things harmlessly and most gently. Similarly, the Good is itself spread everywhere, and it soothes and entices al things. It does not work by compulsion, but through the love which accompanies it, like heat. This love allures all objects so that they freely embrace the Good...

Ficcio, as Copernicus, was a sun worshipper.

But the reaction of the Church to the ideas of these two men, Causanus and Ficcio, could not be more sharply contrasted to that of Giordano Bruno (1548-1600). If we look at his ideas, they do not appear to be all that different from his predecessors. For Bruno, the universe was an infinity of worlds within worlds, where no one universe had a special place than any other. The world has no center, he tells us. But, he goes one logical step further: there could be parallel universes.

Yet this seemingly innocent and small step was of enormous theological consequences for the Catholic Church, for it denied the uniqueness of Christ and implied the impotence of the Catholic Church in saving souls. Were individuals in other universes inherently condemned? What implications did these views have for Church dogma? Bruno makes the strategic error of moving from Protestant Europe to Italy, where he is captured by the Venetian Inquisition and handed over to Roman authorities. Eight years later he is burned alive—but not for advocating Copernicanism, as it is never mentioned in his accusation.

As Herbert Butterfield points out, that it took fifty years for the Church to realize that Copernicus led to Voltaire.

The main philosophical argument for heliocentrism rested on its alleged simplicity. The Ptolemaic modeling of universe had become overly complicated. Copernicus's own personal belief was that his ideas were a defense of Pythagorism; he was merely taking a conservative position by restoring European cosmology back to the authenticity of the Greek originals from which it had strayed.

It is certain to say, however, that quasi-religious factors also played a role in his thinking. As with Ficcio, the Sun held a special place in the universe for our conservative revolutionary.

In the middle of all sits the Sun enthroned. How could we place this luminary in any better position in this most beautiful temple form which to illuminate the whole at once? He is rightly called the Lamp, the Mind, the Ruler of the Universe...So the Sun sits as upon a royal throne ruling his children the planets which circle around him.

Copernicus's revolution was not just about a "sun at center" model, however. The entire universe exploded in size as a result of it. William Gilbert and John

Digges in England were the first to accept the model and explore its cosmological implications, truly mind boggling at time. Contrary to the traditional Catholic world view, the universe had not been made for man as he was merely a spec on enormous space some 120 million miles across, 1000 times larger than previously thought.

Many of the participants of Scientific Revolution were partly affected by what today might be referred to as "mysticism." During the era, each rigorous science had its irrational counterpart or "pseudoscience": alchemy to chemistry, astrology to astronomy, and natural magic to natural philosophy. When Tycho Brahe observes a new star in 1572, he predicts a "New Age" therewith. All would be peaceful and the world would be without wars, but one would have to wait fifty years to enjoy the fruits of this prediction. Brahe in fact was first drawn into science by studying alchemy, and in his castle he had an enormous alchemical laboratory whose assistants prepared diverse alchemical experiments.

Of all the key participants of the Scientific Revolution, Johannes Kepler was perhaps the most moved by mysticism, or at least visibly so. It is so pervasive in his work, that he is often depicted as an oddity, but was fairly normal when seen from the lens of his era. As Brahe, Kepler also issued his own series of astrological predictions. In 1595, for example, he predicted there would be famine, peasant uprising, and an invasion by the Turks. Because these, rather frequent, events occurred, he was regarded with some esteem and prestige for his ability to predict the future.

Galileo Galilei, a figure so symbolic of ultimate rationalism of science, fighting many scholars and institutions in its defense, also tended to issue personal astrologies, as that to the Grand Duke. As a keen political player, however, he likely provided these predictions to gain favor with the Court, a practice common at time, rather than a testament of his concern with the otherworldly. (In spite of his prediction of a long life for the Medici lord, the grandfather dies few days later.) It should be noted that Galileo detested mystical claims, attacking suggestions as "action at a distance" as being unwarranted illusive and unscientific ghosts. A follower of Deepak Chopra he would have not been.

The very same Isaac Newton, who culminated the Scientific Revolution and was glorified as a paragon of rationality, was deeply obsessed by alchemy and religious issues. Nearly a fifth of his entire corpus is dedicated to the study of alchemy. His famous *Mathematical Principles of Natural Philosophy* was written only after he had dedicated many years to the study of 'pseudo-chemistry.' He viewed science as a way of knowing the mind of God. In spite of these facts, this aspect of his life is rejected as a stain upon our hero.

The traits previously mention all point to an overriding theme: The Renaissance was a period of transition.

The period is so called because it was literally a "rebirth" of Greek culture in that it was realized that Greek culture was so much more than Aristotle, made up of other rival views equally as impressive and rigorous. It was as if aliens had suddenly descended from outer space with vast scientific systems ready-made. The period is truly a transition between the medieval world of magic-mysticism

82

and man's industrial conquest of nature. Historically speaking, however, we have to distinguish the 'scientific renaissance' of 1450-1650 AD from its formal counterpart a hundred years earlier (1350-1550).

As expected, its personalities showed a curious mix of experimentalism and mysticism. While William Gilbert in *De Magnete* (1600) concluded that the Earth was a giant magnet, he viewed its forces as a 'living spirit'; the Earth is thus endowed by him with mystical properties. In his work, we can clearly see the seed of the modern sciences, surrounded by the hazy mist of mysticism in which it flourished.

At its heart is also a source of healthy debate in the field. Although 'erasing' magic from the historical perspective we can obtain a 'clean' image of the Scientific Revolution, it is all too easily turned into a mythologized view of science and scientific development. On the other hand, keeping mysticism and other 'irrational' and 'non-scientific' intellectual features make for a much more complicated story, which is not as coherent and clear cut and more difficult to romanticize. It goes without saying that its incorporation provides a much more accurate representation of the Scientific Revolution, while making it more difficult to understand.

What tilted men away from a quasi-religious world view to a secular one; how did the process occur? Ironically, observing the process is somewhat like seeing a magic trick performed before one's very eyes. Three features of its history will be discussed: humanism, print, and hermeticism.

Humanism

The Renaissance's Humanism is typically associated with Florence and the elite fine arts. The majestic and sublime works of Michelangelo (1475-1564) and Leonardo da Vinci (1452-1519) point to elevated standards of culture. Both the sculpture "David" and the painting on the Sistine Chapel ceiling do suggest the presence of God by their utter perfection. Da Vinci's Mona Lisa is actually quite small, but true to life, as were all of his anatomical drawings. Both were preceded by Leon Battista Alberti who first applied geometry to painting, leading to the art of drawing in perspective, a feature taken for granted today with the pervasiveness of our mobile cellphones and tablets. Renaissance works of art are all characterized by an impressive realism, so life-like that our minds are temporary fooled into thinking its subject are still alive—as well as to ignore the difficulty with which they were created.

A similar trend occurred in map-making and the pictorial representation of nature. Broadly speaking, we move from a cartoonist style to photorealism in the depiction of nature. Early mapmaking was very crude. Roman road maps often only used predominant markers, such as a church, tree, which were then filled by 'lines connecting the dots.' The TO map of the world also reflected this elementary style of representation, principally in order to present a Catholic cross over the world's globe. There certainly existed more sophisticated portolan maps, based on angles to determine position, but its labels appeared only on the on a costal land side.

The rediscovery of Ptolemy's' *Geografia* helped revolutionize European mapmaking, as he had solved the mathematical problems of flattening a sphere. Gross distortions could not be avoided, unless one stretched and cut the globe, creating lines of longitude / latitude onto which points on Earth are plotted. Jacobo Argelo had retrieved a great deal of Greek material from Constantinople, only to be shipwrecked off the coast of Naples. The only book he managed to save was *Geografia*. Its translation in 1410 led to more accurate European maps. Ptolemy's surviving work did NOT have original maps or images, but the author had wisely provided instruction in the book on how to recreate them—a tactic that other Greek authors had also implemented.

One of the most pervasive problems was that of longitude, making for the sailing of the oceans all that much more difficult and dangerous than it already was. Its solution required clock technologies of high quality and rigor over long periods of time—absent until the 18th century. There were consequently many attempts using other approaches by astronomers to solve this issue, but were inherently difficult because of the instability of ships at sea. One crafty inventor had even devised a chair on a gimble, but as everything else would rock along with the boat, the mechanism was not a viable one.

We find a similar refinement in the acuity of biological drawings. Up to the period, biology was mainly natural history and medicine, and as a formal science it would not emerge until the 19th century. The emphasis was thus placed on botany, in particular 'herbals' or books which described the medicinal properties of plants. By definition, these tended to be limited in number, as few plants have medical benefits. Because they were copied by hand, their drawings would gradually deteriorate over time. The descriptions of animals suffered many of these same problems, often depicting what in essence were fictional creatures. The lamina was a lion with human feminine anatomy, the bishop fish was thus named for its shape. Similarly, animals from exotic lands, as the rhinoceros, were incorrectly drawn with armored plates. These cases perhaps provide the best evidence for role of cognition in perception, as noted by Al-Hazen: we tend to see only what we think we are seeing, rather than 'seeing the facts' for what they are.

The broad trend during Renaissance is thus a drastic change towards realism, drawing nature "as is" in an attempt to obtain a faithful and accurate reproduction of the objects observed. Leonard Fuch's *History of Plants* (1542) is an encyclopedic botanical work, very detailed in its depictions. The shapes of animals also become more faithful and accurate; rhinoceros appear without plates. Although influenced by Pliny and Aristotle, Renaissance authors noticed that animals they were observing were actually not the same as those which appeared in Greek works, often the case for marine creatures. Believing at first they had committed some type of error, it was then realized that the classical authors were merely describing only the creatures observed in their own part of the Mediterranean. The Renaissance eye becomes more accurate, and begins to look at nature for what it was, rather than for what was desired—a trait that can also be found in social analysis as Machiavelli's *The Prince* (1513). Its aim to describe men as they are rather than as one would like them to be.

The use of the term "humanism" to characterize the era was first identified by Jacob Burkhart. The aim had been not only to recuperate Greek learning but to imitate them as well. The valorization of the human form, the ideal of self-perfection, and the inherent pantheism, defining humanity as divinity incarnated, were commonly internalized. The Greek legacy, in short, was used to establish new cultural standards by which to judge oneself and all works of human creation.

Da Vinci, for example, had initially planned a work on human art, focusing mainly on surface drawing. However, this project underwent a radical turn upon seeing the work of Galen and his detailed dissections. The pattern of inspiration by the ancients is so common, that it would require its own book. Translation of Greek works often would lead to an original contribution, or aided to the contribution of others. Alphonsene planetary tables had become offset by more than a month, and the efforts of Georg von Peuerbach and his disciple Reggiomontantus sought to help remedy by translating all Greek works in Italy. While their plan was too ambitious, their work led to the first printed copy of Ptolemy's *Almagest* in 1528.

Peuerbach dies on the way to Rome for calendrical reform.

The Printing Press

The European search for Greek works began in the 12th century. There was an early phase which occurred in premodern Spain after the *Reconquista*. The Cordova library was an important source of works, with a total of 440,000 texts. Toledo had been an Arabic intellectual center, was conquered in 1085, and Archbishop Raymond began school of translation in that city. One of the best translators was Gerard of Cremona who translated from Arab to Latin some 70 works during 1160 and 1187, many belonging to Aristotle as well as others such as Ptolemy's *Almagest* in 1175. Willem of Murbek (1215-1286), one of few who knew Greek, wrote some of the most accurate translations directly from that language into Latin.

There was a large variability in the quality of translations, the lower quality often occurring at the beginning of the period. A second phase begins roughly at end of 14th century (1390s), drawing mainly on works from Constantinople. However, when Constantinople falls to the Turks in 1453, it is seen by many scholars as a tragic event for having closed the doors to Greek legacy. By this time, however, most of works had already been obtained. The important task was of improving the quality of their translations, but also shows to some degree the contingent nature of this acquisition. Such was the magnitude of new learning that it naturally represented a cultural challenge which could only be gradually discussed, understood, and assimilated. The transfer of Greek learning would not occur "overnight" and would require educational institutions.

The institutionalization of higher learning actually dates back to the middle of the Medieval period. Charlemagne after 800 AD realized that a large state required a bureaucracy to run, and hence drew priests who could read and write. His emphasis on education led to the standardization of writing and the creation of multiple schools throughout empire, leaving a lasting legacy. These stimulate a shift away from papyrus to pergamins in monasteries, which help preserve the Roman legacy; 90% of roman texts were preserved in pergamins. These, however were very costly to produce, requiring an entire flock of sheep to make only a single book.

His minister of education, Alcuin, created what are now known as the liberal arts: the *trivium* (grammar, rhetoric, logic) and the *quadrivium* (geometry, arithmetic, astronomy and music). Its purpose was to establish the mastery of the self and the world through logic and harmony. Mathematics was not then the specialized field of today, but rather included a whole host of disciplines that are no longer today directly associated with it, as astronomy and engineering. In spite of their contributions, monasteries generally acted as shielded fortresses, blind to the external world beyond its walls. All that was read was 'old,' and no new knowledge was produced.

The first European university was Bologna 1158, principally as center for jurisprudence where complicated Roman law was deciphered and taught in order to deal with new complex urban demands. This intense educational activity in itself became a demographic magnet, drawing hundreds students from all over Europe to their center, and clearly a viable economic activity. Its wide success led to ample imitators: Oxford (1167), Paris (1170), Padua (1222), Naples (1224), Toulouse (1229), Cambridge (1231), and Sorbonne (1253). These also tended to develope specialties at an institutional level.

Scholars at the University of Paris began widely applying critical analysis. Aristotle for example had a procedure of binary analysis: determine if true or false and then break down into logical categories. Peter Abelard prepares a list of 168 quotes by Church fathers, placing them side by side and noticing their mutually contradicting allegations. As James Burke noted: "if you don't understand it, don't believe it." It goes without saying that the Catholic Church did not like this type of analysis, but they did not have much to worry about. Although universities would turn conservative, their initial period was surprisingly characterized by a critical analysis of Catholic hegemony.

The printing press, which emerges during the period, had an even more drastic effect. First invented by Johannes Gutenberg, 'movable type' in 1453 led to an explosion of information. Each letter would have its own individual block, of metal and uniform size. As one might imagine, these would be placed in a holder to create the textual content. The ink was spread over it, and a sheet would be pressed upon it to make exact and identical duplicate copies, time after time. The pressing mechanism was adapted from the wine making industry. In 1455 he prints what is now referred to as the *Gutenberg Bible*, a copy of which sits at the Huntington Collection, becoming an immediate hit.

Gutenberg, however, ends up bankrupt as his is invention so successful yet so easily constructed, that it was widely pirated. The term *Incunabula* refers to works

86

printed prior to 1501, an arbitrary definition, as does not reveal any technical change in printing per se but is rather a cultural term still used today to refer to the earliest printed works.

It goes without saying that the printing press led to revolutionary sociocultural changes. Manuscripts had previously been prepared by monks. A monk would read the manuscript's text out loud, and a room full of peers would copy what they heard by hand on parchment. As one might expect, this was a very slow and tedious, and thus costly. Worst of all, due to inevitable human error, inaccuracies and distortions would accumulate over time, particularly so when monks did not understand what they were hearing. This process in itself, the decay of texts over time, also contributed to notion of decay of information and the broader notion of historical social decline. It suggested a fall from paradise of higher intellectual achievement, which was properly coined by in the term *non plus ultra*: we can never reach beyond gates of Greek achievement.

The benefits of the printing press are obvious, in hindsight. It drastically reduced cost of printing, allowing for a much wider audience in a much shorter amount of time—one of the key reasons for the rapid spread of Copernican ideas. With the efficiency of the printing press, the delay between creation of a work and its diffusion was drastically reduced. The absence of manuscript errors also implied that an author would be able to accurately make his ideas known.

The role of communication in science cannot be underestimated. A great discovery not published is a great discovery that never existed. One of the best examples of this dynamic was Leonardo da Vinci. While nobody can dispute his originality and genius, da Vinci never published his works but rather quite the contrary; he tried making them as inaccessible as possible. Using a secret lingo of words written in reverse to prepare his manuscripts, made them hard to disentangle and much less read. As a result of his secrecy, da Vinci is but a footnote in the formal history of science, as he had no direct lasting influence.

The same can be observed of alchemy. By searching for the philosopher's stone, so as to turn any metal into gold, alchemists were by nature very secretive. The practice of alchemy ambiguously alluded to terms and objects whose genuine identity was hard to distinguish. But this process also meant that their discoveries could not be verified and authenticated. Many princes and kings would be robbed by it; if they could only grant money, enormous quantities of gold were promised in return. The duplicity was not well received, and its practitioners were often killed as a result. One curious case is that of Alexander Seton (d 1604) whose claim of transmutation of iron by a red powder led the Saxony king to request its recipe. When Seton refuses to reveal his secret, he is imprisoned, tortured and killed. Prior to his death, a confession suggests that he had discovered an actual chemical procedure.

As natural philosophy and chemistry advanced, their books eventually make alchemy look quaint and foolish. The new sciences were providing natural explanations for phenomena which had been previously believed to have been magic. The publication of practical works took the mysticism out of alchemy and natural magic. Examples include Georgius Agricola's *De Re Metalica* (1556),

Vannoccio Biringuicco's *Pirotechnica* (1540), and Antonio Neri's *Art of Glass* (1612).

Vesalius

Andreas Vesalius (1514-64) is a good case study of the impact of both humanism and the printing press on scientific activity. His *De Fabrica* appears in 1543, the same year that Copernicus publishes *De Revolucionibus*. Both men curiously shared many traits. Both valued the works of the Greek originals, Galen and Ptolemy respectively. Both did not necessarily see themselves as revolutionaries, but rather merely regarded themselves as extending the work of their predecessors. Paradoxically, such faithfulness to the Ancients ultimately led to their obsolescence.

Up to that time the principal medical textbook had been *Anatomy* (1361) written by Mondino de Liuzzi. It was a well respected work, mainly used for forensic autopsies so as to determine whether the cause of death had been natural or artificially induced. Assassinations are a more common historical activity than we might care to imagine. One would naturally have been hard pressed to identify the cause of mortality. Yet because bodily organs were the first to decay, Mondino's method was to get at these quickly during the beginning of an autopsy, proceeding from the 'inside' to the 'outside' of the corpse. It should be emphasized that this was not a study of 'anatomy' proper. Students were only expected to do one anatomy in the course of their training, and in that, they did not even perform it but rather watched the instructor's assistant perform it at a distance. The lecturer would read from book in podium, the surgeon would cut the body up, and the students would watch.

As there was no actual practice, anatomical studies had decayed greatly since the time of Galen. Galen himself learned the benefit of doing dissection himself; with it he would detect things others had not. Oddly enough, it is a debate that still lingered in mid twentieth century Puerto Rico when its first medical school was created.

Vesalius studied in Paris under Johannes Guinter. Guinter had been a medical humanist who dedicated a large portion of life to the translation of Greek medical works. Vesalius, who unfairly criticized Guinter, in fact owed a lot to him, who had initially hired Vesalius to create illustrations for such translations. Vesalius obtains a medical degree in spite of being quite young, and quickly receives an academic appointment. After the publication of *De Fabrica* and *Epitome*, he sends copies of his works to the Emperor Charles V, who names Vesalius royal physician. After a second printing in 1555, Vesalius is again named physician to his son Phillip II (Spain), a position which does not go as well as he had hoped. After a pilgrimage to Jerusalem in 1564, Vesalius dies on route to Padua where he had planned to take up an academic appointment.

De Fabrica is truly a majestic piece of work, rich in anatomical illustrations of humans in daily settings. These are depicted in situations of ordinary life prior to their deaths: faithfully praying to God, arduously digging a hole, whistfully overlooking a city. One can spot immediate differences with Mondino's work in that Vesalius begins anatomical procedures by scraping away the exterior of the body and thereby slowly revealing its interior layers of musculature, as if the human body were an onion. He thus progresses, gradually moving from skin to muscle, then onto bone structure and finally the organs. Vesalius likened bones to the walls of a house; both were the underlying framework on which all rested else rested. Whereas Mondino proceeded from the inside-out, Vesalius moved from the outside-in.

Vesalius closely imitated the structure of Galen's work, literally breaking up his book into the same organization and chapter divisions of his predecesors. Since Mondino had not known of Galen's masterpiece, this helped raise Vesalius's fame. Physically of short stature, Vesalius wrote a treatise as comprehensive and encyclopedic as Galen's, superceding him in various ways, specifically in his description of the structure of the heart. Vesalius would begin to correct Galen's most egregious errors, opening the door to a revolution in medicine.

His first edition of 1543 notes that he could not detect the intraventricular cavities of the septum, where blood swooshed from one side of the heart to the other. The septum is characterized by an irregular surface, acknowledged in first edition. Vesalius however, only claims that he could not see the structures, and does go so far as to claim they did not exist. By 1555 edition, his interpretation had changed. Trusting the data before his eyes, Vesalius finally acknowledges that the intraventricular openings in septum did not in fact exist—a claim more revolutionary than might appear at first glace.

To understand its importance, we first have to turn to the prior model of blood flow. The human body was divided into two basic structures: veins, which originated in the liver as served as a nutrition distribution network, and the arterial system, originating in the lungs and providing the body with its vitalistic life force. The inability to find ventricular openings in the heart forced a reconceptualization of human anatomy by revealing that the traditional pathways in the body were entirely mistaken. It would require an entire remapping of the human body's arterial and venal highways; in short an entire reimagining of the way the human body worked.

To what extent do we trust our senses, and to what degree are we seeing what we want to see? This question is a common issue in the history of science. We often only see what they are taught to see, and do not accept the reality before their own eyes. Preconceived notions tend to cognitively blind us.

Somewhat ironically, Vesalius made many of the same errors that Galen had committed, greatly due to the fact that he was still using animals as the principal dissection subjects. For example, while liver in dogs showed lobes, these are non-existent in humans. The position of human kidneys in the body is not vertically symmetrical, in contrast to other mammals. These and other errors were actually not correctly identified for many decades, until post-mortem human dissections became the norm.

The work of William Harvey (1578-1657) also contains strong Renaissance strains in its character. While he is recognized for the discovery of 'circulatory system' in *De Motu Cordis* (1628), he personally did not perceive himself as going beyond Galen. In fact, just the opposite, as his entire life he viewed himself as a Galenist, and was actually a direct institutional successor to Vesalius in Padua. Harvey was a student of the student of the student: Vesalius > Realdo Columbo > Gabriel Fallopio > Hiermonus Fabricius > William Harvey. He returns to London in 1602 to become a doctor at St. Bartholomew's Hospital, and later joins the prestigious Royal College of Physicians.

As in the case of Copernicus, others had made similar claims as Harvey's—and suffered greatly for their views. In 1546, Michael Servetus published a religious tract in Paris titled *Resuscitation of Christianity* (1553). It was principally a theological argument, and as Giordano Bruno a direct critique of Calvinism, to whom a copy is sent. Servetus argued that blood actually flowed through the body, via the lungs; in the lungs, blood was imbued with a vital spirit as a result of mixture of blood and air, accounting for its change of color of blood. Realizing that his life might be in danger, Servetus attempts to leave Paris by way of Geneva, but is captured and burnt at stake in 1553.

Yet, as Vesalius, Servetus had also been Guinter's student in Padua. Harvey's own work had also been preceded by his professor Fabricius, who identified valves of veins. However, the elder professor argued that valves existed so as to prevent harmful 'dilations' in the veins, so should not be mistaken for the same argument or discovery.

If blood did not ooze out from heart's septum, what happened? Harvey traces the different paths ingeniously, specifically via the quantification of blood flow. He notes that the left ventricle holds 2 ounces; since the heart beats 72 times a minute, which would amount to 540 lbs of blood in 1 hour. It was impossible, Harvey pointed out, for the liver to create so much blood in such a short amount of time; he had hit the nail on the head. The heart was like a bellow, Harvey argues, which pumped blood in a single direction; the network of valves made sure of this.

Surprisingly, Harvey's work did not actually get a positive reception, but rather had the opposite effect, greatly injuring his reputation during the rest of his professional life. Those who openly accepted and supported his views were Parcelsian mystics as Robert Fludd, who viewed biological processes in microcosm-macrocosm parallelisms. The majority of academic men simply dimissed it. There were too many lingering questions in Harvey's model, such as how exactly blood flowed through the lungs. Pulmonary capillaries had not yet been detected, and it would only be towards the very end of his life, that the microscope would verify Harvey's claims of blood circulation. Leeuwenhoek used the microscope for first time and identifies the hard-to-see capillary veins of the lungs. Malpighi would provide further reinforcing evidence with a much stronger microscope, conclusively establishing their existence.

Although Harvey ultimately died knowing he had been right all along, he still suffered greatly for his work. Reason and logic had not been enough to convince his contemporaries of truths that today we take for granted. In this regard, the case of Fludd hints at the positive role that 'magic' and 'mysticism' could play in the development of science; philosophies as hermeticism would create a 'space for scientific activity' unconstrained by the tyranny of long-held assumptions.

Hermeticsim

While the notion that a mystical religion as hermeticism would have a significant impact on development of science appears odd, its contributions were significant and diverse. Not only did it create a space for interpretative freedom, but it also placed greater emphasis on direct evidence and interaction with nature, while also stimulating intellectual activity outside of traditional conservative institutions as universities. Hermeticism, in short, freed Europeans from the intellectual and social hegemony of Aristotelian scholasticism; as a mangrove forest, it became the nursing ground of the Scientific Revolution.

Hermeticism is a body of religious texts that emerged in Alexandria during Greco-Roman in 300-700 AD. It was a syncretic philosophy, where the Greek god of communication Hermes and the Egyptian god of wisdom Throth were allegedly integrated by Hermes Trisegsistrus. It was characterized by three main areas: astrology, alchemy and natural magic, codified in the *Corpus hermeticum*. Although initially accepted by the Catholic Church, it was later rejected.

One should not place too much focus on the details of its pseudoscientific ideology, but rather concentrate on the broad conceptions which proved to be so stimulating to thinkers of the early modern period. Specifically the hermeticists believed in the microcosm-macrocosm analogy, suggestive of a broad parallelism in the universe. The world was in man, as man was in the world, literally speaking; its philosophy is well exemplified in the work of Paracelsus, who enormously contributed to medicine and chemistry. Its importance can only be gleamed by considering the broader intellectual context in which it existed.

Although scholasticism was based on a rather loose interpretation of Aristotle and the early Church authorities, through the medieval period it had devolved to become a hegemonic ideology. This is striking as Aristotle had initially been banned by the Catholic Church in 1210 as a result of Peter of Abelard's rigorous critiques of Church doctrine. However, Aristotle was so widely read, that Church leaders were ultimately forced to formally reinstituted his works 45 years later (1255).

The Catholic Church's principal interpretation of the relationship between philosophy (science) and religion had been established by St. Thomas Aquinas. A student of Albert Magnus (1206-80), Aquinas argued that Greek thought could be usefully incorporated by the Church, and establishes the tone for the rest of medieval period. While faith in God would always be the primary source of revelation or truth, God had given humanity the power of reason and hence the ability to discover that which had not been provided by revelation. For Aquinas, there could be no inherent contradiction between the two, and basically sought to compartmentalize the worlds of philosophy and religion. He effectively locks medieval knowledge within theological chains, placing severe constraints on its ultimate development.

This intellectual confinement is perhaps aptly demonstrated in the university dissertation process. A thesis would be proposed by originator and debated with experts, and is well characterized by Pete of Abelard's *'sic et non'* process: thesis, objections, solution. One could expect this to have introduced original investigation, but in actual practice the system devolved into mere repetition. "Proofs" were reduced to allusions to canon, in that only those ideas which already existed in the Catholic canon were accepted. The entire process of accepting new theses in higher education was reduced to encyclopedic referencing—a system which by definition was inevitably doomed to stagnate and helped reinforce an already conservative and repressive environment.

There are always the exceptions, thankfully, as in the case of Jean Buridan and Nicole Oresme who in 1250 further developed Philoponus's original notion of impetus.

The Aristotelian conception of motion had distinct features. Uniform motion was the norm, in that there was little variation in speed, except with regard to the medium. All motion was considered with regard to its destination; as objects moved closer to their natural place, they gained speed. Finally, there was the constant requirement of an external mover, as no unnatural motion could exist without a direct external force. There were many problems with this system, the main one of which was that of falling bodies. Falling bodies showed non uniform motion (acceleration). Also, if projectiles were pushed by air, in theory these should never stop. Many of these difficulties could be resolved by the reinterpretation motion as a result of impetus.

Impetus could be defined as a phenomenon akin to heat, then regarded as a substance that was transferred to an object but which gradually wore away. As with Giordano Bruno, this was a small but more significant step than might appear at face value. Not only did it refocus the understanding of motion by placing attention on the originating source of that motion, but it also eliminated the 'ghost in the machine' phenomenon; the constant presence of 'external movers of the Aristotelian cosmology were no longer needed. While the development of impetus did not represent a radical change in methodology, as its thought experiments could just as easily have been conducted by Greek thinkers, it did represent a substantial departure from the traditional understanding of movement, or what today we regard as the science of physics.

The opposition to scholasticism can also be regarded as one of the principal features of Renaissance natural philosophy, as shown by Jean van Helmont, Tomasino de Campanella, Francis Bacon, and Paracelsus.

Four case studies

Jean Baptiste van Helmont (1577-1644) is today considered one of the founding fathers of chemistry. Perhaps because of his anti-scholastic attitude, van Helmont was lured into a debate with a Jesuit over a rather trivial topic. The Jesuit argued that one could cure a person with a gunshot by medically treating the gun that had been used to shoot the person. Van Helmont naturally rejected such absurd claims. To read the book of nature, one had to turn directly to it rather than to secondary representations, he noted. Such was his disdain of traditional scholasticism, that he rejected the medical degree awarded upon graduation, as he had not learnt anything during the course of his university studies.

As one would imagine, van Helmont was denounced by his faculty of medicine to the Inquisition in 1623, and was imprisoned shortly thereafter. Upon his release in 1636, van Helmont is kept under house arrest until his death in 1644. His son faithfully publishes his father's posthumous writings in *Ortus Medicinae* (1648), which obtain a great deal of popularity. By 1707, twelve editions in five languages of the work had been published. van Helmont's arrest would be ironic as he is formally charged for comments made about Robert Fludd who had criticized Kepler's perfect solids. Tragically, Fludd had also been a hermeticist chemist, whom would also suffer arrest and witness the destruction of his entire book collection in England.

Tomasino Campanella is an open alchemical hermeticist who openly called for an overthrow of scholastic institutions in order to achieve a scientific utopia. His plot to overthrow the Spanish crown in Naples leads to his consequent imprisonment and torture for 27 years. While in jail, Campanella writes *City of the Sun*, which is published posthumously. He calls for a city ruled by three hermetic priests, bounded by concentric walls with writing and samples on wall. It took an enormous deal of courage to be an intellectual at the time, as public opinions could obviously lead to one's state-sanctioned assassination.

Perhaps the best examples of the influence of hermeticism in the development of science was Paracelsus (1493-1541). Born a bastard son to a nobleman and peasant girl, his original name of "Bombast von Hohenheimlast" was gradually expanded to become the horrifically long "Philippus Aureolus Throphrastus Bombastus von Hohenheim." His father eventually moves to a Frugger mining village, Jakob Frugger being one of the wealthiest men in Europe by his use of the double accounting procedure, allowing him to better track of his funds.

Paracelsus's first treatise is in the field today called 'occupational medicine,' in that he focuses on the maladies specific to miners. Some of these included toxic poisoning, and would later come to include the 'bends'—the same affliction scuba divers get when doing too rapid underwater ascents or depressurization. As when a soda can is opened, the sudden change in body pressure gasifies the blood's nitrogen, forming bubbles which can dangerously lodge in the brain. The more overweight the person, the greater the propensity towards the affliction.

His personality is oddly similar to that of Giordano Bruno's, but while Paracelsus routinely insults people, he was also good at obtaining followers. His insults are rather humorous: 'my beard has more knowledge than your head' or 'my neck hairs have more wisdom than your colleges.' Underlying these insults, however, is a strong sense of his philosophical stance.

During his wide ranging travels throughout Europe, he becomes one of the main proponents of anti-scholasticism. While at a university, he burns medical books in a public bonfire, an act which is not well received by its administration and for which he is dismissed.

Paracelsus calls for a complete overhaul of European academia, launching a direct attack on Aristotelian scholasticism. Educational institutions needed to be demolished and built back up from the ground. The animosity towards scholasticism cannot be emphasized enough.

> Do you think that because you have Avicenna and Savonarola, Valescus, and Vigo that you therefore know everything? That is but a beginning.... That which Pliny, Dioscorides, etc. have written of herbs they have not tested, they have learned it from noble persons who knew much about their virtues and then with their smooth chatter have made books about it.... Test it and it is true. But you do not know it is true,--you cannot carry it out, you cannot put to proof your author's writings. You who boast yourselves *Doctores* are but beginners.

To receive proper medical training, Paracelsus argued that one needed to leave the artificial environment of the university, burn all scholastic books owned, and travel throughout Europe to explore its natural resources, such as the herbal properties of local plants or the chemical effects of local minerals. A direct interaction with nature would provide more true knowledge than any philosopher's book. Many, as Peter Severinus, physician to the king of Denmark, follow his advice.

Paracelsus viewed Aristotle's notions of disease with utter disdain. Rather than an imbalance of the four bodily humors, disease for Paracelsus originated from an external seed, inevitably drawn from air, food, or drink, which lodged itself in an organ and festered. His medical vision of the body alluded to the 'triad' as the key components of the universe and man: sulfur, mercury, and salt, more akin to physical states than actual substances.

At the core of his medical treatment was the doctrine of signatures, an inherently hermeticist doctrine. There was a correspondence between men and universe, a microcosm/ macrocosm parallel, whereby the objects in universe revealed features which would best suited for the respective diseases. Nature was itself a manual, and all the physician had to do was identify these signatures to locate the most beneficial medicine for a patient. Paracelsus attacks the common scholastic practice of bloodletting, arguing that blood contained vital spirits—one of the first to call into question this practice. The practice of draining the patients' blood was tantamount to assassination, according to Paracelsus. It is somewhat hard to see how revolutionary this claim was at the time, as today we agree with its presumptions; in 16th century Europe, however, it went against the grain of the most widely used medical treatments.

In contrast to Giordano Bruno or Michel Servetus, Paracelsus is never captured or killed by the Catholic Church, as his journeys never took him outside of northern Europe. He remained safely guarded, deep within Protestant lands. Unfortunately, he also never formally published while alive in spite of actually

94

having written a great deal of material. Had it remained that way, his impact would have been severely restricted as da Vinci's. His greatest influence is hence a result of his followers' perseverance, the 'Paracelsians.' His extant writings are published posthumously in 1550.

Yet Paracelsus was no armchair philosopher, and contributed directly to medicine by his active incorporation of inorganic chemical materials, partly in an attempt to cure one of the most ravaging diseases from the Americas: syphilis.

Syphilis is a sexually transmitted disease that reaches Europe as a result of its encounter with the Americas. As with the first expansion of any illness, its impact and morality rates were horrific: 30% of victims—a rate which matched the first entry of the bubonic plague during 14th century Europe in the period known as the Black Death. The illness spread particularly quickly in Naples during the 1494 French invasion, hence acquiring the nickname 'French disease.' As soldiers left the conflict, they spread it throughout the rest of Europe, creating a wide-ranging health crises. Syphilis was no mere 'itch.' As its state progressed, it eventually began affecting the cardiovascular and nervous systems of its victim. The consequent bodily deformations, blindness, dementia were horrific to witness, and its sufferers were more often than not socially ostracized, further compounding the pain of the disease.

One early treatment was the use of guaiacum wood from the Dominican Republic, but which had no medicinal properties whatsoever. By contrast, Paracelsus called for the use of non-traditional medicines, influenced also by Germanic folk beliefs. Whereas scholasticism advocated the principle that 'opposites cure,' German folk systems called for a 'likes cure' approach. Violent diseases themselves called for violent treatments, and Paracelsus beings promoting the use of mercury as a treatment for syphilis. Although not entirely original in his claim, first used in Arabic medicine, his systematization of this approach popularized its use. The treatment is not as toxic as it might actually appear to us today; it is certainly the case that the cure was often as worse as the disease. Paracelsus also used sulfuric acid in his treatments—extremely diluted quantities of sulfuric acid which otherwise would have killed his patients.

The main benefit of this approach was ultimately its impact on European medical philosophy. While the extremity of his measures were often attacked by medics of the era, there eventually emerged a compromise between the two. The use of nonherbals was recognized to have positive medical effects on the human body. The European materia medica was thus greatly expanded as a result, and the active exploration of substances not typically considered to be medicinal was initiated. This substantive change in medical philosophy helps establish chemistry as a recognized academic field and profession, becoming a requisite for medical degrees. A chemical chair thereafter also becomes common to scientific institutions as the *Jardin dur Rois* in 1626.

Yet Paracelsus had been so critical of traditional doctors, that a systematic attempt to ruin his reputation was undertaken. Since he could not be hung, his persona would be crucified. Claims began appearing that Paracelsus had committed plagiarism. The alleged writing of a Benedictine monk Basilius Valentinus were published by Johan Tholde. So close was Valentinus's approach to that of Paracelsus, that Paracelsus historically obtains the reputation of being a plagiarist. The claim was actually held until 1885, when the historian H. Kopp detects the existence of anachronisms in the writings of "Valentinus," such as the use of term "French disease" which could not have appeared prior to the age of Columbus. Noxious financial interests and disingenuous religious attacks are not unique to the modern era.

Francis Bacon (1561-1626) has not received a favorable historical interpretation as of late. Viewed only in context of his position as Grand Inquisitor to the British crown where he tortured people in order to obtain information. Consequently, his call for 'knowledge is power' is negatively contextualized as the 'torturing' nature into revealing her secrets—information which is then used to take advantage of her. This feminist distortion of his views lack the nuance with which Bacon regarded knowledge; his *Essays*, as those of Michel de Montaigne, make for worthwhile reading.

Bacon made an appeal for going beyond the Greeks, as evidence in his time suggested that Europeans had already crossed the '*Ne Plus Ultra*' door. The Greeks lacked inventions as the compass, gunpowder or the printing press of his day, and he claimed these boundaries could be pushed even further. Bacon was well aware of hermetic texts, and actually criticizes alchemical practitioners for their secretiveness. Yet he saw alchemy as a useful science whose knowledge of metals was of an inherent social value. He also criticizes William Gilbert for making a religion out of the lodestone in the work, *New Philosophy of the Sublunary World* (1651). While these critiques might suggest Bacon's opposition to hermetic philosophy, it was of greater influence than might be apparent.

In 1614 and 1615, two texts appeared which were a part of the *Rosicrucian manifesto*, specifically *Fama fraternitas* (1614) and *Confesso* (1615). While we may discard the superfluous purpose of the manifesto, the treatise contained models of scientific societies which influence Bacon in his calls for scientific reform. They described travels to the Middle East (Fez) where communities of scholars had been working together for a common purpose. The author noted the impressive outcomes which could be achieved from such a collaboration of minds, while emphasizing the need to replace universities with new institution as well as abandon Aristotle and Galen.

In *The Advancement of Learning* (1605), *Novum Organon* (1620), and *New Atlantis* (1627), Bacon makes a similar appeal for a society of natural philosophers—a body of work which would directly influence the formation of Royal Society of London in 1662. The Royal Society would rise to a preeminent position in scientific practice, publishing one of the first and longest running scientific journals, the *Philosophical Transactions* (*Phil Trans* for short). The *Phil Trans* reads almost like one of today's scientific journals, printed in easy to read type.

For Bacon, the father of inductivism, all men could contribute to the scientific enterprise, even if they varied enormously in their abilities and aptitudes. Science was to be built up by associations of men who went out to the world and compiled facts. He provides an interesting comparison to the world of insects which is rather illustrative of his intentions. He did not propose an ant model as this suggested a mere compilation of incoherent data. He equated spiders to scholasticism and its scholars who mindlessly built complex systems without going out into the world. His chosen model was that of the bees. It was a true model of science in that external information was first gathered and compiled, to be later processed by the few who had the ability so understand and synthesize this data to create new knowledge.

This model is strikingly similar to that of Campanella's, who proposed there would be a reduced group of scholars onto which all information would be fed. It would be these chosen few who would be able to provide a good interpretation of the data. Although hierarchical, both models were also inclusive in that they did not entirely restrict participation in the activity we now call science.

Part III: Revolution

Copernicus

Ingrate Revolutionary

COPERNICUS FORMALLY INITIATES THE SCIENTIFIC Revolution. His *De Revolucionibus Orbitum Coelestum* in 1543 can be easily sloganed: the sun rests at the center of the universe. But, why then not use the phrase *Orbitum Mundi* in the title, as it would have been truer to its topic? The word "coelestum" suggests 'skies' as opposed to the more accurate term 'world.'

It goes without saying that Copernicus initiates a profound shift in the Western world view. That Earth and man, however, were no longer at the center of the universe was afflicted by many troubling issues, making the notion problematic and difficult to accept. Why don't objects fly off the Earth? The resolution of this disparity between the physics of the Earth and the cosmology of the universe would ultimately result in a 'Scientific Revolution,' ushering a radical a change in the way we understand nature, less ambitious in scope but truer to its subject

Johannes Kepler was one of Copernicus's most ardent defenders, intensely arguing that the Copernican model was not just a model, but rather the way the universe was actually shaped. Kepler had briefly worked for Tycho Brahe, who produced the best unaided observations at time, and his three 'laws' set the foundation to Newton's work. Both Kepler and Newton are more closely scientifically related that one might suppose in that Kepler could have become 'Newton.' Like his British counterpart, he had substantive mathematical abilities and had the discipline to continually work on a single problem.

It would be ideal to characterize the protagonists of the Scientific Revolution as heroes and provide a mythologized history of our characters making noble and profound pronouncements. History can have a propagandistic character; it leaves an indelible impression on students which is rarely explored or questioned. The 'tyranny of the default' in education is thus a similar phenomenon to that which occurs in the world of computers; few bother to explore or change their computer's default settings.

Copernicus was actually contemporary with renown figures of the Renaissance: Leonardo da Vinci, Desiderius Erasmus, and Niccolo Machiavelli. Erasmus had been the great Renaissance humanist who, upon producing a translation of an original Greek Bible, showed how corrupted over time it had become.

But, in truth, men are as complex as their ideas. Copernicus's model still retained many Ptolemaic devices as epicycles and eccentrics, being more of a theoretical approximation. Copernicus the man was also not a heroic figure at all. Timid and shy, averse to social interaction, he is a Catholic official with a mistress "Ana." He also betrays some of his most loyal supporters, and abandons his brother when he was most needed. Copernicus is no hero.

Nonetheless, some individuals cannot be understood for the life they led. The character of the man must be separated from his intellectual contribution, as one is not necessarily a good indication of the other. Imperfect men can achieve greatness in a subset of their lives; Nobel Prize winners are not necessarily saints. This is not to excuse vice in our historical actors, but simply to separate their achievements from their failures. As Einstein noted, the focus of their lives should reside with what was written on the page rather than that which occurred outside of it.

Life of Copernicus

The life of Copernicus would be carried out almost exclusively in the Prussia district of Ermland (Varnia), today Poland, in the town of Frauenburg. He was not an adventurous fellow, making only a few trips outside of the region to do his university studies and help out his uncle. With a few exceptions, he tended to be a 'stay at home guy.' Born in 1473 as "Nicolas Koppernigk," his father migrated to region as a copper wholesaler seeking better business opportunities. The family consisted of two girls and two boys; the elder of the two being his brother Andreaus. His father dies shortly after their arrival, and Copernicus the boy and his three siblings are promptly adopted by his maternal uncle, "Uncle Lucas."

Copernicus was his uncle's favorite nephew, possibly because they had opposite personalities. The uncle was very outgoing, extremely so. Lucas marries off one of Copernicus' sisters, and sends the other one to a nunnery. Uncle Lucas thereafter plays an important role in Copernicus's life. As Bishop of Ermland, he was an important figure, and would have been a prince or ruler had he lived in the region now called Italy. He looks over Copernicus, making sure of his well-being. As a canon in Ermland was dying, Lucas immediately sends for Copernicus and his brother at the University of Cracow so they could obtain its benefits. A canonship guaranteed a person's lifelong income, serving as a local administrator with judicial and legislative authorities. However, the canon dies too early and the brothers briefly lose their chance of economic comfort. But, just as unexpectedly, another canon soon dies and both quickly take advantage of the situation.

Yet the importance of Uncle Lucas went much further. He helped to keep Ermland an autonomous region, in spite of being geographically surrounded by two powerful forces. On one side of the lands resided the Teutonic Knights, a legacy of medieval period, with whom Lucas was constantly in conflict. It is said that Lucas helps end the order, tending to support the Polish king and playing off one powerful group against the other. Lucas also tries to establish a university in his homeland.

While attending the University of Cracow (Poland), Nicolas begins his astronomical studies, acquiring a good reputation in the field. After his canonship, he is given a grant to further continue his studies with the intended purpose of becoming physician to his uncle. Copernicus was actually known in England as a physician rather than as an astronomer.

The personalities of the two brothers greatly diverged. While Nicolas is quiet and serious in his studies, his older brother could have easily become a fraternity president today. When asked why he should be allowed into the university, Andreas replies that he should be admitted for having the capacity to continue his studies. "I can do the work," he seems to declare, lacking any visible goal or vocation. Andreas's temperament is closer to that of his uncle's than his brother's. Nicolas is thus caught between two overbearing figures, and astronomy becomes his source of personal liberty.

Both siblings in Italy attend the University of Bologna as well as the famous University of Padua. Nicolas's university education would be general well-rounded, including mathematics, astronomy, Greek, and medicine. While at Padua, his astronomy professor views him more as an assistant than a student, recognizing Nicolas's skills and knowledge level. He then obtains a law degree at a nearby university in order to avoid the lavish parties he was expected to give upon law school graduation in Padua. After ten years abroad, Nicolas returns home.

First living in Heiselberg Castle, his official position is that of physician to Uncle Lucas. In fact, Nicolas serves more as his uncle's diplomat and bureaucrat, assisting in the preparation of letters and documents or by accompanying him to foreign locations. While at Heiselberg Castle, Copernicus makes no astronomical observations.

Copernicus is actually a very poor observer. In spite of having a good income at his disposal, Copernicus makes all of his instruments himself, such as a "Jacobs ladder" to measure star positions. His instruments are so crude, that his measurements contained a margin of error of 10 minutes of arc, or one third the size of moon. To make things worse, he defines his observational location as Danzing, 42 miles to the east of his actual position. Copernicus is also not a persistent observer, making a total of only 70 observations in his career. *De Revolucionibus* would use only 27 of the author's own observations. Worst still, when he observes a planet, he does not observe it across the entire path. He saw himself more as a mathematician, basing his data principally on the records of ancients Chaldeans, Greeks, and Arabs. These impertinent facts make his revolutionary role an unexpected one.

Nicolas does help his uncle out with two important efforts. The uncle tries to get the Teutonic Knights to go on a Crusade against the Turks, but is unable to do so. So great was the constant conflict with them, that their ill will towards him was widely well known. Recognizing the powerful role educational institutions play in national life and culture, Uncle Lucas also tries to get a Prussian University established, but is also unsuccessful in this attempt. He dies under strange circumstances. After going to Cracow for the wedding of a Polish prince, Uncle Lucas gets food poisoning and dies on route back to Ermland in 1512, at the age of 65. Although relatively young, Copernicus officially assumes his role as Canon of Ermland. His uncle's sudden death might have contributed to Copernicus's secretive and guarded character.

His brother Andreaus also falls victim to illness; during his time in Italy he contracts a disease. It is not known whether the affliction was syphilis or leprosy, but in either case the consequent disfigurement led to his social ostracism. The local Church, which had so recently issued him a canonship, now wished to retract its offer. Andreas naturally opposed the loss of income, and would stand disfigured at church so as to gain sympathy and put political pressure on local religious authorities. While they ultimately agreed to continue his annuity, it is only granted on the condition of exile; Andreas would have to leave Ermland for good. It is known that he goes to Rome, where he dies in obscurity.

Nicolas never defends his brother.

During his lifetime, Copernicus does not publish many books: a translation of Theophylactus's *Epistles* in 1509, his *Commentarialus* of 1514, and *De Revolucionibus*. His first work is a translation of a Byzantine work of Greek origin, following the humanist scholarly code which insisted that all good humanists translate some arcane piece of work regardless of its importance. Copernicus dedicates his translation to Lucas, noting that since the piece was both serious and lighthearted, it would be of broad general appeal. Like a doctor who combines sugar to ease the swallowing of a bitter pill, the book's low humor would ease the acquisition of its bitter lessons. While the translation of was little importance, it helped mark his "humanist" credentials at home and abroad.

His privately written *Commentariolus*, by contrast was original and important, and was distributed internally amongst acquaintances. It laid out his basic ideas on the structure of the universe that would be fastidiously elaborated in his third work. The manuscript contained seven principle tenets, one of which was that the eight sphere of the universe was very distant, making the distance between Earth and sun so miniscule so as to make it appear as a point.

Commentariolus surprisingly had a significant impact as knowledge of its ideas were first informally diffused throughout Europe, making it one of the most poorly kept secrets in the history of science. Upon hearing of the ideas, Martin Luther casually dismisses them over dinner. Pope Leo X's personal secretary presents a brief summary of Copernicus's ideas before a small audience.

Although *De Revolucionibus Orbitum Coelestum* is published in 1543, it was actually first written in 1530 but was locked away. Upon the draft's completion, it was only shown to a few select individuals during the author's lifetime.

103

Copernicus's later years were rather uneventful, again making him an 'anti-hero' figure of sorts. He had developed a good friendship with Tiedemann Giese, who although seven years younger had a more outgoing personality. Giese recognized his friend's ingenuity and brightness, and was one of the first to try to convince him to publish his important work—but fails in the attempt.

In 1517 Luther publishes his 97 thesis, which spreads like wild fire in 5 years and drastically alters Europe's politico-religious landscape and the Church's sphere of influence. The Teutonic Knights adopt the new Protestant religion, whereas the Polish royalty retain their old allegiance. The Reformation sets church against church, and town against town. Giese himself publishes a tract allegedly attacking Protestant claim, but whose principal aim was the preaching of toleration. "I will not battle," writes Giese. While one can detect a certain amount of anxiety in Copernicus over the issue, Giese's entry into the era's polemical battles is politically successful. The work helps propel him to a bishopric in Kulmas, ascending within the Church hierarchy.

Copernicus for his part kept quietly busy in his role as a canon, serving in that improper mixture of judicial and legislative authorities as was wont to happen at the time. His position leads him to attend to peasant complaints, and Copernicus is actually very meticulous at his hearings, writing down every agreement established with the peasants, apparently conscientious towards all in the region, regardless of social rank or wealth. Towards the end of a bureaucratic and boring life, something unexpectedly happens.

He is visited by Georg Joaquim Rheticus in 1539.

Publication of *De Revolucionibus*

Rheticus, of all individuals, finally gets Copernicus to publish his magnum opus. However, the book appears in 1543, the year Copernicus dies. It is said that the book was placed in his hands before tragically dying—melodramatizing and distorting the incidents that occurred.

Copernicus had been in poor physical condition during his later years. As he tended to prefer solace, he did not have many acquaintances to aid him in the weakened state that naturally comes with old age. Copernicus has a brain hemorrhage which paralyzes him on one side of body. Giese, worried about his old friend, is able to procure an assistant for him. It is somewhat unclear that he actually ever saw the book.

Part of the implicit tension in the melodramatic characterization is the suggestion that it might have caused his death. While Copernicus believed his system to represent the organization of the universe as it actually existed, the preface had been altered to suggest otherwise. There can be no doubt that, had he read the preface, he would have strongly disagreed with it. To add further pain to the injury, the preface was presented as if it had actually been written by Copernicus himself. Heliocentrism, the preface read, should only be seen as a mathematical tool and not as the way the cosmos actually existed.

However, none of this should have been unexpected to Copernicus. The real question that emerges is why Copernicus took some 30 years to publish his ideas. Given the state of insalubrity and short lifespan of his period, this delay meant that it would have likely never seen the light of day upon the author's early death.

We can turn to the author to help obtain some idea of his motivations. A *Letter from Lysis* of (1499) is mentioned in the *De Revolucionibus's* preface. It was a text of Hellenistic origins, combining Greek and Egyptian ideas, which had been picked up by Copernicus during his days as a student at Padua. The text presents the ideas of Pythagorean ritualism associated with intellectual discovery.

One important message of Lysis was that one should not share knowledge which had been costly to obtain, as third parties were unaware of the effort involved and would not appreciate it or understand its significance. Quite the contrary, implying that it would likely either be misunderstood or mocked; this atmosphere, in turn, would detrimentally affect the circumstances needed for mild contemplation.

> For it is not proper to divulge to all and sundry what we have acquired with such great effort... Let us remember how long it took us to purify our minds of their stains... Thick and dark forests cover the minds and hearts of those who have not become initiated in the proper manner, and disturb the mild contemplation of ideas.

Copernicus was overwhelmed by fear of ridicule. The ignorant in his view would not understand his efforts. These worries, however, had been greatly exaggerated. Arthur Koestler shows that there had been only three 'attacks' on the work of Copernicus. The first was Martin Luther's criticism, which had been mentioned only in a private setting; no formal critique was ever published by him. The second was a minor carnival performance mocking Copernicus, of little scientific merit. Lastly, the third was a private letter written by Melchantor which was never released to the public.

There were other reasons why Copernicus should have been concerned, however.

As we have noted, book publication was new to Europe, and implied a rapid spread of ideas, more so than had been previously experienced in world history. Publication meant 'instant' exposure, akin to the use of Twitter or Facebook in our day. Copernicus was aware that most would not understand his modification of the world system, tending to unthinkingly criticize that which wasn't understood. We might also note that Copernicus was a man of some stature in Ermland (Poland). Holding a position of power, influence, and guaranteed income, Copernicus therefore potentially had a lot to lose. Finally, the social and cultural environment in Europe had drastically changed as a result of the Reformation. When Copernicus had been a young university student, Europe was at the height of the Renaissance, and Copernicus himself imbibed its ideas at elite Italian centers. However, after the Protestant Revolution, the intellectual atmosphere had radically changed during his lifetime, becoming more tense and restrictive.

Whatever the reasons for its delay, *De Revolucionibus's* eventual publication was ultimately due to Rheticus. An outstanding student of Melcantor at Wittenberg, he quickly becomes a professor of mathematics and physics in the same university. Upon being made aware of the existence of the Copernican model, Rheticus decides to try his luck and travels deep into 'enemy territory' to get it published. He visits Ermland in the summer of 1539. Initially planning on spending only 2 weeks, he ultimately remains for two years.

There had actually been two Catholic prohibitions issued in the region, prior to and during Rheticus's visit. The first was issued in March, calling for all Protestants to leave the district. The second and more aggressive one called for the burning of Protestant books. In spite of these state sanctions, Rheticus actually never had a problem moving about the town, and does eventually thank the town for its enormous generosity, claiming that it would become the next Athens for its open-mindedness.

When Rheticus finally meets Copernicus for first time, Copernicus was already 66 years old, and would have only a few more years to live. Although Giese is no longer his student, living far away in his own bishopric, Rheticus's visit must have been of some relief to Nicolas. Rheticus, 25 years old at the time, becomes the only student that Copernicus ever had. Naturally, Rheticus makes a pressing argument for publication. In his reticence, Copernicus decides to delay, arguing that he would need to consult with Giese before making any decisions. He might have felt uncomfortable in the company of Rheticus, a Protestant homosexual.

Both leave and visit Giese. The three, Giese, Rheticus and Copernicus, openly talk about the issues and concerns. Recognizing the historical importance of these deliberations, Giese writes and preserves a detailed description of the events.

Copernicus at first was only willing to publish new planetary tables without the mathematical modeling underlying the new figures. Giese clearly lets him know that the suggestion was a bad idea, emphasizing that his friend needed to do both. Finally, an agreement is reached in that Rheticus would publish a summary of the work, allowing for Copernicus to evaluate its reception before issuing a more comprehensive piece. Whereas Rheticus sticks out his head, Copernicus cowers in a shell.

Rheticus spends the next ten weeks with Copernicus, working on a summary manuscript of the larger work. It becomes a very exciting but trying time. Rheticus needed to make sense of *De Revolucionibus*, and does not merely publish a bullet summary. It is said that he briefly goes mad when studying Copernicus's analysis of Mars, at one point banging his head against a wall and falling unconscious. The *Narratio Prima* is finally published in 1540, becoming an immediate hit—giving Copernicus a notion of the positive reception his book would theoretically obtain. Copernicus finally decides to go ahead with the publication.

To this end, Rheticus begins copying by hand Copernicus's manuscript, but is forced to return to Wittenberg to resume his teaching duties. Upon his return, he is made Dean of his faculty. When the semester ends, he heads back to Nuremberg to oversee the publication of the work. Rumors in Wittenberg had begun circulating about his aberrant sexual proclivities, which lead its faculty to 'free' him for a position at Leipzig, to which Rheticus had previously solicited. Rheticus's sudden departure from Nuremberg meant that the responsibility for Copernicus's text would fall on the hands of Andreas Osiander, a Lutheran mathematician.

Osiander commits two errors. He places a preface arguing that book was only a mathematical instrument for calculation, and was not meant to represent the order of the cosmos. This preface, however, is not published under Osiander's name, giving the incorrect impression that it had been written by Copernicus himself. This error would stand until an 1854 publication, although Kepler became aware of it in 1609 when he reads Copernicus's letters before they were destroyed. Osiander also changes the title of the book from *"Mundi"* to *"Coelestum,"* marking a distinct shift in the tone of the original.

However, these changes should not have surprised Copernicus at all. Osiander was not unknown to Copernicus, with whom he had previously consulted about the very book itself, and whom had offered the very same observations prior to its publication. Osiander believed that such changes would reduce the severity of attacks upon the work by diminishing its abrupt revolutionary character, and hence serving as a psychological transition which would allow the community to gradually come to Copernicus's side.

There can be no doubt, however, that Copernicus did not agree with this strategy and point of view, which is why the final result does have an air of tragedy about it.

One would naturally assume for Rheticus to have become some sort of intellectual 'heir' to Copernicus, akin to the various teacher-student pairs in history, as Socrates and Plato. In fact, no such transition came close to happening. Rheticus actually becomes an anti-Copernican, rejecting the very work to which he had sacrificed so many years of his life. Rheticus oddly began preaching a return to Ptolemy. Why?

While Rheticus had been overseeing publication of the work, Copernicus unwisely sends him a copy of the acknowledgements to be included. As is often the case, the author dedicates the work to the friends who had promoted the publication, in particular the Cardinal of Capua as well as his close friend Geise. Yet Copernicus never mentions the very person who had been the central force behind its publication in the first place: Rheticus.

This must have been a tremendous letdown. In fact, Rheticus's life after these events takes a turn for the worse. His homosexual proclivities come to haunt him in Protestant Europe, and he is eventually forced to leave for Rome to study medicine. Fortunately, as his experience with Copernicus, his work is rediscovered by a young student Valentine Otho, a recognition which briefly offers Rheticus some happiness in his later years.

It was thus how one of the world's most important revolution began: a homosexual scholar publishing the secretive works of a neurotic Catholic from Poland in the heart of Protestant Germany during in the middle of the emerging European religious wars. The relations between individuals are not necessarily defined by the categories or groups to which they belong.

Heliocentrism

Copernicus's main argument in his magnum opus is a conservative one: Ptolemy had violated the cardinal rule of Greek uniformity of motion, particularly so by his use of the equant. As with all Renaissance men, Copernicus sought to restore astronomy back to its authentic Greek origins based on the perfect circle. He was also influenced by the Pythagorean notion of a central fire, recognizing the proper role held by the sun as a universal life giver. Ptolemy's work had become much too complicated a model.

At its core, different 'models' existed for each planet, growing overly complicated and inaccurate, with 60-70 spheres in total. One of the key 'selling points' of the Copernican model was its simplicity, by reducing the total required sphere to 34. By contrast to prior heliocentric notions, Copernicus works out the full mathematical details of the structure, which is his fundamental contribution. That he had been an able mathematician and actually well regarded by his peers added weight to this proposal. As Vesalius with Galen, Copernicus follows the same chapter division that Ptolemy had followed in the *Almagest*, again suggestive of its conservative purpose.

However, the Copernican model also grew in complexity, being forced to incorporate a greater number of Ptolemaic mechanisms that he had originally expected. The final model is thus actually different from what one might imagine. Ptolemy's later models had placed Venus and Mars rotating around the sun, as these were never seen in the middle of the night or day. The Copernican model had the advantage of explaining why this pattern was not anomalous at all, but rather the norm.

Yet the Copernican model was not without its own problems, the biggest issue being that of stellar parallax. As the Earth rotates around sun, according to the model, one should see a shift in the position of stars akin to the displacement of objects when we alternate closing our eyes—which incidentally is what provides depth and perspective to vision.

It was for this reason that Copernicus makes the supposition wherein the distance between the Earth and sun are as that to a fixed point relative to distance of stars: stellar parallax could not be seen, and its absence could only be accounted for by presuming that stellar objects were immensely distant form the Earth. This proposition, in turn, led to an enormous increase in the entire size of universe, from 2,000 to 20,000 terrestrial radia. The notion of such as vast universe was to some as preposterous as it was inconceivable.

Another problem implicit in the Copernican model was that of motion. Why did objects not spin off or fly off the Earth if it was moving so quickly? One might counter this question with another one: how could entire universe spin around the Earth at such enormous speeds? In essence, the underlying debates pertained to the factors which made motion 'natural' in both cases.

The eight circle of the world was that of stars, which rotate freely as there was no friction. Made up of aether, natural motion was circular, and no effort was required to make the universe spin—with the exception of the unmoved mover.

God had only to turn once and all else would follow afterwards. Yet, if the crystalline spheres literally held up the planets, what would keep these in place when shown to be non-existent?

Copernicus's solution to these suggestions emerged naturally from his 'geometer's eye.' For him, everything tended towards a sphere, as water which formed droplets. Gravity, in turn, was the natural outcome of the sphericity of objects in that all would tend to move towards each other. The natural motion of spheres was rotation.

These replies were obviously not satisfactory, but one cannot help but notice how conservative they were. Again, Copernicus is seeking to reform astronomy by a return to the perfect forms of Greek philosophy. Oddly, Copernicus's geometer's arguments seem to have a strange counterpart to contemporary notions of space-time, whereby forces (gravity) acquire their characteristics because of the 'geometrical' quality of space. This particular answer, however, was not accepted at the time, not even by his own followers, which raises the question: how well was Copernicus's last publication received?

Reception of Copernicanism

The reception of his heliocentric model in England is rather favorable. Publicly, Copernicus is supported by Thomas Digges, who issues a formal acceptance of the theory as a result of his macrocosm and microcosm hermetic ideals.

While it has been argued that Francis Bacon rejected Copernicanism, this over-simplifies his reaction. It is true that Bacon did not formally believe in the exact model, preferring instead that of Brahe's hybrid modell: although it maintained the Earth at the center of the universe, all other planets revolved around the Sun, which in turn was revolving around the Earth. In fact, the Brahean model would be the predominant cosmology during first half of the seventeenth century, suggesting that one could have one's cake and eat it too. One of Brahe's key contributions of was his rejection of the crystalline spheres. His popular model, however, had many problems as the orbits of Mars and Venus around the sun would collide with the Earth's. The issue, however, was temporarily ignored.

Bacon's actual stance, however, was reserved. He recognized that it was yet to be determined whether stellar objects are "as so many islands in an immense sea," and thus remained relatively open to whatever model was the more appropriate one.

One of his contemporaries, which Bacon tended to lightly mock, was William Gilbert who not only supported Copernicanism but also helped to provide a mechanism by which universe did not collapse in on itself. For him, planetary bodies were magnets influencing each other at a distance, and did not require efficient motion for their operation. Magnetism provided a temporary celestial scaffolding for the Copernican universe. Although criticized by Bacon for making a religion out of his philosophy, it was pragmatically adopted by leading contributors as Kepler and Galileo as an intermediate mechanism of celestial physics, allowed for the continued productive elaboration of Copernican model.

Although initially rejected by Protestantism, receiving the accusation of "bibliolatry" by Martin's adherents, its gradual acceptance of Copernicanism over the long run would actually modify the character of Protestantism. The sect would become more elastic and liberal in temperament over time, almost the exact opposite of the dynamic which occurs in Catholicism. The two principal Christian divisions were mirror images of each other.

The Catholic Church did not place *De Revolucionibus* in its Index for 72 years (1615) as it was not initially perceived as a threat that would lead to a questioning of its theology. As previously suggested, the Catholic Church had imbibed the spirit of the Renaissance during the sixteenth century, and a long list of Popes actually stimulated the discussion of new ideas. It is also important to remember that Copernicus had been a Catholic official. Pope Leo X's personal secretary presented ideas of Copernicus to a small group in 1522. His successor, Pope Clement VII also encouraged open attitude. Although the next Pope Paul III continued this trend, his attitude contained a slightly more conservative bent: as long as the views did not directly question the Church, these could be discussed in public. Calendrical reform actually provides an important stimulus to the Catholic study of astronomy, given that the never-ending problem of splitting up year into correct units proved to be a very difficult task. Under uneven chronological units, however minor, any calendar will gradually 'sink out of time.' Copernicus had even been asked to participate in the 1514 calendrical reform, but rejected the invitation arguing that his model needed to be improved first before could change. While this might be defined as an impolite dismissal of the Church, Copernicus commonly refused the participate in 'public events' throughout his life.

As long as power of Catholic Church was not directly menaced, the institution perceived intellectual innovations as benign. All institutions, once created, become 'political entities' with an inherent interest in preserving their power and continued existence. This is a social dynamic valid for all organizations, religious or otherwise. Corporations, labor unions, government entities can all can be seen as vested political institutions ultimately focused on their self-preservation. The Catholic Church had obtained a set of social benefits, and hence a vested interest that sought to maintain these—even if based on an obsolete rationale.

Tragically, the Catholic Church did not seem to realize that it had built a castle on a pile of sand, using Aristotelian notions to construct its enormous theological edifice. This would turn out to be a mistake; when Aristotleism was undermined, the entire castle upon which it had been based threatened to come tumbling down on itself. Darwinism would later suffer a similar quandary to the Church in that it was also subject to these same intellectual dynamics with regard to geology; improper foundations could lead to its intellectual collapse.

While Aristotleism can be clearly identified in the work of Dante, his model of the universe is not 'geocentric' but rather 'diablocentric': entire universe spun around the devil at center of the Earth, and hence illusion of a cave in popular

110

imagination. The four elements had been endowed with value structures: fire, being the most perfect, ascends to heaven, whereas Earth, the most base, ascends to hell. An unbroken gradient existed between the center and its outer shell, suggesting the notion of 'great chain of being' where everything had its proper place. So 'tight' was this interlinked chain, that that the slightest deviation, any chink in chain, would break entire structure.

The notion of change was inherently negative in this cosmos as it undermined hierarchy, both in the natural world as well as in the social world, along with the place of the Catholic Church in it. Its anti-evolutionist stance would continue throughout the Colonial period, clearly influencing the work of Giambattista Vico. The clash between the Catholic Church and Copernicanism was inevitable even if initially unnoticed.

The period is thus one of general recognition that Aristotelian system needed to change; it's world view had become outdated and obsolete. Most patent was its lack of agreement with many new facts and events. The first was the nova of 1572-4, which although clearly in heaven, it was rationalized by greatly expanding sublunar region. The comets of 1577 had a similar effect, increasing awareness of the disparity. The exploration of the Americas, and the discovery of new creatures not described by Aristotle also further degraded its worldview. Michael Maestlin began calling for a radical reformation of astronomy, and it is generally recognized that events of 1572-4 did more to change public perception than all of Copernicus's work. *De Revolucionibus* in fact could only be understood by few men, and hence its direct appeal was shallow. Fear of its public reception had been overblown by Copernicus.

One of Maestlin's students, Johannes Kepler, and another figure whom Maestlin had also greatly influenced, Tycho Brahe, would take these matters into their own hands.

111

Tycho Brahe And Johannes Kepler

The First Laws

JOHANNES KEPLER (1571-1630) is one of the most important figures of Scientific Revolution, perhaps more so than Newton. Although Newton is typically regarded as the singular figure of the period, in fact much of Newton's work was based on Kepler's significant contributions. Kepler placed astronomy and physics on new ground, rejected the ancient notion of perfect circular motion by choosing an ellipse as the planet's path, combines physics and astronomy, and ultimately rejects the *anima motrix* force (Aristotle's ghosts) by introducing the modern notion of force. As we will see, he so comes very close to Newtonian gravity.

Kepler's work thus represents a profound shift in astronomy from being merely a descriptive science or a 'geography of the stars.' Not being satisfied with description of paths and believing there existed an inherent order to the universe, Kepler identifies its structure and composition. This inherent order was rational, mathematical, placed in the cosmos by God, and understandable to man, thus lying the essential groundwork on which Newton would build—one of the 'giant's shoulders' on which he acknowledges to have rested. Without Kepler, there would have been no Newton.

The originality of his contribution is particularly impressive when considered in light of the enormous obstacles he faced; Kepler made his mark in history in spite of his circumstance rather than because of them.

Life of Kepler

Kepler came from a rather unstable family. His father left the family as a mercenary soldier in Neapolitan armies during Kepler's youth, and thus was inconsistently present in the household. He had actually contracted himself out to fight Protestants, which naturally increased the local village's resentment towards the family. The family sets up a tavern, not the best of environments for a child to be reared in. Various female relatives had been accused of witchcraft, including his mother, grandmother, and aunt. Curiously, a descendant on his mother's line had been mayor of the town of Weil, suggesting a sharp decline in the generational fortunes of the family. His brother was an epileptic, and suffers a horrific life of abuse. Unable to learn a trade, the brother is kicked out of house by his father, leading to a miserable and unstable life of multiple jobs and uncertain income. He finally returns home, only to die at the relatively early age of 42—a tragic life which would haunt Kepler's own dreams.

As a boy, Kepler himself was rather sickly, suffering from a horrific myopia, repeated bouts of illness and even mange, which contributed to his general state of hypochondria. These origins suggest he would not have picked astronomy at all for a career; his eyesight is so poor that at times he sees double images. His greatest fear as an adult had been that of always living so near the edge of poverty, that he would fall into it. Although his actual income suggests this concern to have been unfounded, it is an irrational fear he cannot escape. He was routinely made fun of and betrayed by other students, often finding himself in fights for which he kept a record of his sins—a practice typical of the era. In spite of these petty troubles, his teachers were well aware of his remarkable intelligence. Such was his precociousness, that at times he would correct teachers in the classroom.

Kepler's coming into his own as an adult also coincided with more intense and protracted clashes between Protestant and Catholic factions in Europe. The Reformation and Counter Reformation conflict creates a period of relative chaos where each individual was forced to declare an alliance, or otherwise suffer the loss of his properties. Kepler was somewhat protected in that he had firmly established reputation as an important scholar, and hence could travel more easily than others. However, he was not immune to these threats. When Kepler sought to unsuccessfully recuperate his wife's substantial properties, he is forced to flee Gratz hidden in a diplomat's cart, thus forfeiting the couple's substantial equity. This forfeiture might have contributed to his wife's early death at the age 37.

When he first goes to work for Tycho Brahe, Kepler quickly found himself in constant disagreements and petty squabbles with the Brahe family. The source of Kepler's contributions had been based on Tycho Brahe's precise measurements, the best astronomical data there existed at the time. However, this was jealously guarded by Brahe whom at first would not share any data with Kepler. When Kepler arrives at Jizerou, he is first indignantly told by Brahe that he was to be a guest rather than a formal collaborator as previously agreed, an unexpected insult more likely due to the inner turmoil and chaos within Brahe household.

There was also some family animosity towards Kepler as his mathematical abilities were above even the most learned, including Tycho Brahe himself. When Brahe dies, his eldest son Tengnagel Brahe takes over the instruments and does not allow others, specifically Kepler, to use them. Although these are sold to the Danish government for large sum of money, 20,000 thalers, the instruments are kept under key and soon turn into a pile of scrap for lack of maintenance. Kepler wisely absconds with Brahe's planetary data soon after his employer's death, well knowing their scientific merit and importance. Brahe did not have the mathematical ability to fully use the data himself. Kepler sarcastically remarks that the rich tend to own objects they cannot truly use or fully appreciate.

The publication of Kepler's first book, *Astronomia Nova* (1609) was another troublesome ordeal. It would present Kepler's first two laws: that of the ellipse and the equal area motion sweep. However, prior to its being published, permission was required from the family. In an unusual twist of fate, Tengnagel, who had by now become Kepler's own son in law, would not consent to the publication, requesting that he be made coauthor of Kepler's book. To this, Kepler strangely agrees, but places the condition that Tengnagel pay him a dividend from the prior sale of Brahe's instruments. This the elder son refuses to do and ultimately relents, giving Kepler permission to publish the work and in the process removing his name from one of the most important manuscripts in Western history.

If that were not all, Kepler also required sales permission from the Danish government. Under the new religious conflicts, the government imposed a tax and an order of requisite verification for all of a book's buyers, which would have hindered its diffusion throughout Europe. However, given that the government routinely failed to pay Kepler's salary, Kepler believed the government's moral claims to have been greatly diminished, and so chose to disregard it. He wisely sells the publication to the publisher, who in turn distributes it across Europe, and thus bypassing state sanction.

Yet Kepler's troubles do not end there. If family life affects the scientific output of a person, it must be said that his was not ideal. His first wife could not understand of his work, though it must be said that few would have been able to. His own professor Maestlin viewed Kepler's work above his own, and hence lacked the ability to assist him directly. Yet, the wife had a temperament and seems to have taken everything personally; Kepler quickly learns to withhold the openness of his feelings. Few of their seven children survive, which must have added difficulty to the relationship.

In spite of all of these multiple problems, Kepler places Copernicanism on solid ground over a series of publications that were to form the underlying backbone of the Newtonian revolution.

To understand Kepler, however, we must first turn to Tycho Brahe.

Uraniborg

Although Tycho Brahe (1546-1601) is a secondary participant of the Scientific Revolution, he was nonetheless significant given that he had produced the best planetary observations there existed to date, without which Kepler would not have been able to realize his own contribution. Brahe, as we have seen, also provides an alternative model to the Copernican system which was briefly more popular. While suggestively trying to combine both models, the Brahean cosmology also did not force a reformulation of physics, in that it retained the Earth in the center of the universe. The acceptance of such a model would thus have never led to the Scientific Revolution.

Brahe actually hoped Kepler would undertake the mathematical work required to prove his own cosmological system. In that Kepler sought to use data for his own purposes meant that, as in old comedies, each man sought to use the other for his own ends. It was only due to Brahe's sudden demise that Kepler is able to gain the lead, but Brahe's dominance in the relationship suggested it could easily have been otherwise, showing again the highly contingent character of our story.

Their brief collaboration was equally fortuitous; when one considers the enormous significance of the encounter, one cannot help but be amazed at the outcome. It was a meeting which almost did not happen. The two men collaborated for less than a year and a half. Both had been de facto vagabonds, having been outcast from their respective centers of work.

When they first began their collaboration, Brahe's new princely settings were in disarray. Kepler is not paid and routinely complains about the working conditions. The exact demands provide an idea of the environment and his own character: he desired a separate household, to be allowed to arrive late in the day so as to work throughout the night, to be assigned a formal salary by the government, and to be given fixed portions of food such as bacon. Kepler had not traveled alone, and was with his family; the situation left something to be desired. The debate between the two men lead to Kepler's temporary abandonment, but whom is convinced to return after some cajoling by the amiable Brahe. Nonetheless, when Kepler later returns to Gratz to secure his wife's possessions, he also contacts Maestlin to procure employment elsewhere. Therefore, the actual interaction between Brahe and Kepler was of an even shorter span than the dates would suggest as Kepler spent a substantial amount of time outside of Uraniborg.

Brahe's treatment of Kepler certainly left something to be desired. Brahe had entered into a priority dispute with a former worker, Ursus, who published a work describing a 'Brahean' cosmological 'dual' model. Brahe claimed plagiarism, and in spite of the fact that Ursus had already died, he forces Kepler to write a letter for support for him, which Kepler naturally finds distasteful. Yet, as fate would have it, Brahe dies shortly thereafter as well. Going to a dinner, Brahe 'holds water' longer than he should, and his urinary bladder breaks. As the toxicity in his blood increased, Brahe enters into a delirium. "Let me not have lived in vain," he kept saying, until finally succumbing to the noxious effects akin to renal failure. Dinner parties could be dangerous to one's health in the sixteenth century.

115

Yet, how was Brahe able to produce such detailed and sophisticated celestial maps? Kepler's own criticism of Brahe gives us the answer: one of Brahe's astronomical instrument cost more than Kepler's own lifetime income. Brahe belonged to Denmark's elite, through very fortuitous circumstances.

His father's brother had been an unmarried military attaché. When Brahe's biological father was to have twins, he promises one of his sons to his brother. However, upon birth, the father reneges on the brotherly agreement. Upon the birth of a second child, Tycho is however sequestered by his uncle, leading to a serious sibling rivalry which is peacefully resolved in the end. Tycho's father acquiesces to the prior contract. The outcome would be of enormous consequences for the infant.

His uncle, now adoptive father, had been accompanying the Danish King Frederick II, when the emperor's cart falls into an icy river. The uncle heroically jumps into the freezing water and saves king, but at the cost of his own life. Feeling deeply indebted to his faithful attaché, Tycho receives the full benefit of the king's gratitude. When Tycho as a young adult was about to move to Basle, then a leading intellectual center, King Frederick II offers Brahe his own island, Hven, with all of the benefits implied in being its Lord. Brahe's salary would be the third highest in all of Denmark, living like a feudal king for some 20 years.

Brahe would build the Castle Uraniborg, a scientific research center in chilly northern Europe that became a 'temple' to science. In its basement a comprehensive alchemical laboratory is set up, and its upper floors contained a diverse array of astronomical instruments, including the largest wall mounted quadrant in Europe.

Brahe had curiously planned for all contingencies, and has all of the instruments designed so as to be easily disassembled and relocated. In spite of its size, the enormous wall mounted quadrant could also be split up into pieces, and transported if need be. Upon his later exile in 1597, Brahe obtains another patron in the Holy Roman Emperor Rudolph II in 1599. Brahe is thus quickly able to reestablish himself in similar conditions, fortuitously also located in a castle on an island. Rudolph's delicate state of mind, however, leads to continual conflicts between Brahe and his new patron's state administrators, who believe the expenditure an enormous waste of state funds. If Brahe obtained 50% of his salary in a given month, it would be considered a success. This is a power struggle which Kepler inherits when he obtains Brahe's position as imperial mathematician.

Brahe's entry into the sciences in and of itself was remarkable, as noblemen were expected to go into diplomacy or statecraft rather than pursuing such 'foolish' notions of looking at stars night after night. His uncle even goes to the extent of hiring a private tutor in history to try to dissuade him. Yet because Brahe would continually sneak out and look at the starry night sky, proving him to be incurable of his interest, his history tutor desists and becomes Brahe's lifelong confidant and friend.

Brahe's exact contributions were the result of two particular factors: the size and scale of the instrumentation used as well as the rigor and thoroughness with which astronomical measurements were made. In contrast to Copernicus, who took only irregular and inaccurate spot planetary observations, Brahe and his research team would make continuous observations of planets throughout the entire cycle of their retrograde motions. He was methodical and precise, kept data continuously, abiding by Michael Maestlin's call for a celestial reform.

Using this data, Brahe comes to the conclusion that the Aristotelian crystalline spheres were a fiction. His careful observations of the 1572 nova and the comet which followed five years later indicated these phenomena to have occurred outside the sublunar region. Most certainly, the comet's path would have been interrupted by crystalline spheres, had they existed.

The meeting of the two men must have been a sight to behold, if anything for their opposing personalities and social backgrounds. Kepler was of humble origins, a hypochondriac math genius with a conflicted personality, and poor. Tycho as we have seen was man from the highest economic strata of Denmark, a feudal lord with his own island, and an avid socializer.

Certainly, conflict at some point was bound to occur, due to their different life experiences.

Mars

In contrast to Brahe's most fortunate circumstances, Kepler was determinedly able to raise himself only by his own 'bootstraps.' He directly benefited from the Protestant Revolution which freed education for gifted students as himself, obtaining early academic successes. Initially deciding to become a preacher—a position he was unfit for due to his personality traits—he attends the University of Tubingen to study theology. While his thesis is actually a defense of Copernicanism, he does not consider himself an 'astronomer.' His advisor Michael Maestlin, a mid-grade astronomer, becomes a lifelong father figure to Kepler, evidenced in part by Kepler's habitually long letters to him—even when these continually went unanswered.

When a math teacher in the town of Gratz dies, the school turns to Tubingen for a recommendation; Kepler obtains the position upon Maestlin's recommendation. As a math teacher, Kepler was so far advanced, that during the first year he has few students; by the second year none registered for his classes. In this aspect, he is very much like Newton. However, the school administration recognized Brahe's ability and kept him in the school employment roll. That he had produced 'accurate' astrological prophecies also greatly aided his standing in the community; his astrological tables also provided an additional source of income.

While teaching, Kepler becomes curious over a number of important astronomical issues. Why were there only 6 planets; why not 50 or 100? What determined the number of planets? Similarly, why were their periods not exactly proportional to their distance? Saturn was twice as far away as Jupiter, but had a period that was 2.5 times as long; instead of the expected 24 years, it took Saturn 30 years to revolve around the sun. These questions deeply puzzled Kepler.

As is often the case, after thinking long and hard on the issue he has an unexpected "eureka" moment while providing a math explanation in class one day. He draws a triangle circumscribed between 2 circles, and realized that it matched the Saturn and Jupiter's data. It seemed like all six planetary orbits could be circumscribed by the five perfect solids. The complexity of the universe could be reduced to a few basic forms—or so he thought.

Kepler, however, is rigorously empirical; if the theory does not fit data, he had no qualms in letting go of an idea he had recently placed so much hopes in. Upon closer inspection, he later finds that a triangle would not match the orbit. But he continues working on the problem until he rigorously answers it, resulting in the work *Mysterium Cosmographicum* (1596). There were only 6 planets because their orbits were perfectly circumscribed by the five perfect solids. While today we would find this suggestion odd, after all there are only 8 planets notwithstanding Pluto, it does reflect the Pythagorean ideal of an underlying mathematical unity to the universe. Upon publication, he distributes his book to various recognized scholars in Europe, one of them being Galileo Galieli.

Upon receiving the work, Galileo writes back to Kepler, congratulating him. While it would be ideal to suggest that two key figures from the Scientific Revolution were in extended contact with each other, regrettably they were not, exchanging only a few letters between them. While Kepler is the more solicitous of the two, writing to Galileo on numerous occasions, Galileo's replies are curt and sparse. One of the objects of their discussion was Copernicanism. Although Kepler defended it openly, Galileo recognized its importance but was not yet willing to take the same bold public support.

When Galileo publishes his *Starry Messenger* in 1610, Kepler is one of the very few men to have publicly defended it against its numerous opponents.

Another copy of *Mysterium Cosmographicum* is also sent to Tycho Brahe, and it is this letter which initiates the conversation between the two men. At the time, Brahe was still at Uraniborg, and immediately offers a job to Kepler, recognizing his powerful mathematical abilities. However, Kepler, relatively secure in his teaching position at Gratz, at first turns down Brahe's generous offer. It is not until the authorities in Gratz begin their persecution that Kepler is forced to emigrate out of the city for his and his family's own safety.

Oddly, it is the inopportune clash between Catholic and Protestant factions which brings the two men together for the first time.

When Kepler finally goes to work with Brahe, Brahe had already given the data on Mars to his assistant Legomontanus. Yet Mars was such a tough case, that Legomontanus gives up as it exceeded his abilities. Consequently, the Mars data falls on Kepler's lap, who incorrectly believes he would be able to resolve it within 8 days. It would in fact take him more than six years, filling 900 enormous folio pages with miniscule writing. (It is no wonder Rheticus went briefly mad.) Upon Brahe's untimely death in 1601, Kepler is named imperial mathematician, and hence is able to secure a stable salary and place of work. Because the nature of his work was theoretical, working only with paper and pen, he is able to survive on a far lesser income than that which had been originally promised to his patron (Brahe). It is then when he finds the three laws.

Mars had not been easy to conquer, in part because it has the largest deviation from a circular path of all the planets. His intellectual progress is well described by Arthur Koestler. Kepler begins with the basics, by plotting the path of Mars, a very detailed and excruciatingly tedious work. Strikingly, he initially portrays it in the form of an egg, yet because of certain inconsistencies he discards this notion. He realizes that his first calculations had an error of 1/8 of a degree, but because Brahe's data had only a 1/15 margin of error, he discards this first attempt.

Then he plots the path of Earth as it would be seen from Mars. While this might appear to be redundant, it was an important exercise at obviously Earth was the source of all Martian measurements, and served as a secondary verification process. He discovers his equal area sweep law, which is titled as the second law but in fact first to have been found. The regularity of all planetary motions could be traced by calculating the area beneath their paths and the sun: equal areas would be traveled in equal times. Since the area of a planet closest to the sun would be much smaller than that from its farthest point, it would be traveling much faster—but at a predictable speed.

The second law is perhaps the most puzzling, in that Kepler appears to have had the answer before his very eyes on a number of occasions, but his initial ignorance of conic sections hindered this discovery. Kepler would take years before 'seeing' the elliptical pattern in from of him. At one point he notices a variation of the radius between sides of 0.00428, typical of an ellipse. At another point, he notices that the variation in angle M formed from the farthest and closes point equaled 5°18," again a trait of ellipses. After 6 years of continual work, the answer finally hits him in the head: planets do not travel in perfect circular paths, but rather in slightly compressed circles or ellipses.

With this discovery, Kepler breaks with a millennial Platonic/Aristotelian tradition of using the perfect circle to account for celestial movement; his study also does not pretend to be a mere mathematical exercise but in fact claimed to reveal how the universe actually existed. There was no difference between the model and the physical nature it represented.

The third law would take a decade to be discovered, and addressed his second important question: how could we account for regularity of planetary motion, as it did not follow a simple rule; double the distance did not mean double the time. But it appeared as if there had to be some sort of underlying order. What was it?

In 1619, Kepler publishes *Harmonici Mundi*, which, like his other previous works, is so autobiographical, describing all of the contours of discovery, that the third law only appears hidden in the middle pages of the thick book—something which might have made Galileo shirk from taking a detailed look at Kepler's work, aside from the fact that Kepler was clearly writing from a region with clear Protestant affiliations.

That $t^2/d^3 = k$, where k is a constant might sound trivial, but this is precisely the purpose of science: to render order onto chaos. Kepler in fact had found the 'physics of the planets,' in that there was an identifiable pattern behind the vast arrays of points and numbers in a planets path. This law applied to all planets, regardless of their size or distance; one did not need a different physics, as Ptolemy had one, to account for the motion of each planet. The square of the time traveled divided by the cube of the distance traveled equaled a uniform constant. In his third law, Kepler begins to 'clean up' the picture of the universe, and set the foundation stones for Newton's marvelous construct: the law of gravity.

Somnium

Towards the end of his life when things looked bleak, Kepler writes one of the first works of science fiction: *Somnium* (1608). In it, Kepler applies his emerging notions of force, turning away from Gilbert's *'anima motrix'* to one of *'vis motrix'*: a cold and abstract notion of force that would heavily influence Newton. Kepler realized that the influence of gravity was the result of a mutual interaction between two bodies, but which did not leave a 'shadow' as light did. If the Earth were to suddenly disappear, its water would be pulled into the moon. Inversely, he realizes that the pull of gravity was proportionate to its mass, and hence the smaller a planetary body, the less its influence would be on objects on its surface.

Kepler thus presents us with a scene that would be readily recognizable by the youngest cognizant member in a family. He describes a moon walk as the many moonwalks seen on so many televised occasions: the first moon walk of 1969 during the Kennedy administration, "Aliens" (1986), "Superman" (1978), and so forth.

One needs to pay attention to the fact that Kepler wrote his play more than three hundred years prior to the actual events, truly seeing far into a future that was yet unimaginable to most other Europeans.

But it had been written during a bleak period in his life, and this bleakness is reflected in the drama. In essence, it captured the trials and travails of his brother, so abused and downtrodden as he has been throughout his life. Despised by the mother, the protagonist is kicked out of the home, but ultimately pardoned on one condition: the exile to the moon. Here he meets bizarre moon insects that would fit well along with Phillip K. Dicks' leftist pieces or Robert A. Heinlen conservatively-oriented works of science fiction. As in many such novels, his demise comes about rather unexpectedly.

After the death of his first wife at an early age, Kepler had the fortune of having a circle of friends to present a number of suitable matrimonial substitutes. As imperial mathematician, Kepler was in a position of some standing, and hence a 'good suitor.' He is offered a total of eleven potential female candidates, each one of whom he evaluates with some meticulousness in a bizarre selection process that must have been odd even for his day. Taking too long, one favored candidate ends up marrying another. Kepler finally selects not the daughter of the elite household that had been identified for him, but rather her maid: Susanna Reuttinger, a woman 24 years of age, half his age and from a distinctly different social class. They have six children together.

During the preparation of his last book, Kepler travels throughout weary Europe, noticing the horrific destitution in which the religious wars were leaving individuals of either religion. Kepler in this sense is fortunate in that he did not have to directly suffer the horrors of war. However, the preparation of his last work required constant trudging through harsh winter conditions, which end up killing him at the young age of 59, leaving his wife and children in destitute circumstances before these had reached adulthood.

Galileo Galilei

Prince of Motion

Inquisition

GALILEO'S STORY IS SO COMPELLING and dramatic that we will begin with the end. Why did the Catholic Church come to persecute the founder of modern day physics? This persecution almost destroyed the birth of physics in its cradle, whose repercussions for the future history of science would have been significant.

In November of 1632, Galileo was called to Rome for an Inquisitorial hearing. He had just published *Dialogue concerning Two World Systems,* a discussion of the merits and weaknesses of the Ptolemaic geocentric model and the Copernican heliocentric model of the universe. As with prior works, he had aroused controversy. Was it a defense of Copernicanism?

Galileo, however, was now fairly advanced in age. At 68 years old, he had suffered much of his life from a long and persistent illness; in this particular occasion, it would leave him bedridden for weeks. What specific disease he had or what its cause was is not known. Some as Stillman Drake have argued he suffered from a lifetime of rheumatism. Other noted that he had the gout, rather common for period. Dava Sobel points to a strange event which occurred many years earlier in a Florence cavern sauna chamber, fed by water flowing through volcanic rock. One day, Galileo and two other friends had gone into the chamber and fell asleep, during which they were exposed to noxious fumes for an extended amount of time. His two other friends eventually succumb to the fumes. While Galileo was fortunately able to recuperate, his health was never quite the same there afterwards.

November 1632 had been a particularly bad month. Winter had begun and his weakened condition had not allowed him to set off for Rome. November passed into December, and his absence infuriated the Roman Inquisition. A team

122

of three physicians were sent and inspect Galileo so as to ascertain the validity of his excuses. All three verified that in fact Galileo was very sick, and that he should not leave until he had recuperated. It was a medical evaluation that fell on deaf ears, and the Inquisition orders him to set off immediately for Rome. But that was not all.

The order sent a weakened old man in a debilitated state right into a plague infested land. Ever since the Black Plague arrived three hundred years prior to Galileo's departure, periodic waves of the disease had routinely swept through Europe; although reduced in severity over time, they still resulted in gruesome and deadly epidemics. Families, believing they were protecting their loved ones, lied about the condition, keeping them hidden indoors—only tragically resulting in the death of the entire household. Galileo's first 'wife,' Maria Gamba, and his own son's (Vicenzo) wife also die during this event, leaving three kids behind. By forcing Galileo to travel at such a time and in such a weakened physical condition, the Catholic Church was issuing a veiled intent to assassinate him.

It took Galileo one month to get to Rome, as he became stranded in a town that had wisely imposed quarantine. Although the journey covered only 130 miles, this was the era of the horse and cart; what today takes less than two hours of driving on a highway took much longer back then—for good or ill. He finally arrives in February 1633. Galileo and his daughter Maria Celeste (Virginia) did not really think the trial would last long nor did they fully appreciate its seriousness.

As a young man, Galileo had fathered three children: Virginia, Livia, and Vicenzo. Upon his ascension to the Medici Court, he had taken Virginia and Livia with him to live in Florence, being old enough at 10 years of age to undergo the journey. Vicenzo at 4 years was too young and stayed with his mother. Galileo had not married their mother, Maria Gamboa, whom he had met in Venice. She was of a lower rank, and ultimately even Galileo advised her to marry another man from her respective social rank.

Upon their arrival to Florence, he places his daughters into a convent as a mere educational formality. A woman was not allowed to take religious vows until the age of 16. Sadly, the convent ended up becoming a prison for the two, greatly restraining their movements. While Virginia thrived in this environment, Livia could not tolerate it and was left with permanent psychological scars as an adult. As they matured, Maria Celeste and Galileo developed close ties, as is often the case with a first born. Unfortunately, their enormous correspondence was burnt by a nun at the convent upon discovering it; only the 125 letters kept by Galileo provide evidence of the strong bond of their relationship. His eldest daughter routinely mended his clothes, prepared jams from the fruits of Galileo's garden. Galileo, in his turn, helped the cash-absent convent with routine contributions.

Maria Celeste believed the trial would last less than a month. In actuality, the trial would be tediously drawn out over 4 formal hearings that would not end until June, when a verdict was issued. Six months would pass before a sentence was read, forcing Galileo to undergo a great deal of psychological trauma, being uncertain of an outcome on which his life depended. The human brain has evolved for intense but brief bouts of 'fight or flight' when confronted with danger, rather than the chronic pain of a prolonged duress. Humans can withstand crises for minutes and hours, but not for weeks and months at a time. Heightening his anxiety, Galileo had been unable to see Pope Urban VIII directly, with whom he had previously held a relatively close and intimate relationship.

The two men had a number of social and familial ties linking them together. Galileo had been professor to Matteo Barberini's (Pope Urban VIII) nephew. As a younger man, Barberini had been able to obtain a canonship for Galileo's son, Vicenzo. When he first became pope, Galileo visits Rome and pays his respect to his old friend. Though Barberini, now Pope Urban VIII, had a busy agenda upon his ascension to power, he met with Galileo once a week to walk in the Papal garden and discuss all manner of philosophical issues. When Galileo had entered into squabbles with Church (1616), Cardinal Barberini defended Galileo, creating a strong bond between the two. After the early incident, both maintained routine contact with one other through letter writing.

In light of these preceding events, it would have not been not unrealistic for Galileo to believe he could have some measure of protection from the Pope himself. Nothing seemed to suggest otherwise. Galileo in fact publishes the *Dialogue* shortly after Barberini's ascension to the papacy, likely believing that such a friend in high places would offer a strong degree of protection.

The radical shift in Barberini's treatment of Galileo must have been unsettling, in and of itself.

During the trial, Galileo was under threat of torture and death for two principal purposes: to force him to tell the truth and as a means of punishment. That he could have been burnt at stake at any moment meant that the costs of losing the case were enormous; disfigurement, impairment, or death are not to be so lightly cast aside. Remember that the Inquisitorial processes was very arbitrary, and today would be considered illegal as these violated the most basic rights of the defendant such as *habeas corpus* (speedy trial). The accused languished in prison, which often led to the breakup of entire families and the financial ruin of the accused. Torture need not necessarily imply physical harm to ruin the life of an individual.

Galileo was certainly more fortunate than others because he had become a figure of some celebrity. Upon visiting Rome, he was allowed to stay at the Tuscan embassy in Rome under the care of a faithful diplomat of the Grand Duke. However, the Church placed pressure on Ferdinand, whom ultimately agrees not to pay for Galileo's stay—contrary to what had been the usual practice during Galileo's prior visits to Rome. This certainly worsened the trial's impact on Galileo's personal finances and state of mind. A right of the official mathematician of the Medici court was undermined by the Church's enormous power.

The trial proceedings were themselves aberrant. The Catholic Church presented itself in the voice of a third party, giving the impression of being an absolute and objective source of information and judgment. Certainly, the tone and grammar used in the proceedings reinforced notion that the accuser could not be questioned. An accused as Galileo always stood alone before the judges, testifying in the first person, which, in turn, suggested his inherent liability and subjectivity. The language and tone of encounter was at its core partial and biased. The accused was always on the defensive, and could never question the case of prosecutor, neither on philosophical grounds, logical argumentation, or evidentiary basis. There was an inherent presumption of guilt by the Church's court, and its evaluation was to determine the degree of guilt rather than a question of innocence. The accused had to prove that he was not as guilty as they believed him to be—not a good place for anyone to be in.

At the final hearing of the trial, Galileo as a 69-year-old man was forced to kneel and recant. He is immediately placed under house arrest for the rest of life, which is not as light hearted as it might first sound. He was not allowed initially to go to Arcetri, his home situated next door to the convent where Maria Celeste lived. Galileo was first placed under house arrest in the care of Asciano Piccolomini, Archbishop of Siena. The surrounding circumstances—the length of the trial, its uncertainty and enormous implications, and his inability to return to his home and loved ones—took an enormous toll on him and his family. Piccolomini worried that he would have to tie Galileo's hand to his bed, from the traumatic emotional state he was in. Galileo had hoped to reform the Church, not to become its prisoner. Maria Celeste had been so worried about her father throughout the trial, that she seldom ate or took care of herself. Shortly after the last hearing, Maria Celeste gets dysentery and suddenly dies, buried in an unmarked grave.

The Catholic Church could have "killed" the Scientific Revolution in its beginnings. It is often the case that 'revolutions,' or changes of enormous social consequence, are not detected at the moment they occur; more often than not, these are identified *post facto*, as is the case of the internet in our day. Had Galileo been found guilty or had he died during the strenuous ordeal meant that he would not have been able to publish his greatest work: *Discourse on Two New Sciences* (1638) where he invents the science of motion and materials. It had been published in Holland by Elzevir, outside of Catholic jurisdiction. The publication could have led to Galileo's execution but he denied that it was by his hand.

The original trial had been undertaken to 'send a signal' and make an example out of him. Galileo had certainly become a well-recognized figure in Europe, and the trial did actually have an effect. Descartes was about to publish his magnum opus, *Le Monde*, but immediately stops publication in Catholic France upon hearing of its result.

There can be no doubt that the Church also delayed Galileo's own magnum opus.

As early as *Siderius Nuncius* (1610), Galileo had suggested he was holding onto two ground breaking studies. Throughout his writings, Galileo repeats this message, letting the reader know that he had such a study in his hands. Much of this work began when very young, but was fully developed until he began teaching at the University of Padua in Venice. Galileo referred to his eighteen years in Venice as his happiest years. Any delay or postponement of his work meant that if at any time Galileo had died, the Scientific Revolution would have been seriously hindered.

The question that inevitably has to be asked is: why did the Catholic Church persecute Galileo? Again, the case is more unusual than might be imagined, given that, as previously mentioned, Pope Urban VIII had been a good friend of Galileo. Galileo was also a faithful Catholic. Both of his daughters had become nuns. Galileo had always sought prior approval of the Catholic Church for his book publications. In this sense, he abided by the rules and had never done anything formally wrong. Why then mistreat him so—psychologically tortured, buried in an unmarked grave next to his daughter, or the death of his beloved daughter?

Entry into astronomy

Galileo was born in Pisa in 1564. His name is not as unusual as one might think. It was a typical honor for the male child's first name to be based on the family's last name. What is curious, however, was that the original family name had been "Bonajuti," being modified by a great grandfather to "Galilei" to honor a physician brother of his named "Galileo."

Galileo Galilei had various siblings, including a sister that married and who would require a dowry—a substantial burden for Galileo at the beginning of his career. His brother had promised to share in the financial burden, but when he marries and moves to Germany with his spouse, Galileo's brother forgoes his responsibility. The rise of Protestant-Catholic conflict, however, would irrevocably alter his brother's life. The burden was particularly difficult as Galileo at the time was only making 60 florins, a very small amount of money when compared to the 1,000 florins per year he made while at the Medicci Court.

Galileo's father, Vicenzo Galilei, was of tremendous importance in his life. There can be no doubt that Galileo was his father's child, like a branch from a tree. Vicenzo had been a cloth merchant who moved to Florence looking for better commercial opportunities. More importantly, he was a minor musician and musical theorist who sought to fortify the theory of music. Vicenzo eventually publishes a study, which although not revolutionary, made its own contribution to the field. Of greater importance for our historical character, Vicenzo had workshop was full of musical devices and instruments: pendulum strings, chords, and so forth.

It is to be noted that the character of Vicenzo also had a substantial direct impact on the son. He openly criticized the reasoning of scholastic academicians, noting that one cannot validly argue by citation. Claims needed to be based on concrete proof—a notion that would be integral to Galileo's intellectual style.

Galileo's father's influence can be detected throughout his works. For example, Galileo often uses music and devices in argumentative points, often in surprising and unexpected places. In *The Assayer* (1623), for example, Galileo undertakes an epistemological discussion of knowledge using the story of a man living alone far in forest who one day detects a lovely musical melody. One of the most joyous occasions in life similarly occurs when he also hears a boy playing a flute reproduces same melody, much to the man's surprise. He then decides to undertake a journey around the world to identify and categorize the many ways by which the beautiful harmony could be made. However, throughout his long route he finds so many different means, that he realizes the total number of ways would always be greater than those which he was aware of. It would become a central point to Galileo's philosophy of science.

As a young man, Galileo attends the University of Pisa. His father wanted him to study medicine in order to have a more stable source of income, as is often the case when individuals rationalize their entry into the field. Vicenzo senior well knew the typical economic woes musicians faced. However, while at the university Galileo acquires a poor reputation for constantly criticizing the positions of 'the Ancients,' specifically Aristotle's world view. In today's terms, he would be called a "*quejoso*," and actually took a strong liking to mathematics. In this, Galileo is influenced by the professor Ostilio Ricci, who was a 'practical mathematician' in his use of math for solution to problems. Specialization hides today much of what was previously considered mathematics.

Galileo drops out of the University of Pisa, and it is easy to see why: he was a creative thinker operating in a conservative environment. He then begins to work at home using his father's instruments, doing his first motion studies. As a young academician, he writes a significant study of motion, discovering that pendulum of given length swings at a constant frequency regardless of amplitude. This study is sent to Christopher Calvius in Rome, who in turn shares it with various important people. The chain of events eventually results in Galileo returning to the University of Pisa in 1589 as a professor rather than as a student. As noted before, his low salary of 60 florins are meager scraps from which had to pay his sister's dowry. The year 1591 is a difficult one for him. His father Vicenzo dies and his contract at Pisa ends.

Yet Galileo had also made a hydrostatic balance, a '*balanceta,*' which is seen by an important nobleman who again initiates a new chain of communications which results in Galileo's first long term employment at the University of Padua in 1592. At the age of 28, Galileo thus begins one of most productive periods in his life.

Galileo as a 'scientist' was principally an experimentalist, which is ironic as he was imprisoned for astronomical work. There are two types of astronomers: observational, as Tycho Brahe who sought accurate, exact determination position, angle, elevation, and so forth. The second type is the theoretical one, as Johannes Kepler, who focuses on the discovery of the mathematical structure of the cosmos. Galileo was neither of these. The main portion of his productive life was dedicated to study of motions; astronomy had been only a distraction for him. The importance of his astronomical works are based, however, in his recognition of the significance of its discoveries.

Galileo could very clearly see implications for Catholic theology. Specifically, he realized that Catholic Church had based its worldview on erroneous notions, and the purpose of much of his life is the attempt to reform its underlying (and incorrect) tenets. However, since Galileo also wrote in the vernacular, it meant that his writings were available to the general public and aimed at men of reason and contemplation regardless of their station in life. In this sense, the use of vernacular was certainly a direct threat to Catholic Church whose policies stipulated that Bible could only be understood through a priest's interpretation

One cannot overestimate the significance of *Siderius Nuncius* ("Starry Messenger"), published in 1610. The book catapults Galileo into international fame with such magnitude that it is hard to conceive of even today. The first edition quickly sold out of its 500 printed copies; in spite of the fact that the publisher had promised Galileo 30 copies, he receives only six tomes due to the demand. Galileo soon thereafter receives diverse public recognitions and awards. Federico Cesi makes Galileo honorary member of the *Academia de Lincei* (Lynx academy), a name alluding to sharp eyesight and a keen mind. Newly formed, with only a handful of members, Cesi promised that the institution would publish all of Galileo's works thereafter. It also provided Galileo an honorary title, which the author proudly used in all of his publications. The term "*member of the academy de Lincei*" was a sign of status and prestige, whose social worth was well appreciated by the author.

Kepler publishes formal praise of *Siderius*, continuing one of the rare direct exchanges between members of the Scientific Revolution. Kepler urged Galileo to publicly defend Copernicus, but Galileo responded that he was actually hesitant to do so given the ridicule the work that had received. It is important to remember that Copernicus never faced public scrutiny for his ideas, which only appear tardily on his death. It is in fact Kepler who becomes one of the first scholars to publicly recognize merit of Copernicus's ideas. Galileo, however, was more interested in promoting his career, and uses Kepler's letter of support while negotiating with the Medici court for position and patronage. Kepler's formal title 'Imperial astronomer' has an air about it, and this support was significant. Kepler is to be recognized for putting his reputation on the line as he had not actually seen Jupiter's 'stars.' His act of common defense, however, places Kepler's own position and employment into jeopardy as he begins to be attacked given that few had actually seen the "Mediccean stars" with their own 'eyes.'

These critiques force Kepler to request a 'Galilean telescope' from its maker, and although Galileo writes back stating that would send a model, he apparently never fulfills his obligation to Kepler. The imperial astronomer, however, is able to obtain a telescope via Prussian dignitary whom had received it directly from Galileo.

The fame and publicity generated by *Siderius Nuncius* meant that, in the long run, Galileo would become the public face of the Scientific Revolution—a role which Galileo's' extroverted and confrontational personality suited him. This is certainly the case when contrasted to the personality of his introverted 'colleagues' as Copernicus, Kepler, or Newton. Galileo appears to have greatly enjoyed the fame and attention as well as the debates his works generated. Galileo is formally a very good debater. By contrast, Newton avoided public confrontations whenever he could, and publishes *Optics* after death of Hooke rather than confront Hooke in the court of public opinion.

Galileo also seems to love the company of people; his good friends Salviati and Sagredo were memorialized in *Dialogue of Two World Systems*. He would often stay with Salviati when sick, and Sagredo's sharp mind likely played a good counterpart with insightful commentary, akin perhaps to that of a Michel Besso to Einstein. By contrast, Newton was a loner who rarely socialized, never married, had much less had children.

Galileo is the public face of the Scientific Revolution, much more so than any of his scientific 'co-revolutionaries.'

One also has to recognize that, as proven in *Siderius Nuncius*, Galileo became the first person in all of his history to actually observe the Moon's surface and the moons of Jupiter—a point that cannot be emphasized enough. Galileo was also well aware of the significance of these acts.

He begins his brief piece with description of moon and puzzles at the information gleaned from his new instrument. He obviously notices the moon had irregular surfaces. He looks at a reflection of sunlight from the moon and notices how their shadows align with the sun and concludes that they were craters and mountain peaks. However, if the moon is so irregular why does it appear so smooth? One would naturally expect a jagged and irregular outline, rather than a smooth and regular profile. Galileo comes up with two possible causes for the visual phenomena: a) the role of the Earth's atmosphere or b) the creation of an optical illusion is akin to that of waves on the ocean. In his consideration of the matter, Galileo opts for the second choice.

A thunderous sea full of waves looks flat on the horizon, in spite of being full of crests and valleys; each wave peak cancels the other out, and thus giving the illusory impression of smoothness. Galileo argues that a similar phenomenon on moon was occurring; each crater and irregularity 'cancels' the other out, hence providing the incorrect illusion of perfect circularity from afar. Later, in *The Assayer*, Galileo also points out that it was precisely the irregularity of the moon that allowed it to serve as a 'reflective surface.' If the moon were a perfect and polished sphere, we would not be able to see it from the Earth; it was actually its very roughness that diffused light in many directions, thus allowing a greater portion of it to actually reach the Earth.

The moon had other puzzling features. Galileo notices an enormous variation in heights. If the Earth had the same degree of topographic variability, it would look vastly different; Mt. Everest would be a common phenomenon. In other words, while both Earth and Moon were characterized by craters, the respective magnitudes of each varied enormously. He also notices that there was a 'backlight' on the moon. In other words, Galileo discovers that the Earth was serving as the moon's 'moon,' reflecting the sun's rays back onto its own satellite. The same processes which occurred in the heavens also occurred on Earth, diminishing the previously assumed distinctions between each.

Yet perhaps the most important piece of *Siderius Nuncius* was the discovery of Jupiter's moons. They were difficult to see at the time even with Galileo's powerful telescope, given that the telescope as a technology was still relatively primitive. As old computers, they required a skillful operator to work. There is an embarrassing incident in Rome due to this. Invited guests at a party were unable to see the Galileo's Jovian stars. In fact, when first seen by Galileo, they were tricky to identify. He describes that the 'stars' that tended to stay close to planet, and were relatively faint. Their odd properties gave them away, however, as these would 'disappear' only to reappear later.

He did notice, however that the intensity of light could not have been caused by their rotation around Jupiter, given that the variation in distance to the Earth was not enough to have caused the observed variation in light intensity. Galileo thus leaves the cause of this variation as an open question, unwilling to hypothesize further than the data itself suggested. He, nonetheless, realizes their enormous scientific significance: the Aristotelian world view upon which the Bible was based would have to be modified.

While Galileo had not been a firm Copernican early in his career, he would become one as evidence was continually gathered in its favor. Only Copernicus's model, rather that Ptolemy's, could account for many of the astronomical anomalies Galileo was detecting. The processes of heaven and Earth were not as dissimilar as had once been assumed.

It is also important to point out the scientific character of Galileo's method. A very acute observer who undertakes many comparisons, Galileo does not take evidence at face value, but rather continually verifies the data he has obtained. More importantly, he is unwilling to generalize to conclusions that are not properly supported by the available evidence. It would have been politically ideal, for example, to have used the Jovian satellites' luminosity variation to support the Copernican model, but he does not. Instead, he prudently leaves the answer undetermined.

Note as well that technology begins to form an integral part of scientific change. It had been Galileo's own technological innovations which further extended the sight of man, and personally allowed him to have 'new experience' which had been previously unavailable in world history. However, a valid question did arise at the time with regard to the degree such objects seen by the telescope existed in nature or whether they had been created by the very instruments used to detect them. Could we trust the information provided by technologically assisted senses?

It is certainly the case that there were some problems with the early telescopes, as these often did produce a double image, again requiring skilled observers to differentiate between them. It was on these inconsistencies which the Jesuits publicly remonstrated Galileo, and end up building a better telescope at the Colegio Romano. The Jesuits do eventually see the Jovian satellites, previously criticized as technological constructs. Galileo had been right all along.

He first comes across the idea of a telescope the prior year when a Flemish spectacle maker accidentally discovers the effect by looking simultaneously through concave and convex lenses aligned along the line of sight. The magnification of the very first telescopes was rather small, and Galileo acutely begins to apply mathematical analysis to improve the basic design. He builds a number of models, each with increasing power: the first of 30x while a third variant rose to increase vision a thousand fold.

Galileo also proves to be politically astute as to the implications of the telescope. He takes his first version to the Court in Venice, where all were astonished to see a ship at sea three hours prior to arrival. Its commercial and military implications were enormous. One could obviously detect naval warships long before they arrived and identify new cargo ships with enough preemption to obtain better market prices given the implicit information asymmetry afforded by the telescope. Galileo's salary at the University of Padua in Venice was immediately raised to an astounding 1,000 florins.

Yet, unwisely, Galileo also uses his new technology and this new income as base for negotiations with Medici court. In *Siderius Nuncius*, Galileo names the Jovian moon after Cosimo II, later opting for the term "Mediccean stars."

This would prove to be a fateful mistake.

Galileo, courtier

Galileo uses his newfound recognition to enter the Medici Court. Since 1601 he had been soliciting the position of tutor to Cosimo but the first approach was denied as Cosimo was too young at the time. By 1605, however, Galileo had gained enough of a reputation to finally begin tutoring the young Cosimo, establishing a close bond between the two. The two families, Medici and Galilei, develop closer ties than might be presumed. A member of the Medici who served in Virginia's convent was going to retire, and pushes for the two girls into their vows in spite of being three years too young to become nuns. Although this helps guarantee their financial future, it also proved to be too psychologically limiting and cruel a life for the younger sibling.

By the time *Siderius Nuncius* is published, Cosimo had become Duke. In their negotiations, Galileo requests that he be excluded from having to pay his sister's dowry and that his court income match the 1,000-florin salary just recently awarded by the Venetian university. Galileo's income and social standing drastically increases in a short period of time, moving into a new mansion with servants and a wide array of artisans working for him. The Medici patronage, however, was a double edged sword.

The Venetians naturally felt betrayed by Galileo, particularly as these had just increased his salary by a factor of three. Clearly, financial considerations did not exclusively propel Galileo into his new position. Most surprisingly, Venice had been one of the few political regions declaring governmental autonomy from Rome, and was generally recognized for its atmosphere of free thinking, lively and active discussion and debate. Giovanni Fancesco Sagredo, a diplomat who had been away on duties during Galileo's negotiation with the Medici Court, warned Galileo as to the poor nature of his decision. He writes a letter to Galileo expressing his grave concern, and notes that the Jesuits dominated the religious landscape in Rome, a political reality which did not look positive for Galileo. Both Sagredo and Cesi held poor opinions of the Jesuits, criticizing these for attempting to monopolize truth, as if they were its only bearers and retainers. Galileo unwisely chose to ignore Sagredo's warning.

There can be no doubt that Sagredo was correct: Venice would never have 'handed over' Galileo to the Catholic Church in Rome as occurred in 1632. Galileo would have found absolute protection against any direct personal threat to him for his religious or scientific views. Had Galileo chosen to stay in Venice, there is no doubt that would not have become a 'martyr' to science, and hence, in hindsight, his decision to move ultimately be of critical importance in history.

Yet the issue is of enough importance to merit a closer look. Why did Galileo then push for a position as Courtier in the Medici court at Florence? Was it greed, or desire for social status?

Various reasons have been proposed. Future threats always look vague in the present. In contrast to the looming cloud of the Jesuit menace, Galileo was already being subjected to attacks by academics in his own university. That Venice was a 'free thinking city,' did not necessarily imply that Galileo was free from any type of opposition—as well it should be in any democratic space. It has also been suggested that his move to Florence was a 'return home.'

As Dava Sobel has noted, each municipality was then known for their own distinctive traits and habits. Naples was well regarded for its parades, Rome for its picnics, Venice for its gondolas, and Florence for its plazas. It had been in these very plazas where Galileo was first launched into the public life of the intelligentsia; in these public spaces people would meet and have active discussions of all manner of topics. Florence was hence characterized by a vibrant social life, in contrast to the apparently more subdued character of Venice. One would be amiss to suggest that there was some amount of 'homesickness' in Galileo's decision to return home, where he had thrived as a younger man. One has to also remember that he was a very good debater, a skill which was likely developed and honed within Florence's socially vibrant plazas.

The university atmosphere was another factor. The early modern university was inherently a conservative one, as most universities were subjected to strong religious influences. Even if no direct ties existed to the Catholic Church, these were predominantly populated by Church men. A good example would be that of Niccolo Castelli, a former student of Galileo who had also become a close confidant. A brilliant mathematician, Castelli was the first to propose a novel way of looking at the sun's light, by indirectly shining its telescopic image onto a piece of paper, thereby avoiding the retinal damage that would otherwise occur. Castelli had become a Benedictine monk, and hence his name "Benedetto." Benedetto obtains a position at the University of Pisa, taking over prior Galileo's prior position.

It is important to note that as professor, Galileo did not teach Copernicus, but rather stuck to the traditionalist interpretation of Ptolemy. So conservative was the intellectual environment at European universities during this period, that all manner of academies had been formed: a private place where scholars could freely discuss new ideas and experiments without fear of state or religions reprobation. Hundreds of academies formed throughout Europe, some of these being the "*Academia deli Umidi*" (Academy of Moistures) or the "*Academia della Cursca*" (Academy of the Chaff).

Patronage was a way to gain 'freedom' in this restrictive context, but one which itself came with some constraints. The scholar was used mainly to enhance the patron's standing and social position, as in the case of Arabic caliphs. But, as also observed in the Arab world, such positions could be unstable and insecure. Galileo's deal with Cosimo did not last long because Cosimo II dies from tuberculosis at the young age of some 30 years. While Galileo did not lose his courtier position as Cosimo's son retains Galileo's standing upon the ascension of power, it is also the case that the intimacy which Galileo had formed with his former patron was certainly lost. Although Ferdinand II respected Galileo, he lacked the close and intimate emotional ties that had been formed between his father and Galileo.

Courtiers by definition play roulette in that their short-lived careers, ultimately depend merely on the good will of men.

Post *Siderius Nuncius*

Galileo certainly played his part as a courtier, engaging as he was wont to in a series of controversies: sun spots in 1613 or comets in 1623. Each of these controversies reveal typical rhetorical tactics used at the time.

One such tactic was the obfuscation of the original author. Galileo at one point is reacting to a publication by "Apelles," a pseudonym for the Jesuit Priest Christopher Shneier. In the debates pertaining to comets, both opponents used pseudonyms. Whereas his opponent was an alleged "Lothario Sarsi," Galileo himself used his student Mario Guiducci in the debate. The rhetorical device allowed its user to claim that they had not personally voiced such comments, and hence evade any possible socio-cultural implications. It also helped gain social support, in that it looked better for Galileo to be defending a young hapless student when in fact the ideas originally expressed (and consequently attacked) had been Galileo's all along. Galileo in *The Assayer* claims that when sick, Guiducci requested permission publish such ideas, and otherwise he was merely defending a young student. It is known that the original manuscript of *The Assayer* was written in Galileo's own hand. In the context of the enormous social and legal power held by the Catholic Church, it was a necessary device where truth was intermixed with falsehood to protect the true debaters from institutional reprobation and sanction.

Ironically, such a rhetorical device was the flame that would ignite the controversy between Galileo and the Church.

Galileo had written a letter to Castelli, expressing his views on the relationship between science and religion. Castelli, a typical mathematician, appears to have been rather naïve and shows the letter to third parties—to the point of even allowing others to copy it. Galileo naturally was furious, and obtains his original letter back from Castelli. In spite of the letter's return, the damage had already been done. Niccolo Lorini, an elder Jesuit who had at previously attacked Galileo but whom had backed down due to his ignorance of astronomical issues, now launches a 'frontal attack.' When Lorini receives a copy of the Castelli letter, he perceived it to be a direct menace to the Catholic Church. Still retaining some amount of influence and authority in the Catholic ranks, he sends it directly to the Inquisition. In his criticism, he does not attack Galileo directly, but alludes to the 'Galileans.'

Such incidents lead Galileo to write *Letter to Grand Duchess Christina* (1613), an important historical document more thoroughly documenting his personal views on the relationship between science and religion.

The sunspots controversy arises that same year when "Apelles" (the Jesuit Shneier) sent a book for publication to Marcus Wesler. Wesler, wanting to verify the validity of the information therein, sends it to Galileo seeking a brief letter of merit on the work. Galileo, however, responds with a lengthy exposition on the subject matter. The case is interesting because it sheds further light into Galileo's scientific mentality. He is not quick to accept or dismiss the claims in the work but rather thinks 'deeply' on the complex issues at hand, as well as obtains evidence before reaching any formal conclusion.

What were sunspots? Shneier claims that they were stars orbiting the sun. Galileo notes that this was unlikely given that stars tended not to change character, shape or luminosity. In fact, their constancy was their unique trait. Sunspots, by contrast, were so subject to change, that it was hard to tell whether it was the same bodies appearing on each rotation. Galileo also observes that celestial bodies did not have to be dark to make spots appear on bodies, using the phenomena of clouds on Earth as an example. It was the relative brightness of objects which rendered shadows. For example, when Galileo observed sunspots on cloudy days, these were still much brighter than the clouds themselves.

All the evidence seemed to suggest to him that sunspots were that were on surface of the sun, but he could not quite tell. As we now know, Galileo's observations hit the mark. The case is interesting because it suggested that Apelles was a Galileo 'wanna be,' implying in his book the notion of that bodies revolving around others were not unique to the Earth. Appelle's claim, however, was at heart a conservative interpretation which confused the issue rather than clarified it.

The Assayer 1623

Various comets appeared in the night sky, raising the issue with regard to their nature and origin. Guiducci (Galileo) publishes his analysis of the phenomena, but reaches no definitive conclusion. The piece is then savagely attacked by "Sarsi." In Galileo's defense of his student, three key points were taken up: 1) Sarsi identified comet distance by size/luminosity, 2) that the comet was generating its own light, as though a small star, and 3) that its tail was generated by friction/heat as it moved through the medium of space. Galileo's wit and rhetorical genius comes full force in the response. Sarsi had misrepresented the certainty of Guiducci's claims, as well as ignored common experience of light.

He validly observes that one cannot determine distance by a star's apparent size or luminosity. If such a claim were true, it would have revolutionized astronomy long ago in that the distance to all stars could have been immediately calculated. However, it is well known that stars vary in luminosity, and thus the claim is incorrect. With his dry wit, he points out that distance is relative: Naples to Rome might appear to be a short to a courier, while a walk from a household to a nearby church will appear distant to an elderly woman.

It was equally hard to tell whether objects shined by their own light or whether they were merely reflecting the light of other celestial objects. Sarsi would mistake a million pebbles lying on the shore of a beach as stars according to these criteria. Galileo provides a curious anecdote, where he observes the diffraction of light on windy day. As one would presume, if there had been no wind, the lake would appear as a mirror, with the sun perfectly reflected on its surface. On a windy day, however, the sun's image was broken up. However, if Galileo traveled far away enough, he would again observe a coherent image from a distance. He uses the anecdote simply to point out the difficulty of interpretation caused by reflections of objects.

However, in a third refutation, Galileo demolishes Sarsi and launches a direct frontal assault on Aristotelian physics with his acerbic wit. Sarsi was ignoring everyday evidence by arguing that a comet's tail was created by friction. When one rides a horse, one's face does not get hotter, nor does one's temperature increase the faster the horse gallops. Just the opposite, as anyone can attest. It is said that Babylonians cook eggs by swinging them in a circle. However, if one conducts the experiment, Galileo observes that it will not work. Given that conditions are the same, it would then mean that the eggs were cooked only because they were swung by Babylonians. (Not being a Babylonian, Galileo could thus not cook his eggs in this manner). Finally, Galileo proposes an experiment: take a compound steel bow and shoot an arrow with the strongest force. If Sarsi were to notice any browning in the feathers of the arrow, going at the fastest speed possible, Galileo would concede the point—particularly so if the arrow caught on fire, which it obviously would not.

Galileo argues with such an acute wit, it is not difficult to understand why he would want to enter the fray of public debates common to Florentine society. An intellectual heavyweight can only show his ability by comparison in a fight. Galileo would win a lot of debates in this manner, but at the same time make a lot of enemies—particularly within the Jesuit division of the Church.

Pretrial interactions with the Catholic Church

The year 1616 was an important one for Galileo. Although it was not the year of his most well-known trial, evidence gathered at the time would later be used against him in spite of fact that 1632 case allegedly concerned only the *Dialogue concerning Two World Systems*. As mentioned previously, Lorini sent a letter of complaint to the Inquisitorial body, but he had not been the only one to do so.

Tomacco Caccini also issues a protest, and Galileo is thus called forth to Rome for his first informal "inquisitorial hearing." These accusations would immediately fall flat on the ground, perhaps suggesting to Galileo that he was immune from repression by the Catholic Church and its Inquisitorial bodies.

It is in this occasion when he first meets Cardinal Matteo Barberini, future Pope Urban VIII. Barberini, of a humble background, actually defends Galileo at the time, leading to a close relationship between the two men. Galileo spent three months in Rome, but without any undue concern that should have worried him. Galileo does briefly fall sick, and he does receive an anonymous letter warning him there had been a meeting of Jesuit priests who wanted to damage his reputation by any means necessary.

A key figure in these discussions was Cardinal Robert Belarmine, a powerful figure behind the scenes whom had actually been offered the Papacy but rejected it. Pope Paul V would also be involved in these early direct interactions. However, Paul V greatly respected Galileo and admired much of the work he had done. Pope Paul V actually informs Galileo that none of the false rumors that were flying about would be listened to as long as he remained alive. It would thus suggest to Galileo that, so far as he could tell, the possibility of persecution was actually relatively minor.

It is paradoxical that men who greatly respected Galileo were at the same time some of the men responsible for the worst abuses of the Catholic Church. Pope Paul V initiated the Inquisition at the Council of Trent, which also drew up definitions for persecution as well as the *Index of Prohibited Books*. Barberini himself (Urban VIII) would canonize San Ignacio de Loyola, and hence validating the Jesuit Order. Up to the time, those who had most attacked him belonged to the Dominican orders.

It would have been hard to charge Galileo in 1616, as most of the allegations did not rest on any formal publications but were rather based on private letters that never reached the public eye. One of these had been a letter Galileo had written to Castelli, which Galileo later ended up expanding into a much longer and comprehensive treaties, now referred to as "Letter to the Duchess Christina of Tuscany." The Duchess was the mother to Cosimo.

Castelli, was invited to dinner at the Medici palace. One of the dinner guests was Cosimo Boscaglia, whose name might suggest him to be a member of Medici family but whom actually was not. At the dinner many abstract theological issues were discussed. Castelli dominated the discussion, and believed himself to have done quite well in the informal theological debate—or so he believes. Right after the dinner had ended and as Castelli was walking out the door, he is again called back into the now relatively empty room. A minor session was then initiated with few people, amongst which were the Duchess and Boscaglia. Throughout the whole impromptu session, Boscaglia never said a word, although intensely looked at Castelli. Christina asks point black whether Galileo's observations did not in fact imply a confrontation between religion and science. It is clear that Boscaglia had been manipulating Christina.

At its core, the *Letter to Duchess Christina* is an appeal to the recognition of a boundary between science and religion. Akin to the well-known statement

136

"render onto Cesar what is Cesar's; to God what is God's," Galileo had begun reading theology so to address issues his many discoveries were raising. His discoveries were in fact so unusual, that Galileo himself could not believe their implications. He would thus spend much time talking and debating among friends to hash out his own ideas and resolve conflicts. Most of the citations in Galileo's letter were based on St. Augustine's *De Genesi ad literam*, a well-known theological treatise. Galileo in essence argues that the truth of God is so complex, that it had to be simplified in order for common man to understand it. Religion was not to be confused for science.

It is theologically well known, Galileo points out, that the Bible was not to be taken literally. If taken at face value, it could be shown to be full of contradictions. None of the planets, with the exception of Venus, are ever mentioned in the holy book. Even then, Venus is referred to as "Lucifer." Much of it was written for common man; the image of God as a man is used so that the simple minded would abide by it. Galileo also notes that as God had given man higher faculties, it was injurious to now deny the use of such faculties. At its core, it was a bad idea to deny 'science,' actually using the word "geometry," in that its practice demonstrated truths that were undeniable. Galileo is very prescient by suggesting that Church would be injurious to its self interest over the long run by denying the veracity of basic truths so discovered.

Surprisingly, Galileo does explicitly defend Copernicanism, using the case of the Book of Joshua. In it, the Sun is described as standing still in middle of day. Galileo explores the implications of this statement.

For the sun to stand still in a geocentric universe, it would have to travel 360X in opposite direction from west to east, something hard to believe given the drastic change in speeds. The allegation also implied that entire universe had to stop and then suddenly begin its entire cyclical motion once more—an incredible assertion that was hard to believe. It was more feasible and much more reasonable to suppose, Galileo argues, that only one planet ceased its motion, specifically the planet Earth, rather than the rest of the entire universe. Even the very odd planetary behavior of Biblical imagery suggested the validity of the heliocentric universe.

Galileo made similar arguments in other writings. Presume for a second that Aristotle is correct in that a Prime Mover at edge of the universe: a very powerful entity that makes the world turn. However, the Book of Joshua's allegation would not make sense even in this world order. The Prime Mover has the power to move the entire universe but not to move a single planet? It defied logic to believe such a situation to be the case, given the vast power in the Prime Mover's 'hands.'

During the course of the 1616 informal hearings, the allegations against Galileo were modified, due to the help of Barberini. The term of "heresy" in the complaint against him is removed, and as well as the strict prohibition to teach Copernicanism. Barberini had himself been an astronomer, and was concerned how such a strong charge would in turn negatively affect his own practice: he would have not been able to study Copernicus without concern of any sort of persecution by the very institution to which he belonged.

Ultimately, Galileo reaches an agreement with Cardinal Belarmine, formally stating the terms of this agreement in a very important document. While Galileo would not be allowed to teach Copernicus as a faithful representation of the cosmos, he could nonetheless discuss Copernicus's ideas and astronomy theoretically. Galileo would keep this document very close to heart, later using it in this 1632 defense against the Inquisition.

But by that year, however, his two principal Catholic supporters had passed away and the political environment had significantly changed.

Trial 1632/33

As we have mentioned, the 1632/3 trial based its accusations on Galileo's *Dialogue of Two World Systems,* appearing the previous year. By August 1632, the book had already been placed on hold by Inquisition. Galileo had begun formally working on the manuscript at the age of 60—a date which is more important than might appear at first glance. In 1623, Galileo's *Assayer* had appeared, receiving a warm welcome from Barberini as new Pope Urban VIII. It goes without saying that this positive reaction likely suggested to Galileo that the cultural climate had further improved in his favor, and was likely a stimulus for publication. With a friend now as Catholic Pope, it might appear that the road was now clear to venture forth.

Dialogue of Two World Systems is a conversation over 5 days between Salvietti, who argues in favor of Coperniansim, Sagredo, a 'moderator' favorable to Copernican arguments, and Simplicito, taking a blind obedience to Aristotle's views rather than the facts presented. Galileo constantly criticized that had Aristotle been alive, he would have changed opinion, instead of ignoring all of the evidence that had accumulated: a) phases of Venus, b) the sun's spots, c) Jupiter's moons and d) the moon's imperfections. One should not be chained to the pages of old texts, but rather should follow the book of nature.

The piece is presented as a debate on two different philosophies, weighing the strengths and weaknesses of each. This formal characterization allowed Galileo to publicly claim not to be in favor of any one particular theory. However, there really can be no denying that tone of the work is wholly in favor of Copernicanism—as it should be. The notion that the entire universe should spin around the Earth for the sake of humanity was patently absurd. Galileo likely believed that the rhetorical device of a debate would allow Church officials to 'save face,' so that its internal reformers could more easily battle opposition within its walls. The Catholic Church is not a homogenous body, but rather was composed of various camps, and presenting too overtly a view could have had negative political repercussions. The Carmelite Paolo Antonio Foscarini had published a public defense of Copernicus, and the book was immediately condemned outright. The publisher was forced to closed down and Foscarini suddenly dies under mysterious circumstances at the young age of 36.

It is also important to review the Church's position. Belarmine in 1616 had counter argued that, if on ship sailing away from the sea shore, we do not suppose that shore is moving away from us but rather that we are moving away from shore. A similar argument could be made with regard to the universe. Looked at from point of view of Earth, it was difficult to believe the implications of motion. It is obvious that the Catholic perspective was not giving due weight to implication for the rest of the universe: that millions of stars should be moving at incredible speeds. As the scale of the universe increased, the stars now being much more distant, this huge circle would imply ever more accelerated theoretical speeds of stars. It is likely that Galileo's sensitivity to motion meant that he was more attune to these gross contradictions than others.

Yet, was his persecution strictly an issue concerning theology?

It is patently clear that the debate between Galileo and the Catholic Church was not strictly a 'scientific debate' due to the many non-scientific issues involved: political, psychological, and sociological. One should never undermine the totality of factors at play in a debate; a debater could hypothetically be "ultra-rational" and logically "win," while being declared its looser in the end. Galileo's case is such an example: one afflicted by strong psychological components which were also tied to politics.

Galileo had in fact traveled to Rome in 1631 to obtain Inquisitorial permission to publish his *Dialogue*, going by its rules and protocols. He was fortunate to some extent in that the person overlooking permission had strong ties to the Medici court. Even then, it took months to evaluate, and Galileo eventually is forced to resort the help of Ferdinand II in order to secure a decision. Ricci requests only a few changes, modification of the title and a reworking of preface and epilogue, and finally approves it for publication.

By 1632 Barberini (Pope Urban VIII) was a very changed man, however. He had become one of the younger popes, more akin to a general often seen riding his horse as well as being a commanding figure. However, the Thirty Years War weighed heavily on him; he had studied at Jesuit school and supported the French. However, he was being attacked internally for not taking defending catholic theology, placing him in a delicate political position within the Church.

One can easily obtain a notion of his debilitated mental state from a number of anecdotes from the time. Barberini ordered all birds from the garden at the Vatican to be killed because their noise would not let him sleep. Although Galileo arrives to Rome in February, his trial is not held until May. The pathetic reason for the delay was that Barberini had been vacationing with nephew by the coast. Barberini believed he was constantly surrounded by Spanish spies, and was in such a paranoid emotional state, that he dared not talk out loud in public, choosing to whisper every important conversation.

But there is more.

A key turning point in trial was the evidence used against Galileo. This evidence strongly indicates that the 'audience' of the trial were not just its 10 cardinals, three of whom ended up siding with Galileo, but rather Pope Urban VIII (Barberini). Its sole aim was to undermine the close personal relation that Barberini had with Galileo.

Given that the Catholic Church was so centralized in the figure of the Pope, any changes of opinion by the person holding the Papal seat would have enormous political implications. If you could make the Pope distrust a figure, as in the case of Galileo, that figure would lose all papal sanction, hence making him much easier to attack. Inquisitors seem to have "played a trick" on Barberini to make him believe Galileo had been taking advantage of him. The "minutes" of the 1616 discussions were presented as evidence of proof of Galileo's betrayal of the Pope.

Notice that Galileo had never seen or heard of such minutes, which easily could have been falsified, and today would never be admitted as evidence in a modern court of law. The minutes allegedly registered that the agreement between the parties was contrary to the formal document Cardinal Belarmine had written and handed to Galileo. The minutes 'evidenced' that Galileo had been prohibited from discussing, teaching or writing about Copernican ideas at all— even hypothetically.

This false proof turned out to be the 'killer blow.' Galileo was then asked if he had ever informed Pope Urban VIII of this agreement, to which Galileo naturally said he had not, as he had never seen the alleged minutes before.

This single incident likely did more to damage the 'case of Galileo' than any other single proof or evidence. Barberini was made to look it as if he had been played the fool. Barberini would never again speak to Galileo. The wave of support Galileo presumed to have been riding on suddenly collapsed in on itself.

Galileo was placed under house arrest for the rest of his natural life. Even though he had an assistant, the able mathematician Vicenzo Vicentia, Galileo ultimately lost sight in both eyes. Galileo died five years later in 1642 before three close assistants: his son Vicenzo, now a lawyer, his assistant Vicenzo, and Evarista Torricelli.

Dialogue of Two World Systems was banned for centuries in Catholic countries. Galileo would not be 'pardoned' until 1979 by Pope John Paul II, but even then would only be a partial pardon.

Newton and the Plenum:

The Rise of Mechanical Philosophy

IMAGINE, IF YOU WILL, THE FOLLOWING. You first go to college, deciding to buy with your meager funds the latest books on diverse topics that look interesting. You get a humiliating part time job as a maid of sorts, which you detest but don't resign as it pays for your schooling. On the third year, a bad bacteria hits your university town, forcing you to leave the campus for fear of succumbing to the disease. At first you were not going to do anything, but your mother made such a desperate plea that you move just for her sake. When you go home, you invent the calculus, a conceptual reformulation of Euclidian geometry, and come close to the law of gravity. When you return to college, you never again work on mathematics for the rest of your life.

Such was the genius of Isaac Newton.

It goes without saying that the importance of Newton cannot be understated in the history of science, a figure defined as the culmination of the Scientific Revolution. "If I have seen farther, it is because I stood on the shoulder of giants," he used to say. There can be no doubt that the world of science after Newton was radically different from that which preceded him, an aspect which encouraged the use of the term 'revolution' to characterize this broader shift. Many of our modern presumptions, such as the breadth and limitations of scientific knowledge, come from Newton. The notion that a mathematical formula (law of gravity) could explain the world becomes a 'trope' in science: the dream of a final theory. Many physicists afterwards would dedicate their entire lives to find another such unifying formula. Newton was to the 17th century what Einstein is to us in the 20th.

141

Newton creates the mythical image of the scientist, that of the absent minded professor. He was distinctly anti-social, never marrying and tended to avoid contact with people early on in life—for which he likely suffered various nervous crises. He recuperated from his mental illnesses due to good friends as John Locke. Later on in life, he becomes the President of the Royal Society of London, attending every single meeting, and commenting on the experiments performed, something which no other president had so steadfastly done before. He imbues the Royal Society with a distinct culture and social dynamic, continuing to determine its course until his dying days.

Yet Newton might have never come to be. He suffered greatly as a young boy; when born, he was so frail and weak that many did not think he would survive. His tortured personality is both cause and symptom of his scientific greatness, two sides of the same coin. To understand him, we have to look at both sides of him, both what is 'good' and that which is 'negative'–as with anybody else.

Newton, however, does not emerge in a vacuum. Important scientific and philosophical changes during the preceding period establish the predominance of mechanical philosophy, setting the intellectual context in which his innovations would be grounded. Newton could be defined as the culmination of the mechanical treatment of particles in collision, the stoic 'plenum' of the Greeks. However, while we might presume that early mechanical philosophy was tailor made to go along with quantification, the two actually clashed during the early modern period. More oddly still, mechanical philosophy also did not coincide with experimentalism. Again, while today it is taken for granted that the two go together, such was not the case in their origins.

One of the key figures in the rise of mechanical philosophy was Rene Descartes, one of the most cited names in Newton's now-famous college notebook he referred to as his *Waste Book*, a blank notebook he had inherited from his detested stepfather Barnabus Smith. While Newton would likely have chosen not to use it, paper was then too expensive to be so casually discarded. Smith had forced Newton's mother to abandon him at the age of three, in order to begin a new family with her.

Newton's tortured genius is both the cause and symptom of his greatness.

Renaissance Naturalism

The origins of experimentalism began with various thinkers whose different approaches would be ultimately blended into physics. What we today conceive of as 'science' is not necessarily how it was seen during the seventeenth century. One of these strands was Renaissance naturalism; William Gilbert and Jan van Helmont are key figures in this movement.

Gilbert undertook eighteen years of systematic experimentation on the lodestone (magnet) culminating in *De Magnete* (1600), a work unparalleled in its thoroughness. He was actually a physician who was initially interested in finding practical ways for improving sea navigation. Gilbert systematically analyzed magnets under diverse conditions to test their properties, thus producing the most comprehensive treatise that existed up to that time.

Some of his experiments would later enthrall Einstein into a career in physics. Gilbert placed a lodestone of a circular shape floating on water, which would always point due north, regardless of how the object was rotated. The magnet's properties were not like those of light, in that they did not cast shadows, and the dip of a compass needle varied with latitude. He looked at the behavior of magnets when cut at the poles, and eventually came to his one 'fundamental conclusion": The Earth was an enormous magnet.

Gilbert's magnet would come to temporarily substitute the 'world engine' role previously played by Aristotle's crystalline spheres, until the advent of Newtonian gravity. Gilbert's magnet, as Newton's gravity, held action at a distance, not requiring the mechanism of efficient cause that stood so closely at the heart of Aristotelian cosmology.

Van Helmont is perhaps more well known for his important contributions to chemistry, specifically his famous ten-year willow tree study. Controlling for water, air and soil, van Helmont concluded that its 200-pound growth could only be accounted for by the water it had regularly received, as all other factors had remained the same. While we would not agree with his conclusions, his persistent weight measurements are a modern methodology. He demonstrates that, contrary to popular opinion, trees did not grow by 'eating' soil, and is the first person to introduce the term "gas." These discoveries and ideas are published posthumously in *Ortus Medicinae* (1648) after conflicts with the Catholic Church's Inquisition.

Yet van Helmont's role was more important that this widely cited experiment. His action must be understood within his unique world view—one which was surprisingly similar to that of Gilbert's. Both believed in a prevailing 'life force' in the universe, 'archeus' for van Helmont, 'magnet' for Gilbert. More importantly, this living spirit had a 'personality,' which meant that the only way to truly understand nature was via experimentalism. As there were no consistent patterns, one had to 'listen' to nature's messages as one would in a dialogue with another fellow citizen.

Contrary to what one might suppose, Renaissance naturalism's 'organic viewpoint' thus helped lay the foundation of modern experimentalism, which later would be ironically tied to its antithesis, mechanical philosophy. Furthermore, the rationalist viewpoint held that, as Descartes believed, since the world is the embodiment of mathematics, one did not need to perform experimentation as all such testing would only serve as a superfluous demonstration. All one needed to do were the calculations, which would reveal nature's truths. Galileo in fact took a similar approach, arguing that experiments were only rhetorical devices to show his ideas to his opponents. Most of the actual work was done as thought experiments or mathematics. Only when publicly attacked was Galileo allegedly forced to do "experiments" to demonstrate the truth or falsity in a public forum. While this is not an entirely valid characterization, Galileo did truly experiment, there is a grain of truth of its characterization. Quantification, experimentation, and mathematical analysis did not necessarily go hand in hand to early modern philosophers.

Brahe was the first to detect the cosmological implications of comets, such that crystalline spheres did not exist, but also was led to the issue of a 'mechanism' required to keep planets in place. What prevented them from flying off? Such was the power of Gilbert's work, that Kepler incorporates it directly into his own: The Sun had an *'anima motrix'* that caused planetary revolution. It is also important to note the inherently 'conservative nature' of the concept. It was still abiding by criteria of Aristotelian "efficient cause," of an original mover or a 'ghost' behind every violent motion. However, Kepler further develops this notion, changing it from Gilbert's *'anima motrix'* to something akin to our modern notion of 'force.'

It is striking that none of its participants fully understood the implications of the changes they were undertaking: our revolutionaries apparently were not fully aware of the intellectual revolution they were starting. Galileo, father of modern day physics, couched his studies within context of Aristotelian natural motion. Kepler, for his part, did not presume a universal motion common to both planetary and stellar phenomena. Intellectual revolutions are not demarcated by clear and coherent developments, but are full of bizarre and strange twists along the way. If it were otherwise, they could be easily plotted on a graph and statistically predicted, lacking that trait of 'magical change' which is so characteristic of such revolutionary periods.

Father of Mechanical Philosophy

The importance of Rene Descartes (1596-1650) to the Scientific Revolution cannot be underestimated. As the 'father' of mechanical philosophy, Descartes established the key questions of the era: how do particles collide? Like a catalyst, however, the inconsistent answers put forth by him were soon superseded by other authors, serving to 'erase' the enormous importance of his original research paradigm. While his name would be one of the most commonly cited in Newton's notebooks, Newton would not be the only one influenced by him.

Christian Huygens, whose father was a friend to Descartes, so thoroughly develops Cartesian 'science' that he comes tantalizingly close to the law of gravity. The first publication by Gottfried Wilhelm Leibniz, who had been a student of Huygens, crushes Descartes' analysis of collisions, showing its many contradictions. Leibniz would also be known as one of the creators of the calculus; and although we give recognition to Newton for that, it is Leibniz's expressions which are today used in its modern form.

As a young man, Descartes had also invented the new math of analytical geometry: algebraic descriptions of geometrical forms which would allow for the rigorous analysis of particles along the now well-known x, y, z axes. Some even suggest that he came very close to creating an "Einsteinian space-time," 400 years 'ahead' of its time.

Descartes asked the right questions, even if provided very incorrect solutions, proving the adage that it is not the answers that are important so much as to the questions that are first raised. Yet, who was this important figure?

As a youth, Descartes attended a Jesuit school. While expecting to find 'enlightenment,' the answers provided were so contradictory and confusing, that he quickly realized that the state of knowledge had not progressed much in 2,000 years. Like Galileo, Rene showed early mathematical genius and begins studying math on his own, quickly dominating much that had been done to date. In a fateful decision, he decided to 'start from scratch,' building a new knowledge system from the ground up. He would reject any ideas that were not clear and distinct, thus laying the first brick on his important edifice.

He briefly enters the military, and did not actually fight but rather used his employment as an opportunity to travel throughout Europe. His *Le Monde* was about to be published in 1632 when he hears of Galileo's Inquisitorial persecution following the *Dialogue of Two World Systems*. Wanting to avoid such persecution, he cancels its publication, and proceeds to instead write up more detailed works on specific topics, which help lay the groundwork of his magnum opus—published posthumously in 1664. The *Le Monde* creates a new world view which replaces Aristotle's, who had so dominated Western thought for a millennia. Descartes' influence is comparatively short lived, however, becoming the main view in Europe until it is overturned by Newton's *Principia* (1687).

It is somewhat amazing to consider that that two important works from the Scientific Revolution could have been published in the very same year, that of Galileo and Descartes, suggesting the existence of a thriving environment of intellectual innovation that had been squashed by Catholic church.

By the end of his life, Descartes was internationally recognized and had become a 'famous' philosopher. At a relatively young age, he is hired by the Queen of Sweden whose habit of waking up early in the morning forced Descartes to walk through cold to give his respective lessons on a daily basis. As one might imagine, this exposure to the extreme temperatures apparently led to Descartes' early death at the age of 54 from pneumonia. Smart men can be quite stupid at times; knowledge in one area does not imply expertise in all others. It is to Cartesian philosophy which we must turn to appreciate his original mind.

His most famous phrase: *"Cogito ergo sum"* (I think therefore I am) serves as the foundation stone for his world structure. To him, the mind was *'res cognita'* whereas the world was characterized of *'res extensa'* whose primary qualities included weight, extension, body, and motion. Descartes thus creates a distinct boundary line between the two, so much so that it created a new problem as to how they interacted, a solution Thomas Hobbes would try to answer, himself laying the cornerstone of modern psychology. Descartes's shallow proposal of the 'pituitary gland' as the seat of the mind was a weak ad hoc attempt to give cohesion to philosophy; however it allowed him to move on to more important aspects.

The key point of his stance was the following: the only thing that he truly knew was his thinking. The world could have been created by an evil demon, constantly attempting to fool and trick him. Hence Descartes could not trust the evidence received by the senses as valid. The notion looks odd to us moderns, as its enormous importance is not immediately obvious. In essence, it implied that the world could be something other than that which our sense experience revealed; what we believed the world to be might be entirely different from that which it actually was. Curiously, Galileo had postulated a similar notion, in his distinction of 'primary' and 'secondary' qualities. Secondary qualities as color and smell could not be used as valid evidence, as these resided within the mind and were not necessarily situated in the object perceived. For Galileo, it was only the primary qualities that mattered: weight, motion, number; these could be measured and counted.

Cogito's demon was of enormous importance in that it allows Descartes to build up a mental construct of the world entirely different to what early modern Europeans presumed it to be. This philosophical state constituted a complete break with medieval philosophy, where the world of the senses is what mattered most. The world had been made for the sake and pleasure of man, and hence to be interpreted at face value for its religious symbolism.

From this base, Descartes proceeds to build the world anew—atop the stoic plenum. For him, the world was full of particles of three different sizes, some of which were so small they could not be seen. Given that a void could not exist, every action had a counteraction; in the absence of a void, all surrounding particles would be affected, and in turn affect those near the neighboring localities, as in a chain reaction. These motions in turn created innumerable vortices and eddies, the total quantity of motion in the world being fixed by God at its creation. All natural phenomena was hence to be explained on the interaction of these particles: solar system, gravity, light, and magnetism.

For example, as the sun moved, it created a countermotion that kept the planets on their orbits. Similarly, as a planet moved through the plenum, particles were moved in such a way as to create vortices which perpendicularly pushed down the objects on its surface, creating the phenomenon of gravity. Light was the result of pressure waves from objects reaching the eye. Descartes's explanation of magnetism was truly ingenious, using screw like particles whose motion was dependent on its orientation, thereby producing both attraction and repulsion.

In the end, however, Descartes had created a nonsensical world that would not have stood up to rigorous experimental verification, and did not readily coincide with Kepler's or Galileo's work. Ironically, it did not lend itself immediately to mathematical analysis given its contradictory nature. The Cartesian world view provided a new space for analysis so utterly lacking in Aristotelian scholasticism. It certainly was intriguing, and one might suggest that the European intellectual landscape was ready for a change.

The Cartesian new world view did have one crucial implication: the role to which it attributed the impact of particles became the central focus of scientific explanation. Since the universe was a plenum of particles in constant contact, the outcome of the impacting particles ultimately determined the nature and shape of the universe, and hence received the most emphasis. Ironically, in spite of having created analytical geometry, Descartes the rationalist did not undertake mathematical analysis of his own ideas.

In his 'mechanical philosophy,' Descartes postulated seven case studies of those impacts, with variations of direction, speed, size of particles. Various scenarios were postulated to create models from which to understand particle collisions. Yet his modeling proved to be intellectually inconsistent.

Imagine having two particles of different sizes, one 'large' and the other 'small.' Descartes postulated that when a small particle hit the large one, it would bounce back with the same speed from which it came; however, if the same large particle hit the smaller one at the same speed and direction, it would imbue on the smaller one with some of its motion, accelerating the smaller one in the same direction while proportionally losing speed. Descartes's model is thus incongruent in that it postulated different outcomes for the same events. He had committed the same error as Zeno in his turtle - Achilles paradox: using varying frames frame of reference for the same events, Zeno 'proved' Achilles could never catch up to the turtle.

Resolving these obvious contradictions, however, ultimately led to key advancements resulting in the Scientific Revolution. Descartes sets the questions to be asked and hence the intellectual and philosophical space in which Newton operated. Descartes establishes a new context from which to look at the universe: the world as a totality of atoms in collision. Anyone who could create a formula for these collisions would ultimately discover the 'keys to the kingdom.'

Quantification

Evarista Torricelli had been one of Galileo's best disciples, sitting at his bedside during his final days. Torricelli would further extend Galileo's work, like his mentor using a combination of imaginary thought experiments and concrete experimental analysis. Vacuum pumps at time had been limited to pulling up water to only 32 feet of height. Why?

Torricelli inverted test tube in water, lifted it up, and an empty space would form at top of tube; what was the nature of this empty space. Was it being created by vacuum or by air? When you suck at a straw for milkshake, are you 'pulling' the milkshake into mouth? Torricelli noticed that if he used heavier liquids as mercury, he could minimize the space to only 4 inches high rather than 32 feet, making its analysis that much easier to conduct. Torricelli came to realize that the vacuum was being created by the 'weight of air.' Air in the atmosphere has weight and creates pressure, depending on column of air above a given object. When one drinks a milkshake, the mouth is only reducing its internal pressure, allowing the atmosphere above to 'push' the milkshake into the drinker's mouth. We are all sitting at the bottom of an enormous 'pool' of air which, although much lighter than water, still excerpts pressure on the objects beneath it. Torricelli had invented the world's first barometer.

Torricelli's findings were famously tested in 1648 by Blaise Pascal and his brother in law at the Puy de Dome, a mountain several thousand feet high. In a

147

French region known for wine and glass making, Pascal took various measurements at different elevations. Unlike Galileo, however, Pascal's experiments were not meant as a rhetorical stratagem only to demonstrate an idea, but rather to show nature in her true form. He found definite proof by showing that each elevation had diff. sized columns of air, and hence different air pressures. The same technique was independently discovered by Francisco Jose de Caldas in New Grenada, believing he had come up with a new device to measure elevation.

Torricelli had also worked with 'impetus' and motion, coming up with an original analysis of momentum. He noticed that the impact of a falling 10-pound rock varied greatly; if it fell from 10 feet, its impact was negligible, but the opposite was true when it fell from 100 feet. Although the observation had already long been established, Torricelli explores it in further detail.

He characterizes the force of falling bodies within the context of a water fountain. A fountain's water flows at a certain rate; at every instance of fall, a given amount of water will be produced by the fountain which could be weighed and measured. For Torricelli, 'impetus' is similarly accumulated over time, which he began referring to as 'momentum.' Momentum, as the fountain's water, increased with each corresponding instance of time. The longer an object fell, the more momentum was accumulated, each instance of fall adding a small amount of momentum to its total.

For example, in the case of huge ship and small tug boat at port, a man pulling on a large ship will take hours to drag the large ship in. But when it touched port, the larger ship had a much more substantial blow when compared to the small tugboat. For Torricelli, the 'momentum' generated by the man pulling the rope was accumulated in the large ship, hence accounted for the enormous disparity of impact upon landing. The small boat, pulled easily and quickly, thus hit the port with a negligible force because it had not accumulated momentum.

Christian Huygens and Gottfried Leibniz extended the work of Torricelli and Descartes. Huygens shared some similarities to Brahe in the sense that it was unusual for a member of the economic elite to go into "science" (natural philosophy), more often opting for diplomatic or political service. Huygens however demonstrated early mathematical ability, obtaining early recognition for his work on optics, which allowed for a substantial improvement in the quality of telescopes of the era. It is Huygens who sees the rings of Saturn for the first time. His early renown is such that Huygens is elected in 1666 to the elitist French *Academie des Sciences*. He settles in to critically analyze the work of Descartes via studies of the pendulum—specifically the Cartesian analysis of particle collision.

Huygens differentiates the respective frames of reference of collision which had given his predecessor so much trouble. He begins by taking a man with 2 metal balls swinging at the end of 2 strings, each ball hitting the other, essentially two pendulums colliding. He redefines the collision by giving a metal ball to two men each, and places one man on a small boat which moves parallel to the other with the same speed as the prior case. By this manner, he recreates the same prior

148

collision. This method thus allows Huygens to isolate the varying frames of reference in his analysis.

More importantly, he asks: under what condition would collisions lead to no change of velocity? He realized that what remained constant was the square of velocity (mv^2) and calculated the force needed to keep body within circle of pendulum at a 45° angle, distinguishing centrifugal force (force pushing outwards) from the centripetal force (force pushing inward) which would later be used by Newton. The best measurements of gravity were thus undertaken by Huygens with his conic pendulum, noting that at 45 degrees, the object was balanced between the inward pull of gravity and the outward push of the centrifugal force. Whereas prior measurements yielded 24 and 30 feet per second, Huygens arrives close at to the modern day measurement 32 feet per second.

This aspect of his work was 'culminated' by Gottfried Leibniz, who had gone to Paris to finish his studies with Huygens. Huygens immediately recognized his intelligence and hires the young philosopher. In 1686 Leibniz publishes his famous *A Brief Demonstration of a Memorable Error by Descartes*, clearly and succinctly putting forth the key mistakes made by his predecessor, and setting forth the law of conservation of motion.

Leibniz notices that a four (4) pound ball falling one (1) foot would have the same effect as a one (1) pound ball falling four (4) feet. Imagine, now that our one (1) pound ball bounced back four (4) feet from its original fall. Leibniz realized that if the ball bounced back 5 feet, he would be able to generate work only by 'capturing' the extra foot generated from it. This 'result' would produce more energy than put into the system. Leibniz in effect had shown the impossibility of a perpetual motion machines; each collision had to result in the same amount of energy at the end as at the beginning. If all particles in the world generated more energy than their initial state of collision, the world would have long ago exploded; on the other hand, if each particle collision resulted in a substantial loss of energy, then the universe would have long ago dissipated into a vast heap of motionless particles.

By resolving Descartes' contradictions in the collision of particles, Leibniz comes up with the law of the conservation of motion, which would serve as the foundation of thermodynamics two centuries later.

Biography of Newton

Sir Isaac Newton was born in the cold winter month of December 1642 in Woolsthorpe, incidentally the same year Galileo died. A premature child, maids sent to fetch items took their time on route because they believed it would not survive. That he was baptized a week after birth indicates how frail he must have been as a newborn. His father "Isaac Newton" had died a few months previously, which must have led to a few months of stress for his mother Hannah, possibly causing his premature condition.

His father came from a family that had been slowly gaining economic prosperity and social ascendancy as sheep herders, as the area showed greater land concentration clustering in the Newton family. His mother originated from a patriarchal family "Ayscomb" whose fortunes had been the inverse of her husband's, declining in prosperity. While the mother married 'down' from a social standpoint, her family still retained upper class values—a feature which would be important for the young Isaac as they would place a great deal of emphasis on education.

After three years of life as a single mother, Hannah married a preacher in nearby town, Barnabus Smith. Smith, however, imposed a number of terms and conditions on Hannah, one of which was that her young son could not live with the couple. In exchange, Barnabus would provide a parcel of land to the now abandoned child. Barnabus and Hannah have three more children, but as he was already 63 years old when married, Barnabus dies shortly thereafter, a decade later.

So, at the tender three of age, Newton is traumatically separated from his mother and sent to live with his grandmother—leaving an indelible if undefined impact on his character. Oddly, Newton never mentions his grandmother in his writings, so we do not know what kind of relationship the two had. There can be no doubt, however, about the hatred he felt toward his parents. Naturally feeling rejected by his most intimate circle, Isaac at one point threatens to burn their house down along with its inhabitants. After the death of Smith, at the age of 11 Newton returns to live with Hanna and new kids. They did not seem to get along and a year later Newton is sent to Freeschool in Grantham, which would end up being a very important experience for Newton, crucial in his intellectual formation.

The schoolmaster Henry Stokes had studied at Trinity College and would come to admire and respect Newton. As an educator, Stokes gave his students an unusual curriculum for the time. Aside from traditional learning of Latin and other practical skills, he taught more mathematics than was usually the case. A special relationship develops between Stokes and Newton, and is revealed by various incidents towards the end of Newton's schooling. When Newton leaves to study at Trinity College, Stokes gives a speech in his honor; the last year prior to Trinity, Newton had moved in with Stokes where he receives a more intensive training than the rest of the students.

While at Freeschool, Newton lives with William Clarke, an apothecary whose family would also play an important role in Newton's emotional development. The family consisted of various boys and an older daughter. To be expected, Newton fought with boys, while romantically falling for their sister. There was a big fight between them one day at school, and although Newton was winning the fight, he decides to back down. However, the local minister's son calls Newton a coward for doing so, whereby Newton proceeds to grab the priest's sons ear and bashes his face against the school wall. Newton develops a romantic attachment to the eldest daughter, for whom he was constantly building model doll houses. This fact in and of itself demonstrated that Newton had a great deal of manual dexterity, which benefitted him greatly in life.

There can be no doubt that the environment in which Newton was raised during these years was a nurturing one. Newton obtains a sociable space in which to grow and develop. His room was covered with all sorts of drawings and portrait sketches, showing that Newton developed a degree of artistic skill. He also learns about the basics of chemistry and chemical manipulation ('alchemy') from Clarke. Newton obtains a book by John Bates, *Mysteries of Nature and Art,* which he reads voraciously. Of equal importance, Clarke's wife was friends with a prominent Trinity College professor, Henry Pemberton, who would come to play a decisive role throughout Newton's career. One can easily visualize that Pemberton was informed of Newton's significant achievements as a youth, likely identifying Newton's promising brilliance early on. Newton was fortunate in that he lived within a protective social environment whereby information of his early achievements were diffused to important 'higher up' figures with close affiliations to the family.

Newton certainly developed many impressive abilities on his own. He was constantly building models and sundials, the latter which were extremely accurate. A sundial of 'equal spaces' would lead to incorrect timekeeping as the length of an 'hour' varies throughout the year. Surprisingly, Newton's dials were adjusted to account for this variation. He learned so much about the cycle of the sun, that throughout his life, when asked for the hour, he could always tell the time by looking at a shadow. Newton also built all sorts of wind and watermills. A new windmill was being constructed in town, and Newton builds an exact replica improved in design by combing it with a watermill. It was clear that Newton was a very gifted student. However, as is often the case, he had a hard time making friends. Newton was very competitive, perhaps overly so.

When Newton turned 16, he was initially sent back to the family farm to become the new landlord. That Newton had now become the adult male of the family, which meant he had to assume a new role in life. But Newton was so horrible at it, that all servants were greatly relieved when Hannah finally grants permission for Newton to go off to college. The stories told are funny, typical of the absent minded scientist. He was so focused on his experiments and models that he often ignored his duties and responsibilities. The farm's pigs invaded the neighboring homes because Newton had not mended the fences—for which the local council imposes fines on the young landlord. On various occasions, Newton would forget to remount his horse after a steep hill, walking back absentmindedly to the farm, and arriving hours after the horse had already returned on its own.

Absolute Space and Time

Newton first enters Cambridge University (Trinity College) in 1661. His mother had been very stingy, in spite of recently becoming a wealthy widow. In spite of her annual income of 700£ per year, she provides her son only with 10£ annually to go to the university. With these meager funds, Isaac buys a chamber pot, pen, ink, and a lock for his desk drawer. Newton also brings his *Waste Book*; contrary to its name, some of the most important notes ever made in the history of science would be written in that notebook.

Newton obtains the position as a 'subzicar' at the university, a low rated part time employment to help sustain himself financially. Although having recently served as a well-off landlord, he was relegated at the university to the lowest rung on the social hierarchy. As a class of students, subsizars were expected to serve the rest of the students, and were not to interact directly with the nobles to whom they served food. Each group even had different vestments to easily differentiate amongst themselves. The experience was humiliating and hard for him. One day, while moping about in the yard about his circumstances, he made one of his few friends in life: John Wickins. Both agree to support each other throughout the harsh ordeal.

Yet, Newton soon ran across many of the university's limitations. Upon arriving, Newton better understood Sanderson's *Logic* than the professor who was supposed to tutor him. Needless to say, they did not meet much after their first encounter. Cambridge University as a whole had a curriculum that was typical for time. Its outdated scholastic requirements for a diploma were seldom met, and local prohibitions such as visiting the town's cafés were just as widely ignored.

However, the city had one reigning glory: its bookstores. Cambridge was home to a thriving book trade, and it is through these bookstores where Newton obtains his advanced education. He begins buying books that are now considered classics of the Scientific Revolution: Galileo, Kepler, Descartes, etc. The best way to understand Newton is not by his associates but by the books in his library.

When he first came across Descartes' *Analytical Geometry*, Newton could not make heads or tails out of it. Consequently, he began to slowly read it, sentence by sentence, paragraph by paragraph, page by page until he gradually came to master it on his own. When taking the required examination, given by Isaac Barrow, Newton actually does rather poorly. Barrow, who held the new Lucasian Chair of Mathematics, tested Newton on Euclid, of whom Newton had never heard. Although the student passes the exam, Barrow instructs him to learn Euclid, presuming that Newton could not possibly know Cartesian mathematics as the former was a prerequisite for the latter. Needless to say, Newton quickly 'catches' up to speed. When Barrow later finds out about Newton's self-studies in analytical geometry, he is astonished, and becomes a mentor to the student. Newton continues buying books in mathematics, and soon catches up to the leading edge of mathematical research in his day.

Yet Cambridge was not all that bad. From its professors Isaac Barrow and Henry More, Newton obtains his key metaphysical constructs of absolute time and absolute space.

Henry More was a colleague of Barrow's at Trinity. This might suggest a mutual influence, yet More was not a mathematician but rather a theologian. He had been initially attracted to the Cartesian world view, but the more he thought on it, the more he came to realize that Descartes' plenum left God completely out of the universe. He was the first to come up with the notion of the universe as a clockwork, first formed by God who then set it in motion to go on its own. Even though a theologian, More also begins to critically analyze various assumptions

152

about Descartes's worldview, specifically that of space in which his vortice model operated.

Descartes rejected the existence of a void, pointing out that it was relative space or the relation between particles (interactions) what mattered. More took a critical look at this notion, to argue that that it inherently implied the notion of absolute space. Three particles vertically aligned with each other will shift relatively to one another, but within the context of an absolute grid work. More also imbues "space" with religious/theological connotations. For him, space existed before and after the formation of Universe. More becomes philosophically aware of the immense magnitude of space, and comes tantalizingly close to equating it with God, but does not take that step. He attributes 20 important qualities to space, including the traits of being uncircumcised, incorporeal, and omnipresent. Space had no boundary, no beginning, and no end—all traits that could be used to characterize a deity. The universe presumes the prior existence of God and angels.

In other words, More believed space preceded the existence of the universe, a notion with many similarities to Barrow's own ideas of time. Yet Barrow, as a mathematician, wanted to imbue time with a more 'concrete' and 'specific' meaning that would render it mathematically useful and practical.

What is time? If there is no change in the universe, did this mean that time ceased to exist? It was clear that time is a measurement of change and motion, without which it is impossible to detect; the patterns of the sun and moon are used for calendrical purposes for this very reason. However, to truly measure time, one needed to have consistently uniform repetitive change. Barrow well recognized that 'regular' patterns in astronomy were in fact irregular over long periods of time, ultimately undermining man's measurement of time and change in the universe. One could not know whether time in one century had been the same as that in another century because of this inconsistency. Barrow thus concludes that if local perfectly uniform and consistent rhythms could be created, one would be able to detect 'universal time.' Barrow, in essence, had pointed the way to an atomic clock.

For both More and Barrow, time and space are absolute, becoming Newton's principal metaphysical notions. Both existed before and after end of universe, and equally recognize human limitations in that one can only determine time and space relatively, using a common unit of comparative measurement. The exact definition of this unit was an arbitrary decision; rather, it was the proportions between the respective groups which is important.

Proportionality as expressed by ratios for the Greeks is the same as our modern 'number.' However, because measurements could not be not accurately taken, these would ultimately affect the work of Newton. One could not know with absolute certainly whether a law was valid or not; rather one had to issue an approximate 'good enough' confirmation.

In 1665 the plague hits Cambridge. At first it was barely detectable, with only a few individuals falling victims, but it quickly spreads throughout the city. King Charles II and most of its inhabitants leave the city while the epidemic died down. Incidentally, a few students stayed during entire ordeal at the University of

153

Cambridge and were relatively safe, as the low population density made its contagion difficult. The only letter that was kept of Newton's mother is one written to her son during this crisis, concerned as she was about his wellbeing and pleading for him to return home. Newton's two years back at Woolsthorpe would be some of his most productive period in life.

The role of disease in the development and outcome of the history of science is more significant than commonly presumed.

Four years later, Barrow is offered a political position by the King Charles II, leading to the fortunate circumstance that Newton, recognized for his abilities, is awarded the Lucasian Chair at the early age of 29. The incident would guarantee his income for the rest of his life, and certainly no longer be under any concern that he would have to return to manage the family estate, far away as it was from a thriving intellectual center

Yet Newton's income was also guaranteed from another source. Hannah had inherited all of Smith's properties, and had made Newton the sole heir, ignoring all of her other kids. Upon her death, Newton would never have to worry again about financial issues for the rest of his life. Curiously, this financial security also characterizes many of his peers: Copernicus (Catholic canon), Brahe (Imperial Astronomer), Kepler (Imperial Astronomer), Galileo (Philosopher to Mediccis), and Descartes (philosopher to the Queen). Most of its participants had relatively comfortable financial situations from which to undertake their revolutionary endeavors.

Optics

In 1671 Newton publishes his first formal paper, a treatise on optics. It was 'revolutionary' given the chaotic number of interpretations surrounding the phenomenon of light. What is light, after all? Aristotelian notions suggested that light emanated from the eye. Some argued for a 'corpuscular theory,' other suggested a 'wave theory.' As Galileo questioned, is color in the object or in the eye that perceives it?

Newton's experiment brilliantly uses simple tools that were readily available, and ultimately leads to a comprehensive treatise of light of 1704. Newton passes light from a pinhole in a window shade through a prism that had been purchased at a fair, onto a screen 22 feet away. Newton's genius is shown by the long distance used between light source and the resulting image. Descartes had performed a similar experiment, but only projects the light ray a small distance, and thus is unable to differentiate the variable refractions. Newton's use of a longer distance allows him to more clearly reveal the full impact of the prism: an oblong circle appeared with its corresponding colors all stretched out according to wavelength. Each color was refracted at different angles; as light progressed from red to blue it had a more acute bending.

Newton then took the light and projected it through a convex lens, which inverted the different colors once again onto a single point. As Newton moved a sheet along the light's path, the colors met once again, creating a spot of white light. Newton writes up report and sends it to Barrow, who then proceeds to submit it to the *Philosophical Transactions* of the Royal Society of London. Light was composed of colors, and these could not be subdivided once separated.

Newton had in fact already been obtaining some renown for his contributions. He had shown Barrow how a given formula was but a subset of a larger infinite series. Barrow sends a copy of the paper to the formula's original author, John Collins, who immediately recognized the brilliance behind the paper's authorship, and pleads with Barrow to obtain the author's name. Although Newton had insisted that his authorship remain anonymous, Barrow leaks it to Collins, and hence initiates the beginning of his formidable scientific reputation, to be distinguished from malicious gossip. Newton in fact had used his fluxions (calculus) to derive the more general law, but did not reveal it in the final piece, keeping it a secret even from Barrow himself. The plagiarism that was common at time eventually forces Newton to begin placing his name on his publications—a tactic that had been wisely used by Barrow to get Newton's name to the public.

However, the publication of Newton's optics article results in a traumatic experience for Newton, which leads him to withdraw from formal participation in what was then the most important scientific institution of England.

Robert Hooke at the time was Secretary of the Royal Society, known for his important *Micrografia* which systematically explored the hidden organic world of the very small using the microscope. Not only does Hooke criticize Newton's experiment but, even worse, argues that Newton favored the corpuscular interpretation. The charge infuriates Newton, as he clearly stated that it had only been a hypothesis. The debate spills out over some 11 letters appearing in the pages of the *Philosophical Transactions*. This must have been a traumatic experience for Newton given the prominence of his rival. At one point, Newton visits the Royal Society, to be warmly greeted by Robert Boyle who recognized and appreciated Newton's genius. Newton however withdraws from the Royal Society and would not again enter its halls or publish his second magnum opus, *Optics*, until the death of Robert Hooke some thirty years later. As we will see, there was some justification for Newton's antipathy towards Hooke.

All telescopes at time were faulty in that their refraction lenses created chromatic aberrations as each color had varying focal points. The only way to properly build a telescope, Newton realized, was by reflection: use a concave mirror to focus light at a point which was then observed by the eye. Newton spends a great deal of time personally grinding down and polishing its mirror, creating an impressive small 4-inch telescope. As usual, he gives a copy of the device to Barrow who again shows it to Royal Society. All were astonished in that it had greater amplification than much larger telescopes, and ultimately establishes the foundation for modern optical astronomy. All modern optical telescopes follow the same basic design principles established by Isaac Newton, leading to a telescope 'space race' during the twentieth century. Unlike refracting ones, reflecting telescopes did not bend under their weight as they increased in power.

Hard to believe as it may be, Hooke adds fuel to the fire of their animosity by again claiming plagiarism, arguing that he had previously built a similar telescope.

The conflict between the two men was also fueled by Henry Oldenberg, at the time Secretary of Royal Society. A naturalized German, Oldenberg did not get along with Hooke, and would publish personal letters by either men so as to make the other look bad—encouraging the animosity between them.

Newton was fortunate in that a lot of 'second rate' men recognized his talents but nonetheless still decide to play an important positive role in his life. Barrow's assistance in this sense was devious but well intentioned. Plagiarism was rife in early modern period. Galileo often wrote anagrams of discovery: jumbled letters of findings, which were then send to trusted others to claim priority should others falsely claimed to have done so. In contrast to Hooke, Barrow uses the threat of plagiarism to help Newton overcome the negative traits of his personality.

Alchemist

The period after conflict with Hooke, 1671-1684, was not a pleasant one for Newton. At Trinity College he was relatively isolated, often lecturing to empty classes. Newton was also isolated from his colleagues, and did not generally interact that much with them. We see him thus entering another 'phase' of work: the study of alchemy.

An enormous amount of alchemical work was done by Newton, who even builds his own laboratory. It consisted of a small closed shack with an oven to process materials, and had no window openings. He had the only key to the place. The absence of windows and a chimney was an indication of the excessive secrecy typical of the practice, but which could have ended up killing him: fumes had nowhere to escape and so would readily concentrate in the room.

Some have suggested that Newton suffered from mercury poisoning, and it is known that one time he fell asleep while processing chemicals, the room filling with its toxic smoke. Newton also had a mild 'mercury elixir' which he regularly drank for 'good health.' At the age of 30, Newton already appeared older than was usual, given his white hair. Visual evidence suggests, however, that he had not suffered from the affliction, as his teeth would have fallen out. It is also to be noted that brain would have inevitably suffered neurological damage, but Newton dies instead as a fully cognizant intellectual at the ripe old age of 84 from a painful kidney stone.

One can obtain an idea of the scope and amount of alchemical work done by Newton by considering that he had a 100-page index of 5,000 individual alchemical works. About one fifth (1/5) of his corpus is related to alchemy. Newton was so proficient at the practice, that he is made Warden of the Mint, later being promoted to Master of the Mint: the person in charge of coinage for all of England.

Determining whether coins had been adulterated required a great deal of chemical knowledge of metals. It is Newton who introduces ridges on the edges of coins, as those found in United States quarters, so as to prevent thin slices from being removed and thus debasing the 'value' of the coin by reducing the total amount of metal in it. The quality of coin in England was unsurpassed during his tenure, and it is for this work, rather than the *Principia* or the *Optics*, that he is knighted by the British Crown. Newton is buried at Westminster Abbey in London, along with Charles Darwin and other world renown British figures.

Newton's alchemical writings were first seriously studied by Betty Jo Dobbs, who at the time was harshly attacked. When she first began studying Newton, the figure had been venerated as an ideal figure, the very paragon of Western rationality. His first biographer in 1844 stated categorically that Newton had not done alchemy, hence setting the context in which Dobbs's work would be received. Part of the reason for the animosity was due to the loss of Newton's papers.

Newton died without a testament or will, and thus his belongings fell on his immediate descendant Catherine Barton, a niece who had taken care of Newton during his elderly years and whom married Newton's former Mint employer. Over the years, the collection was dispersed as they were passed down between generations—divided, fractured as they are wont to do under these circumstances. Eventually these were sold at auction at Christies, a portion of which were bought by John Maynard Keynes, the famous World War II economist who had foreseen the economic calamity brought upon Germany with its reparations after World War I. He contacted other collectors and gathered what is known as the Pemberton Collection. Still today, no single Newton's archives exist, combining all of his original writings.

Yet Keynes, using his newly obtained documents, writes another biography revealing a much different figure of Newton than that which had been traditionally portrayed.

The topic raises an important question: was Newton a mechanical philosopher?

Mechanical philosophy presumes a universe composed only of only dead matter: inert particles blindly reacting to collisions, without any will nor inner directed guidance. This view was seen as ultimately determinist, which so sharply contrasted to Renaissance Naturalism whereby nature was defined as being full of 'spirits' and truly 'alive' in some sense. Therefore, the puzzle arises in that, if Newton was a mechanical philosopher, why did he dabble in alchemy and its implicit metallic sympathies?

It appears that he was a bit of both, again indicating that Newton was more of a transitory figure rather than the apex of the Scientific Revolution. If we judge Newton by his actual work rather than public claims, one will immediately notice that Newton's notion of 'gravity' was attacked precisely because it raised the specter of '*anima motrix.*' Newton's gravity consisted of a mysterious action at a distance, which later scientists as Waterston and Herapath during the 19th century would try to mechanically remediate. Galileo in his own day had very harshly criticized such notions, and might also explain why he might not have taken Kepler's very seriously.

We can also point out that Newton's mechanical philosophy was 'humble' in contrast to the euphoric view held by Descartes and Galileo, both of whom believed that the world was ultimately knowable through mathematics. There can no doubt that Newton was outwardly what today might be regarded as a 'positivist.' He believed that we can only have superficial knowledge of nature, captured in his '*hypothesis non fingo,*' even if he did allow himself room for hypothetical explorations in his *Queries*. It is incorrect, however, to imply the complete absence of metaphysical positions, as E. A. Burtt shows. In practice, Newton's 'positivism' meant that his metaphysics would imperceptibly spread 'uncritically,' as its underlying assumptions were rarely verbalized and rigorously evaluated. They were presumed to be a natural 'default' state of affairs.

Aside his chemistry, Newton was also deeply religious. He had traced down Biblical stories in the Old Testament, plotting family lines, calculating star positions, etc. Towards the end of his life he wished to write a biblical history, and began doing so but was unable to finish before dying. During his London period, Newton met and debated with religious philosophers as Charles Berkeley, who was taken aback at how much religious theory and history Newton actually knew.

Psychologically, Newton undergoes two periods of general depression in 1679 and 1692. Oldenburg died in 1679 and Hooke ascends to presidency of the Royal Society, reinitiating a 'dialogue' with Newton—egging Newton into a public trap which Newton detects. Hooke asks Newton what the path of a falling body through the Earth would look like. The question consisted of a thought experiment whereby a man on tower drops ball, and the ball is hypothetically able to pass through the Earth. Newton sarcastically responds that although he no longer dealt with philosophy, he too would also like to reinitiate a dialogue. Newton makes an error and claims that the ball would radially curve towards the center.

Hooke takes the opportunity to write back and point out that Newton was wrong. The center of the Earth only acted as a center of gravity but did not make up the only point of gravity of the body (Earth), and hence an object would travel in an elliptical curve, using the inverse square law. This response likely came as an enormous shock to Newton, who does not respond. The letter likely took Newton very much by surprise in that Hooke was 'close' to Newton's findings on gravity, an event which forced Newton back to the studies on gravity he had done during college school years at Woolsthorpe.

The incident does raise the perplexing question: how could Hooke ever have bothered to even suggest such a sophisticated answer? Hooke was not a mathematician, and clearly lacked the abilities so predominant in Newton. While we can rule out notion that Hooke discovered the path on his own self, it begs the question of how Hooke obtained such a solution. Did he or his 'allies' break and enter into Newton's office or residence to rummage through his papers? Newton lived a solitary life, and it would not have been too difficult to achieve when he was away from his papers. Today, the issue is difficult if not impossible to determine, but certainly strikes as a very suspicious incident—validating Newton's mistrust and animosity towards Hooke.

In 1692 Newton suffered another bout of depression, but on that was much more severe and degrading than any prior episodes he might have experienced. One might refer to it as a type of 'post-partum depression,' in that it occurred after the *Principia* had been published. Newton was now a well-recognized scholar. Although few people could actually understand his magnum opus, those who could did recognize its brilliance. Some 700 copies of *Princpa* were quickly sold, and the author was already working on a second 1713 edition. Newton added a new section to end of book: *Queries*, where he would state tentative hypothesis and new lines of research, becoming extremely influential during the next two centuries.

Our character at the time had moved from Cambridge to London. Professors at Cambridge University did not actually have to teach, and Newton tended to neglect teaching duties without any negative institutional repercussions. In London, Newton befriends important scholars as John Locke as well as Christian Huygens, whose noble character shines through the anecdotes. Although Huygens strongly disagreed with many of Newton's conclusions, he gives due credit to Newton's brilliance. The story of Newton's nervous breakdown shows him isolating himself from everyone. He then begins writing scathing letters to friends as Locke, but alternates his letters between an overly dominant and an overly submissive attitude. Locke recognizes Newton's' psychological condition, and even tries to help him out by introducing potential female partners. Newton, in typical fashion, rejects these.

There are many studies done on Newton's psychology. Some have suggested Newton suffered from Asperser's syndrome, but again, rigorous evidence for such a claim is lacking.

Without a doubt, Newton was a tragic soul, suffering greatly while giving so much to the world.

Principia, Magnum opus

Newton's *Mathematical Principles of Natural Philosophy* (1687) showed that the world could be analyzed mathematically, providing such exact results that it established a level of unprecedented achievement in the history of science. Newton had solved problems previously believed to be unresolvable, and the importance of his work cannot be overstated. Its principal contribution consisted in the unification of terrestrial physics and planetary cosmology, unquestionably proving the validity of the Copernican model. The treatise also showed how prior work could be derived from Newton's gravity law $F=G(m_1m_2)/R^2$: Kepler's planetary laws, Galileo's law of motion proportional to square of time of fall, etc. While few people could understand it, the work established the paragon scientific achievement for some 200 years—until the rise of kinetic theory.

As might perhaps be expected, upon the publication of the *Principia* Hooke again claimed Newton had plagiarized him, alluding to their prior letter exchange where the inverse of distance between two objects had been discussed.

How did such an important work as the *Principia* ever emerge in the first place?

Newton had taken ample notes on John Wilkins in his *Waste Book*. Wilkins was one of the first 'scientists' in Europe to realize that the Copernican model implied an infinite universe. More specifically, he wondered what prevented universe from devolving into chaos. How do objects know where to go? How do they know where to gravitate to; if each was its center of gravity, what determined who was to be the predominant one? Why should the Moon be a moon of Earth rather than a moon of Mars or Venus? How did moon know to be moon only of Earth? Newton realized that influence of gravity declined with distance this allowed for multiple and distinct 'centers' in the universe in 'harmonious' existence.

The publication of the *Principia* can be attributed to Edmond Halley more than to any other man. Back at the Royal Society, Christopher Wren and Edmund Halley had a similar debate to that between Newton and Hooke—but in a much more cordial fashion. Wren was an architect with scientific interests and had an ongoing bet in the Royal Society with regard to the path of a falling body influenced by the inverse square of a radius $(1/r^2)$. Wren and Halley disagreed as to the conclusive trajectory, and Halley actually visits Newton in 1684 to see if he knew answer.

By this time, Newton's tumultuous emotional state had passed, and Newton himself had better social ties and in a better mood. When Newton receives Halley, he quickly answers the puzzle by stating that the path would clearly be an ellipse. Halley is taken back by Newton's quick response, asking how he knew. Newton responded that he had calculated it early in youth, during the plague years at Woolsthorpe. When Newton then tried looking for his old notes and could not find them, he quickly writes up the proof, leaving Halley astonished. The visitor then asks Newton if he would write up a paper on the topic, and it is this specific request that formally leads Newton into writing the *Principia*.

Newton's most important biographer, Richard Westfall, points out that if Newton at any point had died prior to 1684, most of his work would have been lost to history. As we have seen, Newton had the bad habit of never publishing his discoveries, jealously guarding these and often needing these to be pried from him for scientific consumption. Discoveries only count when published and made 'common knowledge.' Newton would even publish anonymous solutions to certain puzzles that had long concerned European scientists. However, journal editors as Bernoulli were able to recognize Newton's logical style and correctly identify its source.

Newton begins work during an intense period of activity, often becoming so absent minded that he would forget to eat or where he actually was; a level of intense concentration similar to that which Newton had previously shown at Woolsthorpe. Newton then presents a brief article with summary to Halley, which everybody at the Royal Society agreed was important. However, the Society was unwilling to fund it. While it is argued that the institution had recently published an illustrated book on fishes at an enormous financial loss, it is to be noted that Robert Hooke was the Society's president at the time.

Halley again comes to the rescue. He determines that the work was too important to not be published, and decides to pay for the entire publication out of his own pocket. In that the effort constituted a substantial financial burden for Halley, we cannot give enough credit to Halley role in creation of *Principia*.

The book Newton had originally promised grew and expanded greatly beyond its original scope, no longer being merely a consideration of the planets but applying his gravity law to all manner of phenomena: tides, orbit of moon, etc. Newton had truly recreated the universe with his unique conception, vastly extending gravity's applicable range.

It is wrongly suggested that sitting under an apple tree at Woolsthorpe, Newton came up with the idea when an apple hit his head. Yet the myth is not so far from the truth, as Newton was considering the common forces at work for both an apple and the moon. Whereas Huygens had argued about a centrifugal force pushing objects outward, as they spin, Newton began interpreting the same dynamics as centripetal force, or a force that pulled objects inward to the center. Newton realized that, contrary to what was supposed in Aristotelian physics, the natural tendency was for objects to continue moving in a straight line, unless otherwise impeded or acted upon by an external force—what we refer to as inertia.

Just as the moon would naturally float away into space, the centripetal pull of gravity meant that the moon was always falling to the Earth, just as when an apple fell from a tree to the ground; the same forces were at work, but at different magnitudes and distances. All points of mass on the Earth are constantly pulling on each other, creating a common center of gravity between them. It wasn't just that the Earth was pulling the moon, but also that the moon was pulling on the Earth—and likewise that the apple was both pulling and being pulled by the Earth. (The apple's negligible mass meant that the Earth's had predominance in their gravitational interaction.) These dynamics could be exactly represented by Newton's law of gravity $F = G(m_1 m_2)/R^2$.

The formula explained why the ellipse was the actual shape of planetary orbits. All ellipses actually have two focal points—as one can experience when visiting a famous grave near the Greek city of Napflion whose roof is a parabolic ellipse. (When two persons stand at both focal points, they will be able to hear each other distinctly, as if a microphone was beside the opposite individual.) When objects orbit around one another, they create mutual focal centers, thus the natural path of the smaller object is an ellipse. In fact, in the solar system, all planets pull on each other, each influencing the paths of their orbits, but settling into an equilibrium state. Newton even tried to unsuccessfully calculate the result of the gravitational interaction between moon, sun, and Earth, a dynamic referred to as the three body problem. Its complexity forces Newton to abandon his effort.

When finished, the *Principia* was divided into 3 main parts. The first was made up of the 'axioms' or the three general principles from which he would derive the system. These included a: law of inertia (Galileo), b: F=MA – force proportional to mass, and c: For every action an equal and opposite reaction, sounding like Descartes. The second part is a study of fluid dynamics, and is actually a proof against Descartes' vortice model. Oddly, though Newton is able to disprove vortice model, this section of the book is the weakest scientifically speaking. The third section was an application of gravity to various phenomena, for such phenomena as comets, the moons of Jupiter, tides, and so forth.

The *Principia* showed that an immense number of phenomena in nature could be accounted for by his simple formula. The formula's simplicity is deceptive with regard to the genius that was required to create it. What was gravity, however? Newton refused to speculate on the matter. Newton, however, had shown its well-defined existence with a mathematical accuracy so precise so as to allow men to send rockets to Pluto, some four billion miles away from Earth.

All of that from a child that was deemed too weak and puny to survive his first week of life.

Part IV: Adolescence

The Hegemony of the Newtonian Paradigm

"Hypothesis non-fingo"

THE *PRINCIPIA* ENJOYED SIGNIFICANT RECOGNITION in that it established a new criteria of human understanding by showing that a simple law could explain such wide-ranging phenomena. It created a universal physics that applied to both Earth and sky, whereas previously the physics of both was separate, as demonstrated by the work of Galileo (physics of the Earth) and Kepler (physics of the planets and stars). Few could actually read and much less understand the work, but Newton's name would be guaranteed for posterity. Hooke's claims of plagiarism would ultimately fall on deaf ears.

When Newton finally joins the Royal Society after Hooke's death in 1703, he is elected President, ruling with an 'iron fist' during the next two decades. In contrast to prior officials who did not even bother showing up to the presentations of its member's activities, Newton attends every meeting and scientific experiment, helping endow Royal Society with a prestige and social status it previously lacked. He also imbues it with a new set of rituals and cultural practices, setting a greater formalism that is still carried on to this day; in England science would obtain an aura of respectability and prestige lacking elsewhere.

That Newton became an iconic symbol of progress and scientific achievement cannot be emphasized enough.

The seventeenth century had been a very turbulent and chaotic one, and helps explain his fame. King James II in England tried to impose Catholicism, forcing important thinkers as John Locke and others to exile in Holland. There Locke first comes across Newton by reading the *Principia*. The Glorious Revolution emerges, seeking to restore Protestantism, and Charles II is placed back on the throne. "Newtonianism" comes to be perceived as way of establishing a new social order. Incidentally, Newton plays a minor role in the Glorious Revolution at the university, and, strangely enough, is elected to Parliament where he briefly serves. It is certainly hard to consider Newton a 'politician.'

Newton's main influence, however, was through his work. Just as the *Principia* had established an ideal order of the planets working in perfect harmony according to a well-defined law, it was believed that a discovery of the laws of society would help establish the ideal social order. This became a central tenet of the Enlightenment. In science, the Newtonian paradigm was applied across many areas for the next 200 years, further extending its impact and range. At the heart of the Newtonian model was the notion that 'empiricism' and mathematics' could be used to identify key aspects of phenomena; via analysis and synthesis, the truth of nature would be revealed. There was a secondary aspect to his work as well: never draw more conclusions than are warranted by the evidence provided. Deep in the sinews of Newtonianism was a humility before nature absent during the Medieval period. One could never presume to ever 'know it all.'

Newton's scientific influence thus diverged into two distinct lines, mathematics and empiricism. Mathematics began to be more continuously applied to the study of nature, often characterized as abstract and abstruse. A good example of its trend is Pierre Simon Laplace (1749-1821) and his monumental *Mecanique Celeste* (1788) which aimed to solve the remaining problems with Newton's work. The Newtonian universe appeared to be unstable, and Laplace proves its stability. The Bernoulli family contributed to this revolution, consisting of 11 members over 4 generations of physicists and mathematicians. "Bernoulli" was not a single person but rather a multigenerational family who adored mathematics. The last name appears so frequently in the history of science, that it is easy to forget who exactly is being discussed: 4 were named "Nikolaus," 3 "Johann," 2 "Jakob," and 2 "Daniel." As is often the case within families, they fought amongst themselves.

The second line was that of experimental work, as shown in the *Optics* (1704). Newton never postulated a definitive account of the nature of light, which led to a large amount of experimental work in the field. Yet Newton's *Queries* also helps to establish numerous lines of inquiry, suggesting a vast array of topics which could be more fruitfully explored. Charles Coulomb in 1784 finds a 'law' of particle 'attraction' and 'repulsion' similar to gravitational law. Newton's gravity law appeared to be truly universal, applicable to both the very big (stars and planets) and to the very small (small electrified particles). Yet, how truly universal was it?

For 200 years, there was a systematic effort to answer this question, ultimately leading to the kinetic theory which would paradoxically undermine the Newtonian paradigm.

Finally, could there be 'laws of society' and 'humans' as well? Newton's work suggested that if the laws of human nature could be found, a more harmonious social order would be established. The absolutist monarchies typical of the era were social orders, legitimized on a false basis. It was absurd to pretend that King Louis XIV, the 'sun king,' was God's representative on Earth. A new legitimization would come to be rationally based on 'natural right.' John Locke, influenced by Newton, argued that since ownership was based on effort, all products of man were an extension of the human hand that had created them, private property therefore could not be arbitrarily confiscated. Locke in this manner establishes the rational basis of private property as inalienable, inherent, and 'natural.' As Forrest McDonald points out, the United States Constitution sought to create a new social order based on Locke's natural rights; '*novus ordo seclorum*' is imprinted on every single dollar bill, repeating and reminding its founders' wishes time and time again.

Hence the new approaches to the study of social phenomena led to genuine contributions in knowledge. The formation of social sciences in this aspect owe more of their existence to Newton than is commonly supposed: economics, sociology and psychology. Adam Smith's *The Wealth of Nations* (1776), which establishes the foundation of economics, argued that each actor seeking their own self-interest established common patterns that benefited all. Akin to atomic collisions which produced a harmonious totality, autonomous individuals unknowingly produced the broader coherence of 'the invisible hand.' Giambattista Vico's *Sciencia Nuova* (1725) similarly lays the foundation of sociology. The individual actions of each actor seeking power, lead to more complex social forms and the modification of men's minds. Vico pointed to what today is called 'emergent properties,' accounting for how civilization could have emerged from brutes, and actually sends a copy of his work to Newton. Thomas Hobbes would have a similar influence in what today is regarded as psychology.

It is hard to avoid the common pattern between the three principal founders of core social sciences (economics, sociology, psychology): broader orders emerged from the blind actions of individuals, akin to the creation of the universe arising from the influence of gravity on individual particles. The creation of new ways of understanding man and society had been inspired by Newton's revolutionary achievement.

The Enlightenment

The Enlightenment, focused principally in Paris France from 1689 to 1789, was characterized by a certain 'anglophilism' due to Newton's prestige in that region. It is mainly an urban phenomenon, marked by the intelligentsia meeting in cafes and salons, with its lively debates and discussions. Its participants identified themselves as 'reformers' rather than 'revolutionaries,' and tended to be part of 'establishment.' It was composed of a very heterogeneous group; the notion that they represented a 'uniform' way of looking at world is an incorrect one. Whereas Rousseau argued that civilization corrupts, Voltaire put forth that humanity could not exist without it. They were all, nonetheless, well aware of larger social actors as the Catholic Church, and the dangers these posed; when attacked, they came to each other's help. They referred to themselves as a 'family,' even if the formal term used to describe themselves was '*philosophes.*'

The *philosophes* were composed of three main generations, whose main leaders included: 1) Montesquieu and Montaigne, 2) Voltaire, the longest lived, and David Hume, and 3) its extension to the United States with Benjamin Franklin and Thomas Jefferson—as well as Jean Jacques Rousseau in France. The term "illustration" was first used by Immanuel Kant (17214-1804) who noted in 1784 that he did not live in an enlightened age, but rather lived in the Age of Enlightenment. It is Kant who introduced the notion of an absolute moral law, a categorical imperative that one cannot use other humans as means to ends, so clearly reflecting the values of his peers.

The Enlightenment became an examination of the social implications of the Scientific Revolution. Europe was still a very religious society, in which the Catholic Church determined 'truth' and helped establish the social order. Newton's '*hypothesis non fingo*' helped lay the foundation against empty speculation and the many obtuse philosophical 'systems' that were so typical of Medieval Europe: cognitive systems built out of imagination with little empirical verification. Newton did not claim that he could reveal truth in every domain but rather his approach was the best available means to obtaining truth of certain aspects of reality; his work pointed the way to the modern world. There can be no doubt that, by end of period, the Catholic Church would not have the favored epistemological and social standing which it previously held.

The period's key beliefs revolved around the faith in reason and progress. One could use logical mind to determine most reasonable institutions, and in this Thomas Jefferson is perhaps the best exemplar. Jefferson sought to rationalize political processes in the United States. His patent law established a 14-year monopoly, which stimulated innovation by allowing all innovators a defense against plagiarism. His patent law was an important factor that would lead the United States to world power; whereas in 1800 both Mexico and the United States had economies of approximately the same size, by 2000 the latter's economy had outpaced the former's by a factor of 20.

Central to the Enlightenment was the notion of progress: society was not destined to forever remain in an infinite cycle of life and death but rather could be improved over generations. The Abbe de Condillac most vocally expressed the notion of the perfectibility of society; a better world could be obtained via rational changes to social structures. The period was equally marked by the profound impact of secularization. This should not be taken to mean that 'religion ends' but rather amount to a gradual change in the orientation and perspective. Life is not to be seen a symbol of an underlying religious truth; objects were not metaphors, but rather had to be seen for what they were, in and of themselves.

It is certainly the case that Voltaire is most vociferous in his anti-Catholic stance. Ironically, however, the period does see the emergence of pseudo-sciences. As the Scientific Revolution had overturned 'common sense'—the stars did not revolve around the Earth as had commonly been supposed—meant the acceptance of 'wild claims' that would previously have been immediately rejected. The Scientific Revolution' implied a certain degree of suspended judgment, implicit in Cartesian mind/body duality. The world is not as we think it is, and sense perception is not necessarily a good guide to nature.

New technological achievements also 'broke all the rules,' as the case with new hot air balloons in 1783 of Paris now allowing men to fly—an experience which not previously possible in human history. As the limits of human achievement and knowledge were not clearly set, this stimulated rise of absurd claims. One typical example was the notion of a perpetual motion machines. Various secret societies emerge during the period, as the Rosicrucians and the Convulsionists who preached the second coming of Christ. Leaders as Count Alessandro Cagliostro obtained a surprising amount of popularity and influence; his 'Military order of Malta' was not a military organization but rather a magical one.

Newton's influence on the Enlightenment is in part direct. John Locke, called the "Newton of the mind," had been a close friend of Newton. Locke likely developed his notions through his long interaction with Newton, whom closely confided in Locke and was one of few persons Newton actually trusted. We can obtain an indication of their bond in that, during his mental breakdown of 1692/3, Newton writes letters of insult to Locke, seeking to break the friendship, while later writing very apologetic letters to his friend. Locke does now allow these anomalous behaviors affect their relationship, as he truly respected Newton, seeing him as a tortured genius.

Given that Newton's empiricism sharply contrasted to the rationalism of Descartes, we can detect affinities between it and Locke's *tabula rasa* concept of the mind. For Locke, the mind held no preconceived notions, very unlike the ideal forms inherent in the mind to Socrates. (The purpose of philosophy, for Socrates, was to get at the untapped internal ideas.) Locke's views became one of the most prominent 'psychologies.' For him, humanity was purely the result of experience, and hence institutional changes were of enormous consequences in the ultimate character of men. While these notions are placed into question by evolutionary psychology, their sociopolitical impact is undeniable.

Newton's European influence, particularly in France, was largely the result of one man: Francois Marie Arouet (1694-1778), better known as Voltaire. A brilliant writer, Voltaire was one of first to actually earn a living from his writing; specifically, the sales of books helped to sustain his continued intellectual activity—an economic dynamic sadly waning today. He travels to England to learn more of Newton, and is so impressed that he writes *Lettres philosophiques* (1734), a popularization of Newton's work for French audiences with the help of the mathematician Emilie du Chatelet. The *Lettres* thus constitute a cross-national diffusion of scientific ideas; breaking linguistic barriers always had a greater impact than most realize.

168

Perhaps the most important work of the Enlightenment was the *Encyclopedie* (1751-76), prepared by Denis Diderot and Jean D'Alembert. As the title suggests, it was the world's first 'encyclopedia': and ambitious compilation of knowledge consisting of 30 volumes, 2/3 text; 1/3 illustrations. Initially, it caused such a scandal that it was at first banned. D'Alembert had traveled to Geneva and interviewed its citizens, being himself very frank, unaware that his views would be published. An article on "Geneva" stated that the clergy of both Geneva and France held doubts about biblical truths. The scandal leads D'Alembert to resign from the *Encyclopedie* shortly thereafter, and in this sense, Diderot should be seen as its true father. Nonetheless, the compilation was so popular that it became a huge bestseller. It is today hard to believe that such a large multitome collection sold more than 20,000 copies throughout Europe. It also revealed the affluence of the emerging middle class.

Previous attempts at a similar project, as one first tried in England, failed miserably—greatly in part to its principal scheme of publishing 'trade secrets.' Each individual group was unwilling to share its techniques due to the potential commercial implications of its diffusion. Each had developed unique processes, and their publication would give economic competitors an edge in the marketplace. Commercial and manufacturing secrets thus tended to guard these 'close to the chest.' However, as Diderot well recognized, the dynamics was detrimental to progress in that it inhibited the diffusion of new knowledge within the entire community. The craftsman took his secrets to the grave, and were lost forever. If each economic actor shared, the sum total of knowledge in society would multiply exponentially. Diderot's 'virtue' as an editor was in being able to obtain the 'trade secrets' of how European products were made. Today this dynamic, creating a sum greater than its parts, would be referred to in the corporate world as 'synergistic.'

One of the *Ecyclopedie's* most iconic images was its tree of knowledge, establishing a hierarchy of information on an epistemological basis; the Lockean faculties of the mind. The manner in which we learn things became the basis for the new taxonomy of learning; on its seedling grew the tree of knowledge: reason -> memory -> metaphysics -> sciences of man -> sciences of nature. The structure placed an emphasis on analysis, the Baconian notion that "knowledge is power," and a faith in progress. The application of reason would lead to social improvements, instead of an endless repetition of traditional processes. The notion can be illustrated today in the sophisticated world of sports, which have been minutely analyzed to obtain ever smaller improvements in performance. Symmetry of motion and technique has had enormous implications in the world of Olympic swimming and running.

The process of rational analysis could be applied to all areas, as politics, and hence why it was perceived as such a threat to the status quo. The middle classes new claims for political participation directly conflicted with political justification based on religious grounds. The French government naturally felt threatened, and revoked the *Encyclopedie's* permit in 1759. This inherent clash of worldview culminated in French Revolution: a violent overhaul of political system. Even then, the influence of science could be felt. The mathematician Lazare Carnot was made minister of war during the Revolution and Gaspare Monde, a chemist, its secretary of navy. Monde's chemical expertise and knowledge of the production of saltpeter, the basis of all explosives in the era, led to an increase in production twelvefold.

The shape of the Earth

Pierre Simon de Laplace was a faithful revolutionary. He wanted to push the Newtonian research paradigm to all areas of nature, believing that everything was ultimately the result of gravity, the fundamental force of the universe. Laplace was able to achieve this to a substantial degree due to his able political wrangling; he became good friends with Napoleon, was elected senator and became a member of *Academie des Sciences* (1666)—the French version of Royal Society.

In contrast to its British counterpart, the *Academie des Sciences* was an emblem to elite science, where only 42 practitioners in the entire region were chosen to participate. It was an arbitrary social creation that did not democratically open a space for other scientists. Those in the group held a great deal of social power, serving in diverse state committees resolving national issues, such as setting standards and measures, as the unit of length. Whereas prior measurement had been determined on the length of the king's arm, the *Academie des Sciences* opted for a more universal unit. In 1790 a committee set the national standard of a meter to 1/10 millionth of a quadrant of the Earth's surface.

Yet, what was the size of the Earth? Was its shape perfectly circular, and if not, how would this alter the French meter? While Newton argued that its shape was akin to an egg lying on its side (an oblate spheroid), Jacques Cassini took the opposite view of an egg 'standing up' (a prolate spheroid). Various missions were sent throughout world to determine the shape of Earth, and in turn, the universal unit of measurement known as the meter.

One of these was headed by Charles-Marie de La Condamine; other members included Louis Godin and Pierre Bouger. Their voyage took them to Peru, near the equatorial region. The other party was headed by Pierre-Louis Moureau de Maupertuis, who along with Alexis Clairaut headed for the 'north pole' near what is today Tornio in Sweden. While one might assume the tropical exploration to have been easier to conduct, the opposite was the case.

Maupertuis was able to obtain plentiful wood logs from the surrounding forest to establish his calculations via triangulation. The cold temperatures helped in that they created a perfectly flat highway of a frozen river, on which their logs could be easily transported. By 1737, less than two years after the beginning of the expedition, Maupertuis had finished.

It would take La Condamine more than a decade to obtain his data complete his calculations.

La Condamine's expedition was actually led by Godin, who soon entered into a dispute over the type of measurements to be realized. Whereas Godin wanted to measure mountain heights with a barometer, La Condamine objected as it would lead to inconclusive results and would as double the amount of time required for the already-difficult-to-do measurements along mountainous regions. La Condamine argued that measurements at sea level, although not perfect, could be quickly completed and would be good enough for the purposes of the expedition. The team ended up taking three different sets of measurements as no consensus could be agreed upon. Upon completing the task in 1744, their return to France led to the realization that the entire trip had been an unnecessary sacrifice, as enough readings had already been gathered to faithfully complete the calculations.

The expeditions had determined that one degree of arc at the pole equaled 111.094 km (Maupertuis) whereas only 109.92 km at Peru (La Condamine). Newton had been right all along; the Earth is an oblate spheroid

Clairaut, a member of Maupertuis's team, would become an important scientist during the century, hailed in France as the new "Thales." As a child prodigy, Clairaut had mastered Euclid by age of 9, and became the youngest member ever to be elected to *Academie des Sciences* in 1731 at the age of 16. His ability was such, that he assisted Madame du Chatelet with Newton's *Principia*, and became a member of the politically powerful group pushing the Newtonian paradigm in France—a group which naturally included Maupertuis. Clairaut's number of contributions to mathematics are far too many to be discussed here. His *Theorie de la figure de la Terre* (1743) was a theoretical analysis of shape of Earth from hydrostatic point of view, or the interaction of internal liquids beneath what today we call the mantle. Clairaut also worked on the three body problem which had so escaped Newton's able mind.

In Proposition 55, Book 1, of the Principia, Newton suggested a possible solution. It was then believed that, if solved, it would allow sailors to establish longitude at sea, and hence the scientific issue was seen as a matter of enormous economic importance. It is to be noted that problem appears deceptively simple: if one could find an equation for two interactive bodies, it would not be too far-fetched to suppose that adding only another body would not result in too many difficulties. Given Clairault's genius, it was natural for him to attempt to tackle this 'extreme' problem, a Mt. Everest of mathematics. It also would likely establish Clairaut's reputation in history, 'surpassing' the genius of British Newton.

While one might suppose an inter-generational rivalry between the two, in fact Clairaut had developed an intense rivalry with Jean D'Alembert over the issue. Both alleged to have presented solutions to the *Academies des Sciences*. Clairaut's "*Probleme des Trois Corps*" (1752) concluded that Newton's gravity law was wrong. Euler, who evaluated the debate, agreed.

While today we might assume that the acceptance of Newton's gravity law to have been immediate, the incident reveals the many fissures and fractures underlying the emergence and establishment of scientific truths. Clairaut was certainly successful in more accurately predicting the perihelion of Halley's comet than Newton's had been. This success leading to the denomination of Clairaut as the 'new Thales.'

However, in 1887 Henri Poincare actually shows that both men had been wrong all along; there is no solution to the infamous three body problem.

171

By 1793, the elitist *Academie des Sciences* was abolished, and a year later the system of *Ecole Polytechnique* is formed, of a much diminished 'monarchical' tone. These provided three year courses in science and engineering for the specific purpose of creating professionals for the service of the state; meritocracy rather than family ancestry would be its primary criteria of evaluation, and all were run as military schools. All leading French scientists of the period would study in these various Polytechniques, imbuing these with a world view radically different from its institutional predecessor.

When Napoleon takes over as "First Consul" in 1799, he appoints scientists to high government positions: J.A. Chaptal becomes Interior Minister, while Laplace and his good friend Claude Louis Berthollet become senators. Both use their newfound position of power to promote the Newtonian research programs. It is rather ironic that, upon crowing himself emperor 1804, Napoleon invades Spain believing his actions to be justified so as to bring 'progress' and Enlightenment ideas to Spain. In actual practice, however, the war soon becomes a long and protracted civil war, horrific in the scale of its atrocities. Napoleon's armies were not fighting the Spanish army but rather the population itself. The atrocities of the conflict are aptly captured in the Francisco de Goya series of drawing titled *"Desastres de la Guerra"* (Disasters of the War).

Laplace's power and influence help establish Newtonianism as a research paradigm, a combination of qualitative experiments and math theory. His own *Mecanique Celeste* (1799-1825) was an ambitious analysis of planetary and lunar motion. Laplace, who was then called the "Newton of the age," claimed that Newton had been the happiest man ever to exist as there was only one system to the universe, and only Newton had discovered it. His faith in Newton was absolute.

When his good friend Claude Louis Berthollet moves to Arcueil, a Paris suburb, Laplace moves next door, and both set up the "Society of Arcueil" (1807-14). Berthollet's mansion became a research facility that hired Ecole Polytechnique graduates. Because French life so centralized, those in power could easily dominate and establish 'national' research agendas. Inversely, however, the loss of political power also meant the unbearable loss of scientific influence. Both had well-thought out research programs, and establishing competition so as to extend Newtonianism.

Political power? Check. Economic resources? Check. Absolute truth? Maybe.

These competitions showed that results couldn't always be anticipated. Such was the case of the in 1807 in a contest on the refraction of light.

The nature of light

Newton did not claim to know what light was, much as he did not claim to know the nature of gravity. He merely stated that it was composed of colors, and his fight with Hooke rested precisely on this point. Hooke claimed that Newton arguing for light as particle, in what was then called the corpuscular theory. Newton was furious precisely because he recognized the difficulty in differentiating its corpuscular and wave nature; light at times behaved as a particle while at other times as a wave. Descartes had suggested light was made up of waves in the aether; 'pulses' of aether which were horizontal to the line of motion. Ironically, however, Newton's fame and recognition lead to the predominance of a corpuscular theory of light over and above Descartes' pulse aether. It would be a debate that would continue throughout the eighteenth century without any clear and convincing results.

In 1802 Thomas Young presented his famous two slit experiment, whereby light from candle was passed through two slits that were not in the direct line of sight of the candle. Young, a medical physician, was led to startling discovery: evidence of the wave character of light. The screen onto which the whole was projected revealed diffraction bands that could only occur if light were a wave: overlapping bands of dark and white ridges. As ocean waves, light was either canceling or doubling up on itself, creating distinctive patterns. The experiment presented the most concrete and direct evidence against a corpuscular interpretation, offending Newton's followers who then sought to undermine the validity of the experiment.

Laplace thus establishes a contest to 'resolve the issue in 1818. The judges were 'stacked' so as to be biased in favor of the corpuscular theory: Laplace, Jean-Baptiste Simeon, and Simeon Denis Poisson; the only jurors in favor of the wave theory were Francois Arago and Joseph Louis Gay-Lussac. Laplace naturally presumed that the winner would provide a mathematical proof of particle theory and that would be the end of it.

Its principal contestant was Agustin-Jean Fresnel (1788-1827), a shy civil engineer who had graduated from the *Ecole Politechnique*, designing roads and bridges for a living. His scientific interests led him to move to Paris in 1815 where he was befriended by Arago, a second generation scientific 'renegade' opposed to the Laplacian agenda and whom resented Laplace's hegemonic control over French scientific institutions. He recognized Fresnel's brilliance and convinced him to publish early in the *Memoires*.

Fresnel's contest paper presented a rigorous wave model of light, and accounted for the results of Young's two split experiment.

Poisson did not like the results of Fresnel's paper at all. A firm adherent to Newtonian corpuscularism, as well as a very able mathematician, Poisson painstakingly looks through Fresnel's mathematics and is able to deduce a particular and unusual consequence. Drawing exclusively from Fresnel's calculations, there should occur a bright spot in the middle of the shadow of a perfectly illuminated disk. As Poisson had never seen one, he had no reason to believe possible: if light was a particle, it would just be blocked by the solid black circle. To his credit, Fresnel sets up an experiment that is able to precisely show the tiny spot as Poisson's meticulous calculations predicted.

Poisson, and all other opposing members of the jury, were forced to changed their verdict. To his credit, Laplace recognized the validity of Fresnel's work and the prize was ultimately awarded to him.

Unfortunately, our story has a tragic outcome; Fresnel does not become the leader in French science he was destined to be. Obtaining a new job by

Lighthouse Commission of France, he invents the "Fesnel lens" and in 1823 is elected to the *Academie des Sciences* as well as the Royal Society of London. Strangely, a 'dispute' over priority emerges with Young. Fresnel had published unaware of the existence of Young's work, and thus does not give credit to his predecessor in his important paper—to which Young naturally complains. Members of the *Academie des Sciences* visit Young's home to settle the dispute. During the arguments, which had been patiently observed by Young's wife who had all along been quietly knitting nearby, she gets fed up and suddenly goes to Young's office to bring back his notebooks clearly proving Young's precedence.

In 1827, in spite of his emerging recognition and scientific importance, Fresnel dies at age of 39 from tuberculosis.

The history of light, however, is even more complicated than it might appear. In Query 26 of Newton's *Optics* he argues that light was somewhat different from sound as its showed transversal displacement, perpendicular to direction of travel. A study by Etienne Malus in 1808 revealed this to be the case. Malus had accompanied Napoleon in hi scientific expedition to Egypt. While looking at the reflection from a window pane made out of calcite crystal, Malus noticed only a single image, instead of the typical double image that was often seen. He realized that the second reflection was being 'absorbed' in some way by the crystal. His further study of the phenomenon led to the detection of polarization of light.

However, Malus's results proved to be controversial; even Arago, who so pushed for wave theory of light, refused to accept the paper. Although the paper on light polarization was eventually published, it was distinctly clear that the phenomenon of light was still poorly understood—and would remain so for various centuries.

Lavoisier's *experiment crucis*

In his *Queries*, Newton also postulated the existence of intra-molecular forces or forces of attraction, best demonstrated by capillary action. If you took two glass plates with a drop of water between the plates, the droplet would be pulled up beyond what gravity would naturally allow, distinctly showing some type of molecular forces at work. It was also clear that forces of repulsion existed. Boyle's *Spring of Air* showed that a "J curved" test tube filled with air would create a counterforce proportional to the pressure exerted on it; twice the force reduced the content to half the space, four times the force to a quarter of space and so forth, suggesting that corpuscles 'repelled' each other. The world of the small showed many complex and contradictory properties.

In his analysis of fire, Voltaire argued that fire was a substance, an 'object' per se, and the absence of regularity in chemical processes contributed to a degree of ritualism in the practice of alchemy. Given the irregularity of outcomes in mixtures and potions, the only 'control' a practitioner had was his own behavior; alchemists thus placed a great deal of emphasis on ritualized procedures as well as on their own internal frame of mind and spirit to obtain successful alchemical outcomes. Alchemists not only had to cleanse their spirits, but also had to make sure that experiments were not undertaken during 'stepmother periods,' dates which foretold gloom as stepparents were associated with unfortunate outcomes.

This tendency to ritualism was ultimately inhibitory to the development of chemistry, as it did not stimulate a deeper exploration of the causes and correlations, deemed to be unknowable given their erratic and inconsistent outcomes. What did clear the way for chemical advancement, however, were the crafts. The creation of oils, medicines and pigments helped establish basic chemical procedures which became well known, as distillation and sublimation. Plants were crushed, placed in a vessel which could be characterized as an ancient coffee pot; its oils would thus be drawn from the plant.

The Greeks unfortunately had not been good guides in the development of chemistry; the work of Aristotle, so oriented to the biological, actually hindered chemical development. For him, wine was the corpse of vinegar, just as vinegar was the corpse of a grape. Chemistry did share the same traits as biology, in that its objects were tangible and could be directly manipulated, unlike the ever distant objects of astronomy; they were, nonetheless, tantalizingly difficult to understand, and their complexity equally hard to decipher. The Stoics, did provide a guide, in that their atomic pneuma pervaded universe; while one cannot see atoms (or pneuma), one could infer their existence indirectly by their effects. For example, while we cannot see the air that enters a balloon as it is blown up, we can infer the existence of air from the stretching of its membrane and its overall expansion in size.

Generally speaking, however, the practice of chemistry ultimately owes its origins to alchemy, today so wrongly associated with 'greedy get rich quick schemes' of the philosopher's stone: a device which was purported to turn all metals into gold. In fact, alchemical practice contributed a large knowledge base by its exploration of materials and their respective properties. Francis Bacon recognized its value, realizing that alchemy fit his scientific model: go out to nature and accumulate facts, which would then be synthesized by geniuses. Bacon, however, did criticize the strange secrecy that so typically surrounded it. The use of bizarre symbols and diverse terms only confused the study; one substance might go by various names, and hence be unrecognized across multiple works alluding to the same entity.

Alchemists believed that their laboratory experiments recreated phenomenon occurring within the Earth, which was associated to a kind of 'womb.' New metals were created in the 'womb' of Earth; when these metals were mined, the Earth simply made more metal to fill the gap. Metal distributions showed a vein-like character; similar substances were regularly and distributed in specific patterns, sulfur being one of the most commonly found. However, it was believed that when these objects were pulled from womb, metals stopped developing—as with traditional cell cultures in biology prior to HeLa. There were also disparities of timeframes. Whereas the Earth's timeframe occurred on the scale of centuries, humans lived in the framework of days. Alchemists could thus try to accelerate nature's processes but it was a difficult task. A 'liquid' is thus a relative term, tending to flow over periods of centuries, thousands times more viscous than water and what men could visually perceive.

'Base metals' were those that liquefied under heat; 'spirit metals' were those which would volatize (turn to gas), and these were in turn correlated to specific substances, as mercury (quicksilver) for the principle of liquidity, and sulfur relative to the degree to which it glowed red under fire. Paracelsus is perhaps alchemy's best exponent, ultimately contributing to the institutionalization of chemistry in universities through his advocacy of its medicinal benefits.

The slow and gradual growth of chemistry means that the use of the term 'revolution' to characterize its development is inappropriate. That being said, one may identify a point at which its practice emerged from the ashes of mystical obscurity into a more concrete and empirical form; ironically almost as if by magic, the vague problem of 'incorporeals' suddenly becomes transformed into the concrete and systematic analysis of chemical elements.

As is often the case in history, this shift was result of a rivalry between two men: Joseph Priestly (1733-1804) and Antoine Lavoisier (1743-1794).

Joseph Priestly had been librarian to William Petty. As Eratosthenes, his position gave him ample time for scientific activity. In 1786 he publishes his extensive six volume treaties on "airs," *Experiments and Observations on the Different Kinds of Airs* (1774-86). The period, unlike that of Eratosthenes, had developed new public institutions which made knowledge widely available to the general public, as the Bodleian Library or the Ashmoleaun Museum (1683).

Lavoisier at the time had become an important scientific man. His marriage to Anne Pierrette Paulze in 1777 proved to be a wise decision, as their Parisian home would become a focal point of European scientific activity; all leading scientific figures passed through it at one point or other while in France. The informal salon met twice a week, on Tuesdays and Thursdays, and its prominent activity gave a great deal of social power to Lavoisier, whom is able to more easily extend his scientific views.

Priestley, by contrast, had been forced to go into exile in 1794, given his vocal support of the rationalization and democratization of politics, vocally supportive of the political upheavals of his era. As this did not sit well with crown, Priestley is thus forced to 'flee for his life.' During the eventual French Revolution, however, Lavoisier would be less fortunate, being executed by the guillotine for his duties as a tax collector. His wife Paulze would continue to run the salon after his death.

Years before his exile to America, Priestley presented Lavoisier with his new theory of phlogiston in 1774, using it to account for the 'fire' of metals, hoping to convince him of it. Priestly showed he was able to separate phlogiston, pour it into a vase with flame and extinguish it. Priestley's experiments also produced what he called 'dephlogisticated air,' which had the marvelous quality of being "eminently breathable." It made animals more lively and a flame would burn with brighter intensity when in dephlogisticated air. While we might mistake phlogiston for 'carbon dioxide,' it had a diverse set of properties that were incongruous with its modern equivalent.

"Phlogiston" was not exactly original with Priestly. The term 'phlogistos' had first been used by George Ernest Stahl,alluding to the principle of flammability. Priestly argued that metals were composed of calx and phlogiston, such than when metal rusted, it lost phlogiston. Phlogiston, however, had a lot of odd properties as 'negative weight.' During calcination, metals gained weight while losing phlogiston, but there were no clear and consistent patterns in this regard. The many complaints over concept of phlogiston, also tended to be denied by practical chemists as Gabriel Fancois Venel who adopted the notion for its practical utility. He was less interested in the underlying dynamics of substances, and more interested in their chemical behavior.Three years later, in 1777, Lavoisier offers his own interpretation and revolutionizes chemistry. He comes up with the notion of 'oxygen' as an acid former and overturns the chaotic terminology which had so beset chemical practice during the centuries. How exactly was he able to do so?

The foundations of the chemical "revolution" are due in part to new technologies and techniques in the field, allowing for the capture of 'spirits' or 'gases.' Stephen Hale's pneumatic trough' was key in this development. Simply put, it allowed the capture of gases from any experiment. When the original material was heated, the resulting gases were run through a tube, which passed into another chamber holding nothing but water. The unmodified gas was thus collected and isolated in a glass bottle, allowing for further experimentation and testing. Hale afterwards switched to mercury, as he finds it would be less reactive with the gases flowing through it. The importance of the pneumatic trough in allowing for the empirical evaluation of 'incorporeals' cannot be given enough emphasis; it would serve as a cornerstone of Lavoisier's revolutionary experiments. Curiously, Hale's *Vegetable Staticks* (1727) was the last book approved for publication by Newton in his final years as president of the Royal Society.

Another problem which beset chemistry at the time was the large range of names that were inconsistently used and which did not reveal relation between substances. For example, the term "butter of arsenic" was often used, but rather than being a digestible delicacy, it was a poison. Other such terms included, "flowers of zinc," and so forth. With greater accumulation of information on reactions, chemists had been increasing the total number of chemical substances discovered, but as of yet had no common and rational nomenclature. Lavoisier led the way to the reconceptualization of chemistry through his redefinition of its terms; for him chemical language should reflect nature of item described. If accurate, the wording would also further reveal information about the substance under study.

Via his combination of trough and heating of metal calx, Lavoisier is able to achieve results analogous to Newton's *'experiment crucis'* in the *Optics*: via analysis / synthesis, he breaks up and reconstitutes the substance, thus unquestioningly demonstrating its underlying components. Lavoisier places four ounces of mercury in a bell jar, heats it for 12 days. The remaining calx mercury was measured along with changes in the volume of air in bell jar, showing they were reduced by 1/6. Lavoisier then superheats the calx (the red grains of mercury) to once again obtain the same volume of gases lost. It is no wonder that Lavoisier introduced the notion of oxygen as an acid former; as shown by Priestly the new gas was eminently combustible and good for breathing. Lavoisier wrongly

177

believed he had discovered a key component in all acids, and began to systematically test all metals. Some 33 further elements were identified by him, and his persistent terminology would further imbue chemistry with his underlying theories and assumptions—that could not be rejected without also throwing away its terminology.

Lavoisier's influence was not exclusively scientific. He also used the institutional power in his hands to impose his theories via the formation of a new scientific journal: The *Annales des Chimie*. As Director of *Academie des Sciences* in 1785, Lavoisier also removes the phlogistonists from positions of influence in the organization. His *Elements of Chemistry* (1789) was the final *pièce de résistance*: a synthetic work, akin to the *Principia* which brought a 'death blow' to Priestley's phlogiston theory by becoming a common reference textbook.

Lavoisier's influence would remain long after his assassination, due to the enormous institutional influence he had held during his life—ominously pointing to future concerns in scientific practice. Religious or not, institutions used to preserve legacies could also be used to demolish innovative activities

.

The Industrial Revolution:

The Birth of Coal and the Death of Nature

WE WILL BEGIN THIS STORY WITH TWO CASES: Fomento's industrialization program in Puerto Rico and that of Eddie Aikau in Hawaii, a big wave surfer. Oddly enough, both share some common traits.

Fomento's logo of the 1960s had the icon of a man turning a metal wheel, symbolizing the 'self-driven' character of Rexford Tugwell's "Operation Bootstrap". (Tugwell established some of its most important government institutions.) However, if the image is taken literally at face value, we obtain a distorted interpretation of its actual dynamics: men moving machines rather machine moving men. In other words, the logo suggests man to be the 'energy source' of a large engine.

Consider now the case of Eddie Aikau. Eddie Aikau was a big wave surfer from Hawaii. One day while on a trip, the boat's engine fails and becomes adrift in the Pacific Ocean. Some amount of panic naturally arose, and Aikau, a strong swimmer, decided to leave the boat on his large surfboard to seek help. While the boat is eventually found, Aikau's body was nowhere to be seen. It is likely he was killed either by dehydration, extreme sunburns, or sharks.

These vastly different island stories have one theme in common: the absence of a notion of a power revolution on which modern civilization so closely rests—and without which it could not exist. While the Industrial Revolution is a chapter in its story, the 'power revolution' encompasses a more accurate and much broader phenomena over a longer time period with a more diverse cast of nations other than England.

179

The wealth of the modern world is based on conversion of external energy sources, coal, oil, nuclear, solar, and geothermal, into usable forces. From the point of view of work generation, the origin of energy is somewhat irrelevant. The tragic aspect of the Aikau story is the ignorance of this feature of modern life, suggesting that one source of energy (human) was equivalent to the other (endogenous coal or petroleum). If we all had to rely on own our motive power for every daily task, we would be able to accomplish only a fraction of things done in a day.

The Industrial Revolution is perhaps the epitome of this broader change. Occurring in England roughly between 1760 to 1820, it helped establish the nineteenth century as 'Britain's century'; a nation that controlled nearly entire globe via imperialism. It was the first time in world history that goods on a truly massive scale were produced, specifically textiles.

Clothing in and of itself is an essential 'technology,' often taken for granted and not defined as 'technology.' Yet clothing assists humans just as other more sophisticated tools: it protects the human body from the exterior: solar radiation to avoid skin burning, freezing temperatures to maintain the extremities warm, and the rain to keep the body dry so as to not quickly dissipate heat. Other nations sought to emulate Britain during the 19th century, the most successful of which was perhaps Germany, whose chemical industry led to its enormous economic growth during the latter half of the century. Its principal product had been dyes for British textiles.

While the Industrial Revolution is typically associated with 'technological change,' with its development and use of the steam engine to produce all sorts of goods. However, this period of substantive change in Western economies should principally be considered a power revolution. Britain had enormous coal deposits, providing an 'endless' supply of energy for productive ends. Its particular contribution (steam engine) allowed the transfer of this potential energy into usable work for human ends.

The Industrial Revolution, however, also had unexpected precedents.

The first was Leonardo da Vinci, a surprising case as he is often more associated for his sophisticated Renaissance art and secret technological artifacts. However, one of his most important contributions was his vision of a 'power revolution' long before it was ever conceived by others. However, it is important to note that he was equally weary of its consequences. While he was not able to put his most important designs into actual work, he did foresee the consequent ecological destruction that would result from them.

The second predecessor is the Dutch Republic. The most widely recognized iconography of Holland is the story of boy with his finger struck in dike. It is illustrative in that Holland became first nation to actively use external energy sources through its windmills, a practice which was widely imitated in both France and Germany. Dutch power innovations actually emerge under enormous military and economic pressures. Through them, a nation of 3 million people was able to defeat the Spanish Empire which then spread throughout the world and appeared to be insurmountable.

180

Leonardo da Vinci

It is curious that Leonardo da Vinci should have had similar life experiences to Newton, both being rejected early in life by their most intimate family members. Leonardo was illegitimately born in 1452 to a peasant girl and a lawyer, eventually being incorporated into his father's household. The father actually had four marriages, and all of Leonardo's stepmothers rejected him when young. Upon his father's death, his half brothers steal his inheritance, and he is setup for arrest by the police. His inherent distrust of human nature, arising from such hardships, likely helps account for his secrecy and relative isolation. His notebooks reveal a continued fascination with suffering and the evil aspects of humanity, somewhat ironic as Leonardo also worked for a series of tyrannical princes during his life, the worst of which were the Borgias.

Early in his 'career,' Leonardo tried to make a living as a courtier, and was humorously known for telling terrible jokes. However, he amused patrons with skillfully designed inventions and mechanical contrivances. In 1481, the monks of San Scopeto commissioned a painting from him, which Leonardo never completes. His main source of income in life appears to be what today we would call civil engineering. His earliest formal position was with Duke Ludovico 'Il Morro" Sforza in Milan. In a letter of solicitation to the Duke, Leonardo writes that he knew how to construct war machines, civil engineering, and sculpting. 'My work will stand comparison with that of anyone else, whoever he may be'— quite an understatement. More oddly still, Leonardo chose 'routine engineering' jobs, such as inspecting and repairing fortresses. However, these jobs paid well and gave him ample leisure time to experiment and undertake his private projects.

Few of Leonardo's *Notebooks* survive today, one of which was recently bought by Bill Gates. He invents the miter-gate canal lock for Sforza – one of most original designs, that was widely copied and still used today throughout Europe. The traditional canal lock was composed of a heavy door, which moved along a vertical axis, lifted up or down according to need. However, it was slow to move; as it was heavy to lift, boats that came too close when opened would often be pulled into it with the current flow, leading to damages. The traditional canal locks also tended to leak. Leonardo's miter gate design had two doors close at an angle, creating a type of triangular arch structure that made the canal gate incredibly strong but easy to manage. The harder the water pressed onto its sides, the more tightly it shut. More importantly, Leonardo had ingeniously built a lower inner door below the water line, which when opened quickly depressurized the chambers. The canal almost opened up by itself, requiring a fraction of the effort to operate than the traditional design.

As is well known, Leonardo also had a number of military innovations. The traditional muskets used a matchlock that would go out in rain. Leonardo created a spring loaded iron wheel with iron pyrite which set off sparks, and thus working in spite of external atmospheric conditions. Leonardo also created the milling principle for the musket. The space between a 'bullet' and the wall of a musket or cannon was important in that any irregularities reduced the power and speed of the respective bullet; the full force of the explosion would not be transferred directly onto the bullet or cannonball. Thus, exact milling of the interior of musket by Leonardo led to much more powerful and accurate shots, given that the bullets now traveled at much faster speed. Leonardo, however, gave this idea to this his assistant "Tedesco," who then takes it to Germany and becomes a wealthy man. Leonardo also invented a submarine, but destroyed his notes because of his concern of its use to kill innocent lives.

As Edwin Layton points out, Leonardo had enormous moral dilemmas with technology—something rather ironic as one of his main business interests was 'war mongering,' perhaps history's original "Iron Man." From 1502 to 1503, he works for Cesar Borgia, the bastard son of Pope Alexander XV. Both father and son were utterly corrupt, involved in continual murders of political rivals and devious treacheries of all sorts.

Leonardo specifically feared that technology would ultimately destroy nature. All forests would be cut down, mountains demolished to get to its metals, terrors and afflictions would be dealt to every living thing; all those who opposed would be slain. "O monstrous animal, how much better were it for men that thou shouldst go back to hell! Because of this the great forests will be deprived of their trees and an infinity of animals will lose their lives."

Leonardo is one of the first environmentalists in the pages of history.

Creatures shall be seen upon the Earth who will always be fighting one with another, with very great losses and frequent deaths on either side. These shall set no bounds to their malice: by their fierce limbs a great number of the trees in the immense forests of the world shall be laid level with the ground; and when they have crammed themselves with food it shall gratify their desire to deal out death, affliction, labors, terrors, and banishment to every living thing. And in their boundless pride they shall wish to rise towards heaven, but the excessive weight of their limbs shall hold them down. There shall be nothing remaining on the Earth or under the Earth or in the waters that shall not be pursued, carried away, or destroyed .... O Earth! what delays thee to open and hurl them headlong into the deep fissures of thy huge abysses and caverns, and no longer display in the sight of heaven so savage and ruthless a monster? All the animals languish, filling the air with lamentations. The woods fall in ruin. The mountains are torn open, in order to carry away the metals which are produced there. But how can I speak of anything more wicked than (the actions) of those who raise hymns of praise to heaven for those who with greater zeal have injured their country and the human race?

However, Leonardo also abides by a quasi-utopic vision of the potentiality of technology. He realized that canals could be greatly expanded, not just for transportation but also for the production of goods. Canals represented an enormous amount of energy that could be tapped, akin to 'oil' right under their feet.' He began by a study of Greeks writers, likely while working under Sforza in Milan (1482-99): Hero, Archimedes, and Philon. He then undertakes a number of hydrostatic experiments, studying for example the eddies and turbulent flow. His first attempts can be seen at Mantua in 1500. In Milan, he designs flood gate damn, which could be quickly opened to flood invaders: a tactic that was also used by military in Holland. Leonardo also undertakes flood control in Venice.

The project which came closest to achieving his ideal was the Arno Canal, the seed for his 'power revolution' in Florence. The project began due to Niccolo Machiavelli, better known today as the author of *The Prince*. Machiavelli wanted a canal to bypass Pisa, then controlling access to the sea. Leonardo was to design the Arno Canal, to ultimately reduce Florence's dependence on the rival city.

Yet Leonardo designs much more ambitious project, in that the water would serve as an energy source of industrial production, with the potential of revolutionizing Italy's economy. In 1800 for example, France had around 20,000 horsepower from water; Leonardo's plan, by contrast, would have created 40,000 horsepower in 1500. He realized that water traveling from vertical iron pipe could be fed into a small turbine. Given the large fall and change in cross section, his hydraulic turbine would have led to more than a tenfold increase in the generation of horsepower, from 15 to hundreds.

Ultimately, however, the project failed for want of advanced digging technologies, specifically dynamite. In 1500, men could not yet tunnel through mountains.

Dutch windmills

Holland is perhaps one of best cases of the impact of technological change. A small area of 3 million people was able to defeat the powerful and apparently invincible Spanish Armada (Phillip II), surmounting enormous odds. The Dutch went on to establish a thriving economy, creating new forms of social organization and institutions which later become the modern corporation. Its windmills were the child of necessity, whose common presence on the Dutch landscape make them prominent cultural traits for which the Dutch are recognized today. These also establish a model which other European powers would imitate in their own 'power revolutions.'

The Dutch power revolution emerged at a time of great turmoil, 1570-1620, when Holland was in conflict with Spain. Phillip II had been crowned king in 1556, becoming the most powerful man in the world. In 1581, he inherited the crown of Portugal. However, the poor use of power led to the loss of the Portuguese empire, Dutch defeat, ultimately turning Spain into a second rate European power.

In 1568, Phillip introduced the Inquisition into the Netherlands, resulting in an immediate revolt. An army of 50,000 professional soldiers were then sent by Spain under the Duke of Alva, and quickly defeated the Dutch rebels. However, the rampant torture, rape, and theft by Spanish soldiers implied that those who submitted received a worst treatment than those who continued rebelling. All

183

who had surrendered were ransacked, tortured, leading to 'fates worse than death.' The policy led to an enormous change in mood, and in turn to a new revolt in 1572.

While a very dangerous reaction, as the Dutch then lacked a trained army, the alternative was far worse. The Dutch revolt paid off with surprising successes. They destroyed their own dykes, flooding towns, and tended to adopt 'guerilla warfare' in that in the beginning they did not confront the Spanish forces directly. The revolt turned into an eight-year war, where seven provinces were liberated. In their improved defenses, the Dutch built 'Italian style' fortresses: rectangular, sloping walls, diagonal ditches, and narrow shaped bastions. The Dutch were ultimately successful in war because of the thriving economy created by the Dutch East and West India Companies.

In contrast to Spanish economic institutions, the Dutch East and West India Companies were self-governing, placing these on a much more efficient basis. The corporations created successful coffee trade in Java, slavery in Africa, and the conquest of northern Brazil. This economic success paved the way for the construction of a navy superior to Spain's, and ultimately defeated the Spanish Armada in 1607. A truce was established in1609-12. While Philip IV's aggressive advisors broke the truce, the Dutch captured an entire treasure fleet 1628, and ultimately destroy the Spanish Armada in 1639.

To provide an idea of the scale of the Dutch naval activities, by 1620 it had a 16,000 merchant fleet manned by 160,000 men: 80% of the world's total. Holland had more trade than Spain, France, and England combined, becoming one of richest nations of Europe. By contrast, the Spanish treasury went bankrupt by 1640, contributing to Catalonia's revolt. The Dutch success was literally 'David" beating "Goliath."

There are various aspects to the Dutch power revolution. One of these were the educational changes in the engineering curriculum by Simon Stevin (1548-1620), also known as the 'Archimedes of the north.' Stevin was brilliant engineer, patenting an improved windmill in 1586. He supervised the hydraulic works near Delft, serving as quartermasters general to the Dutch army, writing *Statiscs and Hydraulics* (1586), and establishing the earliest curricula for engineers for Prince Maurice of Orange. The key to his reform was the establishment of standardized course of instruction.

The Dutch were favored by local steady and strong winds off the North Sea, present all year round. The first windmills operated principally as pumps given that Holland is under sea level; by pumping the water out, the windmills helped it reclaim the land from the sea. They also develop vertical windmills whose sails rotated about a horizontal axis. The use of airfoil-shaped vanes allowed for an increase in lift and a consequent augment in output by a factor of 11. The windmills further change from small post windmills to giant power tower mills, becoming the 'prime movers' of all sorts of manufacturing activity: sawing wood, grinding oil from seed, making gunpowder, grinding dye pigments, shaping metal, polishing gems, and so forth.

This active external source of power stimulated the importation all sorts of raw materials, which were then transform into consumer goods. The entire process was obviously facilitated by the successful Dutch naval fleet, distributing the raw finished goods which emerged out of the 'windmill factories.' The Dutch, as we have seen, even became leading book publishers of Europe in light of Inquisition. As the Dutch had no censorship, and their paper was cheap to produce, the Dutch print industry surpassed that of Venice. It is to the Dutch that Descartes, Galileo, and other scientific revolutionaries turn to for their publications. The Dutch windmills become an exemplar of what the industrial revolution would turn out to be, allowing for the systematic creation of enormous 'value added' activities.

Dutch windmills were the source of Dutch national power, combining technological and institutional innovations—an example the British would imitate in the nineteenth century.

Industrial Revolution

The Industrial Revolution (IR) is typically defined by James Watt's steam engine. While it does form 'the core' in allowing for the efficient conversion of coal energy to work (lift), the phenomenon was a bit more complex than that, as is usually the case in history. As Edwin Layton has shown, the Industrial Revolution had as much to do with craftsmen innovations as with science driven innovations. The tacit knowledge acquired from working directly with materials led to better production methods, demonstrated in the kiln processes and the making of steel.

It is often asked why France did not have a IR, typically characterized as a 'British' revolution. Yet this is a wrong depiction in that the British borrowed widely from many other European nations: French, Dutch, etc. France did have its own variant of the IR, typified by the many canals built as product highways, which were later imitated by British. It is also to be noted that the IR in France did not displace agricultural workers, as it occurred at a much slower pace and was more 'humane' when contrasted to the British variant where displaced agricultural workers flocked to congested and disease-ridden cities such as London.

The poet William Blake referred to Manchester for its "dark satanic mills" to portray the British IR. Coal residue could be found all over its streets, buildings, and houses, and its enormous social inequality contrasted with the eventual rise of the French Revolution which eliminated the privileges of aristocracy by dividing land among peasants. While violent and chaotic, it ultimately led to a restructuring of French society.

The creators of the IR formed the Lunar Society, a group which met at full moon so as to see where they were walking; constituting a group of 'radicals' of sorts. The Lunar Society's members believed that industrialization would be ultimately beneficial for the working man, and had an alternative view as to what technological changes meant, or should mean, for society. Their aim was to steer the IR to a general improvement of society, but were unable to detect all problems arising from it.

Wealth is ultimately a relationship of goods to population; the greater the population relative to its goods, the less rich a country is. Its problem was foreseen by Robert Malthus. In his dreaded vision, population grew exponentially whereas resources 'arithmetically' (linear), leading to an eventual imbalance between the two. Something naturally has to 'give' in the fateful scenario. The Lunar Society did see many problems and created solutions to these but were seen as too radical at the time. In England, its members were followed by British spies, who ultimately shut the group down. During the French Revolution, many members of the upper classes in England were concerned; could the revolution happen in Britain as well?

The British Industrial Revolution began with the water wheel, which had existed since the 1st century BC (Roman). By 1500, these had become the most important source of motive power; some 20,000 mills existed by 1890. When John Smeaton (1724-1792) was hired to build one, he revolutionizes the industry. Smeaton, oddly enough, was a lawyer that turned to 'mechanical engineering.' He noticed there were a lot of inconsistencies in the literature. There were two main kinds of water wheels. The first was the 'undershot,' also called 'impetus,' whereby water passed under the wheel, pushing the ladder up and thus generating force. The strength and volume of the river current generated the motive force. 'Overshot' water wheels, on the other hand, were the more 'typical.' Water poured from the top of mill. Water would accumulate until gravity pulled the wheel downwards, the mass of water and gravity generating the work.

Commercial interests led to conflicts as to which wheel was better; each side naturally claimed its own design was the most efficient and powerful. Smeaton decided to put the dispute to the test, and undertook detailed quantitative measurement of the amount of work produced by the different variants, or the amount of weight each could lift under the same conditions. Smeaton measured the force of each design's wheel, lifting a determined weight over a set distance, in what was a classic Newtonian experiment which quantified the different elements of each system.

He discovers that the best model was actually neither, but rather the 'breast shot' water wheel. In this model the water entered at the back of wheel, but pushed it downward at a half way location. The breast wheel generated the greatest force with least resistance and turbulence. Smeaton's case showed that the application of science to technology could lead to very useful, if unexpected, results, unquestionably determining which model was the most optimal. Smeaton stood at the beginning of the long history of science-technology relations, one pushing the other and vise versa in a positive feedback loop. New technologies lead to new discoveries, and inversely, new discoveries to new technologies. Yet this was by no means the only dynamic at work.

Craftsmen techniques were quietly incorporated into IR, but were not mentioned anywhere by British observers. One can only obtain information of these by foreign visitors, who are so impressed with the results, that they were compelled to write about them. One of best innovations was the reverberatory furnace, sometimes called an 'air furnace.' In it heated air passed over the brick surrounding the material and did not come into contact directly with the ore. One

186

of key problems with traditional furnace was that sulfur in coal/air would interact with the heated substance, thereby degrading the resulting material. The separation introduced by craftsmen led to a higher quality output. Other contributions rested in the making of crucible steel. One needed sensitive awareness as to the properties of clays. Crucibles made out of different clays reacted differently, and craftsmen adopted the technique of grinding up old crucibles for the production of new ones. Other techniques for thoroughly mixing clays were also incorporated in the making of the famed steel, one of strongest in the world.

More importantly, craftsmen also knew how to 'read' heat. In steal made via crucible, which had no air exposure, temperature control was an important factor. Craftsmen realized that color was an important indicator of temperature and in turn the quality of the finalized output. Again, while they did not understand the relationship or the formal underlying scientific causes between the two, they could judge the process by color differentiation, i.e. a 'bright red heat.' Simple technologies are enormously complex, and appear 'simple' only because complex processes become 'black boxes' to the consumer.

The Steam Engine

Watt did not invent *'de novo'* the steam engine. Although it could be argued that his was the 'final design,' it had been built upon precedents—a fact which helps understand true nature of Watt's exact contribution.

Huygens and his assistant Paupan created an early version of the steam engine, but their main aim had been scientific: prove the notion that work could be generated by the lifting of a piston. At first, they tried to use gunpowder in early models, leading to humorous, if dangerous, explosive results. Yet the pressure generated by this source was too irregular in intensity and force. Huygens eventually gives up, but his assistant continues to develop the experiment, away from gunpowder. In 1690 Paupan substitutes steam, and he is finally able to push the piston to its top position in an even manner; it was then held by a wooden wedge inserted by Paupan. The following day, the steam had condensed, and upon the wedge's removal, the piston fell down, lifting a weight attached to it. Obviously this not a viable working prototype, as it was too slow and inefficient. But the experiment was certainly a 'proof of concept.' Steam could be used to generate lift, and hence, work.

An earlier precursor was Giovanni della Porta (d 1615) who observed that, in an inverted vessel filled with steam, water rose as steam condensed. Similar perhaps to Galileo, della Porta was persecuted for his scientific activities. He had been a part of the 'natural magic' movement, also attempting to form an *Academia Secretorum Naturae* (Academy of the Secrets of Nature). He is imprisoned by the Inquisition, and allegedly took own life by jumping from his prison tower—an interpretation difficult to believe given the circumstances.

By 1699, Thomas Savery had developed a working steam engine, consisting of three parts. The boiler chamber received steam from boiler tube. When cold water was poured on it, the chamber would cool; the steam would condense, thereby lowering the pressure and hence pulling water up the shaft. This early steam engine had no moving parts and worked as a primitive 'pump,' limited in its use. As previously noted, the 'vacuum-suction' principle had been identified by Galileo and Torricelli who showed that, through condensation, a partial vacuum could raise water only to a maximum of 32 feet. The Savery engine did not allow for deeper mine exploration given its limited functionality.

The first stream engine was that designed by Thomas Newcomen in 1712. This turned out to be an immensely successful variant; by 1800, there were 2,000 Newcomen engines but only 400 Boulton-Watt engines. Its success can be attributed to its reliability, in that it could run for decades with little maintenance—something which no digital technology can do today. English mines were routinely affected by flooding, either due to the constant rain or water tables that were close to the surface. Water inundations were thus a constant problem for mining concerns. Newcomen's 'steam engine' solved this by pumping water out of mines, allowing miners to their work.

Another key to its success was the abundant availability of coal in Britain. Newcomen's engine was enormous, not the size relative to one found in a compact car today but rather filling the space of a two story building—exemplars of which can be found in the London Science Museum. Newcomen modified Savery's model, replacing the chamber with a piston attached to an inverted lever. The enormous piston, 4 feet in width, was lifted with steam; atmospheric pressure also helped by pushing the piston down, forcing the other end of the lever to rise. The use of ambient air pressure led to the Newcomen engine's nickname: 'atmospheric engine."

One key problem was that the walls of piston changed temperature. As in the Savery model, these would be cooled down, to then be heated up in a repetitive cyclical process. The chamber, in other words was not consistently hot, and thereby lost much of its efficiency in the changing of temperatures.

James Watt modified the Newcomen model to arrive at the first true 'steam engine'. Watt had been an instrument maker at the College of Glasgow when he was first asked by Prof. John Anderson, a professor of natural philosophy, to repair a small Newcomen model. Watt immediately identified its main problem. The cooling of the chamber walls meant that the engine had to reheat the walls back up again, slowing its cycle down. Watt realized that the chamber walls had to be kept consistently hot so as to make the engine cycle more continuous. He devised multiple innovations, the principal of which was to separate the condenser from the main chamber, leading to a three component system. The creation of a separate condenser chamber allowed the piston chamber to consistently retain its heat, which then cooled in the condenser.

Watt apparently has a 'serendipity' moment when an early model did not work correctly. A laboratory accident led to a leak, which let water into the chamber but created enormous force in the resulting downward piston movement. Watt realized that by spraying cold air directly into the piston chamber, the force of resulting contraction could be greatly increased and thus

188

generating a much greater amount of power than that which was generated by Newcomen engine, and four times as efficient.

Watt entered into partnership with Matthew Boulton, a hard-headed businessman but with sound technical expertise and political acumen who made buttons for a living. Boulton was able to get a 25-year patent on Watt's invention, clearly understanding its potential value as well as Watt's creative genius. Boutlon's wealth allowed Watt space and time for his work. Curiously, Watt also had other great ideas, such as chlorine bleaching of textiles. However, Boulton wanted a working machine and prevented Watt from distracting himself on other projects.

Watt ingeniously creates PV diagrams of the engine's internal thermal dynamics: an analysis of heat-pressure relationships in the piston chamber. This allows him, as Smeaton, to create a detailed scientific analysis of steam engine showing the exact amount of work (force) produced by the engine regardless of engine's actual mechanism. Each type of engine in fact has its own distinctive pressure volume graph, amounting to a unique 'fingerprint': The Otto four stroke engine or the diesel engine can be identified by their unique PV graphs. Watt showed that the area under curve equaled to work done by the engine, and hence a calculation of its power.

By 1800, the Watt engine had been successfully applied to locomotives at very high pressures—much more than original Watt engine. Watt and Boulton became the first technological 'millionaires' of 'Manchester valley.'

Origins of Thermodynamics

Watt's PV diagram of a steam engine is akin to Smeaton's first analysis of the watermill; in both cases, quantitative analysis were undertaken to determine the amount of work done by each respective model. The studies pinpointed the exact location of 'error.' Smeaton was led to the breast shot water wheel whereas Watt the condenser. In both cases a generalizable approach was developed which could be used to identify work done by any engine: a rat's heart is also an 'engine' producing 'work' (pumping of blood) and amenable to comparable quantification. These, in turn, also allowed cross category comparison, including (strange as it might seem) the phenomenon of urbanization recently as shown by Geoffrey West. Both biological and urban structures benefit from scaling effects, a phenomenon that had been detected by Galileo as well.

Yet the thinker who did most to contribute to the science of heat was Sadi Carnot (1796-1832). Whereas both Watt and Smeaton's analysis were technologically driven, with the sole aim of improving a given engine design, Carnot's effort was to more abstractly generalize from the data obtained, laying the foundations of modern thermodynamics. His story, regrettably, is also a tragic one due to his early death as a result of an epidemic of cholera, or untreated sewer water.

Carnot's father had been tied to Directory during the French Revolution, and gains fame by organizing peasants into army, which is able to defeat early attacks.

With the rise of Napoleon, his father retires and sends his son Sadi to the Ecole Polytechnique. In 1824, Sadi publishes his renowned *Reflections on the Motive Power of Heat*, an analysis of principles of steam engine.

Sadi noticed that in order for any engine to work, it required heat differentials; it is the difference in temperature of an engine that is actually establishes the amount of work produced. If all parts of an engine were at the same temperature, no work would be generated by it. Sadi also realized that there is a limit to work, a limit which is also dependent on temperature, and further develops Laplace and Berthollet's notion of a caloric using a theoretical windmill model to describe the thermal dynamics of steam engines.

Caloric, or heat, always wanted to flow from high to low. If in a compressed space, it wanted to expand and cool. Since there was, by definition, a limit on the highest and lowest temperatures an engine could reach, this difference thus set a limit on the total work produced by all engines. Akin to a waterwheel, whose torque is dependent on its size, the steam engine's 'torque' was similarly calculated by the 'height' differential of heat. Although not spelled out exactly, Carnot had alluded to the Law of Conservation of Energy.

Industrialism and Liberalism

It should seem strange to suggest, when compared with the close ties between technology and corporate capitalism of our era, that the core vision of the early industrialists was in fact a liberal and socially egalitarian one. Yet liberalism, so defined, is the most striking and marked trend of the Industrial Revolution's core participants and creators, as exemplified by da Vinci. The Dutch society had been founded upon liberty of religion and speech, the Lunar society in the Britain sought to create new social utopia, and Sadi Carnot was guided by key democratic and meritocratic principles.

Yet this fact—the belief of the early industrialists that technology would lead to a more just and humane world—is particularly perplexing and paradoxical when viewed in light of the ultimate asymmetries of power the Industrial Revolution ends up creating, both at home (social inequality) and abroad (Imperialism).

When Sadi Carnot studied at the Ecole Polytechnique, he internalized the Ecole's core value system of meritocracy. Positions were to be awarded to the candidate who had performed the best, rather than the candidate who had the wealthiest parents. Perhaps as his father's revolutionary activities, this meritocratic criterion was then seen as a 'radical' political view, particularly so in light of traditional hierarchies of French society. Carnot had even been offered a position as government advisor, which he rejects for this very reason as it was given on the basis of nepotism.

More importantly, Carnot believed that all democratic systems were ultimately linked to economic development, and hence became active in the foundation of the new College of Arts and Manufactures, a vocational school for the poor akin to the new Mechanics Institutes in England. So 'liberal' was Carnot in his views, that he was widely followed by the French secret police. An anecdote suggests that in one lecture which had been attended by 100 persons, 75 of its audience consisted of French secret agents. Although exaggerated, it helps provide a notion of Sadi's social views and the reaction it created by his very own government—an experience that would be shared by the Lunar Society.

Dr. William Small, who had been Thomas Jefferson's professor while living in the British American colonies, was a catalyst in the formation of the Lunar Society. After his return home to England due to illness, he introduces James Watt to Boulton, stimulating not only their powerful commercial relationship but laying down the seeds of the Lunar Society. In 1766 Birmingham, its first meeting is held in Boulton's home. Many such societies were actually common to the era, including the American Philosophical Society in Philadelphia and the Manchester Library, but the Lunar Society was more informal and did not publish any papers or 'transactions.' However, its meetings were attended by some of the most important scientific figures of the period, including Joseph Priestly (chemical revolution), Erasmus Darwin (Charles Darwin's grandfather), William Withering (discoverer of digitalis for heart conditions), and Josiah Wedgewood (an important industrialist).

The group had some awareness that a phenomenon such as an 'industrial revolution' was being brought about by them, and viewed it optimistically. While they did foresee some of the problems being created, and did try to remedy these, they were unable to fully conceive of its full nature and impact. They simply believed that overpopulation was its noxious byproduct, a structural cause as the tuberculosis epidemic in overcrowded London. Watt himself believed that the new 'airs' (oxygen) discovered at the time would be able to cure tuberculosis, and establishes oxygen tents for treatment. Watt also forbade work and design of high pressure steam engines, fearing these would ultimately explode under load, creating a negative public view, public outcry and extreme legislative reaction.

The Lunar Society also sought to reduce the social impacts of industrialization as social insecurity and family disintegration; for Watt, the two were intimately linked, and he encouraged workers to organize 'friendly societies' of mutual assistance. These societies became the very first corporate unions. Watt also supported Methodist religious revival as an antidote against drunkenness and other forms of disintegration.

Priestly was also acutely aware of the problem of education, in that children who worked 12 hours per day, 6 days per week were simply unable to receive any education at all. This led to his formation of the Sunday School movement, which was initially not for indoctrination purposes but simply to provide free education to the many children that required it. It consisted of the three R's: reading, writing, and arithmetic—ultimately contributing to the rise of England's literacy levels.

Their common liberal views, as one would expect, made the members of the Lunar Society targets of the British government. Priestly and Wedgwood were overt political radicals. Wedgwood himself openly expressed sympathy with the French Revolution, and was enthusiastic about the possibility of reform and change. The British government feared these views might contribute to the spreading of the French Revolution in England. James Watt, strangely enough, was even elected member of Revolutionary French National Assembly.

The British government used underhanded tactics to suppress the Lunar Society. In particular, they promoted false riots, making Priestly the focus of mob attacks populated by the criminal underclass. During one such attack on his home, Priestly is able to safely escape via a back door; it had been a 'lynch mob' whose actions could have easily ended in his assassination. Priestly chooses to migrate to the United States after this, a country then more amenable to such views. The Lunar Society formally 'closes' its doors in 1791, having no further meetings after that date. Yet why had its members been so blind to the ultimately consequences of the Industrial Revolution?

Coal had been the key ingredient to the British Industrial Revolution; it is an abundant material throughout Europe, situated near population centers which in turn facilitated its use. Coal mining rose from 2.7 M tones (1700) to 250 M (1900). However, its use did lead to serious problems, as the abuse of children. Small children could easily fit through the small mine shafts, leading to their prominent employment in the sector to haul coal. The 1842 Coal Mining study revealed abuses, providing images so shocking, that they helped facilitate the passage of prohibitory laws to this practice.

Yet the ultimate cost of the industrial revolution was never foreseen by its key founders: climate change—an issue now widely recognized as the leading threat to humanity in the near and long-term future. It now appears that Leonardo da Vinci's concern about the use of technology had been on the 'right track' all along.

There is a 'law' in critical literary theory which is applicable to the story. Once an author lets go of his literary creation, he cannot control it in the public eye, specifically its reception and reaction. The only moment an author has the most control is when it is in his hands, prior to publication, where he can mold it to his will and concerns. Once in the public realm, the public begins to read and interpret in their own manner, and the work acquires a life of its own that could not have been predicted by its creator. Similarly, once an inventor releases his technology to the general public, it will be used in ways and manners he did not initially intend or foresee.

We can best account for the inability of its early creators in foreseeing the impact of the technologies they were introducing to the world in this manner; once out of their hands, they could not control the scope, reach and extent of its development and use. Once out of their hands, as Mary Shelly's Frankenstein, the creators lost control over their creature.

However, the 'unintended consequence' of technology can be both a good and a bad thing.

Creative users push the boundaries towards positive ends as well. While there will be no single technological 'fix' to climate change, there is no doubt whatsoever that technological changes—solar, wind and water power generation—will not only help ameliorate changes wrought on by 200 years of industrialization on Earth, but actually help mends its damages to the environment.

192

It, of course, will not be the only factor at work; overpopulation is still a continuing threat to humanity's and the Earth's future prosperity.

Charles Darwin and the Beagle:

Biological Evidence of Millennial Change

THERE ARE PERHAPS TWO KEY MONUMENTAL FIGURES in the history of science who created the fundamental concepts onto which most later science was built. We have already seen the first of the two: Isaac Newton. Newton redefined the agenda in physics for 200 years, and his work is still used today, rather than 'invalidated' by quantum physics as is often heard. The second figure is the focus of this chapter. Charles Darwin, as most already know, established evolution as the fundamental process of all biological life. As with the Newtonian notion of gravity, the Darwinian concept of evolution became the key paradigm in the biological sciences during the next two centuries—the skeleton onto which all other theories would be built.

Contrary to overly vociferous public claims, there is no question whatsoever as to the validity of evolution. The topic is often 'debated' by the religiously inclined who want to imbue their deity or deities with greater agency in the natural world than they merit. Life is defined as 'spiritual' and 'unique.'

Yet Darwinian evolution, as a paradigm, took longer to become established than many other key scientific concepts. When first presented in 1859, the *Origin of Species* was hotly debated. Its key defendant was T.H. Huxley, known as "Darwin's bulldog" for his intellectual acuity and dogmatic ferocity. There was a period of acceptance during the mid-nineteenth century as a result of Huxley, but by the end of the century, there were doubts as to its validity. After all, Darwinian evolution required immense time spans, much longer than had ever been previously conceived in Western intellectual history—so much so that it played the important role in drastically expanding Western conceptions of historical

194

depth, as described by Stephen Toulmin and June Goodfield. Human evolution required millions of years, rather than hundreds, or even thousands.

The human brain is not made to truly grasp the meaning of 'deep history,' and still today most do not fully appreciate its meaning.

Yet the idea of 'evolution' in itself was not strictly new. Darwin's own grandfather Erasmus had previously proposed it. He did not provide a rigorous proof as his grandson did, but nonetheless had the notion. Robert Chambers also suggested it in 1844, but was widely ridiculed—even by T.H. Huxley himself, whom we might presume would have held a more favorable position. The reception of Chambers'' work actually taught Darwin that he needed the most rigorous of proofs before going public with his claim, as he obviously did not want to be needlessly humiliated. Evolution would require a mechanism. How exactly do new species arise?

Being his intellectual descendants, most of us already know some of the answers. Mendelian genetics was the key, culminating in the discovery of the structure of DNA by James Crick and Thomas Watson, for which a 1962 Nobel Prize was awarded. But the combination of 'genetics' and 'natural selection' is much more complicated than presumed in that the combination was not initially obvious. This process occurred roughly during the middle of the twentieth century, in what today is referred to as the "Grand Synthesis." Its key participants included T. H. Morgan, Theodosius Dobjzansky, and Hugo de Vries.

Sadly, this development in science also includes the story of scientific abuse. The data accumulated to determine the structure of DNA was 'stolen' by Watson from Rosalyn Franklin, one of best crystallographers of the time. Her x-rays photos showed without a doubt the molecule's double helix structure. The images, which had been sitting on her desk, were simply taken without permission. One has to wonder how often this phenomenon has occurred in the history of science.

The discovery of humanity's biological history began with Darwin's exciting voyage through Latin America as a young student. It became the adventure of a life time—that almost did not happen.

The Voyage of the Beagle

Darwin's voyage lasted five years, from 1831 to 1836. Although the principal scope of trip was South America, it had also been a 'worldwide' journey crossing the Pacific Ocean and its regions as Tahiti, New Zealand and Australia. It was a voyage around the world undertaken only in a 90-foot boat, 13 feet in depth: "The Beagle." Its 74-person crew was relatively young, with an average age of 25. Captain Robert Fitzroy, at 26 years old, was one of the oldest. Darwin himself at the time was 21 years old. Yet Darwin's expertise was quite advanced for his age; we should not falsely presume from his youth that he was a novice. When it came to identification of species, locales were awed by his ability to distinguish venomous from non-venomous snakes. Darwin had already accumulated a very intimate knowledge of the flora and fauna of England prior to his journey, and was ready to devour more.

For him, the trip was… 'awesome,' as it would be for any young man. He had seen more in one foot of the Brazilian forest than in a mile long walk in England. Latin America's wildlife was also characterized by an enormous diversity of creatures, much more so than back home. Darwin had also been an avid experimentalist; his narrative routinely captures his dissection of animals to examine their diets: fishes, octopi, birds, etc. He finds unexpectedly that the lizards of Galapagos ate seaweed. Although he detested being in the ocean, he unsuccessfully tries to throw one into a boat. Darwin cuts the heads off fireflies to see if they would still light up (they would); he feeds a condor meat wrapped in cloth to determine if it detected its prey by smell or sight (both). He is also very well read. An enormous corpus is cited by him, some of the more common authors included Hans Sloane, Captain Cook, Alexander Humboldt, and Georges Cuvier, amongst others. Darwin had been preceded by a long line of naturalists that had traveled the world.

It is shocking to consider, given the historical significance of the trip, that it almost did not even occur—reminding every historian as to the unpredictable role of contingency in human affairs. Darwin had been the third person asked to fill the position.

Professor John S. Henslow was asked to recommend a scientific companion for Fitzroy, and while Henslow himself considered going on the trip, his new family (a new baby) and his 'old' age (35 years old) made him desist from the offer. Henslow then offers the charge to Leonard Jenyns, a friend, but Jenyns also rejects it as he had just obtained a charge as a religious pastor. In any job and time period, if an individual resigns soon after having obtained the position, their merit as a responsible individual will be placed into question. Interestingly, the request does hint at the close relationship between natural history and religion in England during the 19th century.

Natural history, what today we would regard as biology, was the science of elite. Nature was used as a visible proof of God's order in world. Its presence can also be implicitly detected throughout Sir Hans Sloane's *Voyage to Jamaica* (1707), a detailed description of the island's flora and fauna. An enormous amount of work had been done by Sloane, as if in observing God's hand in the diversity of creature and plants in nature.

Once offered the position, Darwin was ecstatic. As a naturalist in England collecting beetles, he recognized the trip for what it was: the opportunity of a lifetime. Yet Darwin's father at first opposed it. As a physician fortunate enough to have invested in industrial concerns early during the Industrial Revolution, Darwin's father made a fortune from his investments. The father vehemently opposed his son's proposal, criticizing Darwin for spending too much time in nature rather than developing a well-established profession. However, the father makes an important concession: he would agree to the trip if Josiah Wedgewood, his father's brother in law, agreed to it. Wedgwood was an ardent supporter of Darwin, and wrote a long letter in favor of the trip. Darwin's sister also helped convince the father, who seems to finally realize what a unique opportunity the trip would be for his son.

But, even upon the father's final acceptance, other issues emerged.

Darwin had taken too much time in answering the request, and so the slot had already been filled by a fourth nominee. Captain Fitzroy, however, agreed to meet both, so as to evaluate his alternatives. The two young men hit it off, to Darwin's advantage. Yet Fitzroy warns him of the dangers ahead, to which Darwin said he was willing to risk. Curiously, however, Darwin was not chosen for his biological knowledge per se but simply because he would be a good companion for Captain Fitzroy, an aristocratic member of the elite who would be surrounded by sailors of the 'lower class' during the trip. Fitzroy needed company that was appropriate to his social standing and education. There was another naturalist on board, and a rivalry quickly emerges between the two.

Darwin, however, also becomes uncertain of his decision just days prior to journey; apparently his imagination almost 'got the better of him.' He begins having heart palpitations, a classic sign of an anxiety attack, and insists on meeting Fitzroy more often before the trip. But Fitzroy was simply too busy making preparations, and did not have the time. Darwin would have to 'man up' and deal with his anxiety on his own. This momentary relapse proved to be useful, as it was from it that he begins to keep a journal to soothe his emotional state. These journals would form the material from which the *Voyage of the Beagle* (1839) was written.

Darwin's fears were not without due merit. Fitzroy had become captain of The Beagle because the prior captain had committed suicide. The original plans for the trip called for two ships, but Fitzroy was ultimately assigned only to the Beagle for this reason. It is an important fact to consider. There would only be one boat in the open ocean. Should an emergency happen, there would have been no relief at all—a scenario aptly captured in the recent movie "In the Heart of the Sea" (2015) whose crew was forced to resort to cannibalism in order to survive. In fact, Fitzroy also commits suicide later on in life.

The first days of the trip were extremely difficult for Darwin, physically and emotionally. One of the main purposes of the trip had been a cartographic exercise, to trace more detailed contours of Latin America. Darwin never adjusted to life at sea; as a tall man in a small rocking ship, he was constantly vomiting. The confined spaces meant uncomfortable sleeping on a hammock. At one point, Darwin was so nauseated, that he requested to be left on land at the next port of entry; many of the crew did not believe Darwin was going to endure the trip's first week.

Darwin did have a girlfriend, Fanny Owen. As with all partings, the separation between the two was emotionally difficult. However, shortly after Darwin's departure from England, Ms. Owen marries an aristocrat—an act whose suddenness took everyone by surprise, including her own family. Her father writes a note to Darwin expressing that his affection for Darwin would not in any way be diminished by his daughter's new nuptial status. The emotional distress caused by the romantic rupture compounded the physical difficulties faced by

Darwin, who seriously thought of abandoning trip on a first landing. Darwin, as the reader may correctly guess, decided to go on.

Throughout his travels in Brazil, Argentina and Chile, Darwin writes a great deal about both people and animals of the region. He is shocked at the impact of slavery on the character of Brazilians. While talking to a slave, Darwin lifts his hand reaching for something, to which the slave ducks believing he was about to get struck. In Argentina, Darwin is taken aback by the gaucho, who could have enormous wealth with vast amounts of territory, while at the same time live in very humble conditions. Yet the gaucho was prone to using knife at the slightest offence, intended or not. By contrast, while the Chileans were more 'civilized,' there was a greater degree of inequality between them, with a quasi-medieval serfdom dedicated to mining in its lands. The 'serfs' worked seven days a week under horrific conditions, and their positions were passed down between generations.

Generally speaking, Darwin was shocked at the social inequalities as well as the government corruption of Latin America. Obviously Darwin had a different social lens coming from a very different society. Bribery was rampant in the region, and most citizens did not expect justice from the government. Poor men were often arrested for the most trivial of offenses, while the wealthy not only avoided jail for the most horrific of crimes, but often bribed police to arrest others to take their pace the very same evening of the crime. The elite were uneducated, knowing little of geography. Its inhabitants were often suspicious of Darwin, wrongly presuming he was a mining prospector looking for gold given that he was often breaking rocks apart looking for fossils. They were not very curious, failing to inquiry why some springs were cold and others hot; they did not wonder what caused tsunamis or earthquakes and generally tended to accept 'because God said so' explanations. The notion of systematic inquiry, of looking deeply into an issue, was not a common mental habit.

They failed to recognize that a functional social system depended on integrity of its citizens; without a core population who believed in justice, stable governments could not long endure.

Alexander von Humboldt

Darwin had been immensely influenced by Humboldt. Darwin once told his sister that he had committed to memory many parts of Humboldt's works, and could recite entire sections by heart. Another indication of his emotional affinity can also be detected after the trip. If he respected Humboldt before going on his journey, he admired him even more afterwards as he now appreciated the difficulties Humboldt had encountered. Darwin is in disbelief when Humboldt becomes a relatively forgotten figure later in life; such were the curses of heroes, Darwin notes.

Alexander von Humboldt was a German scientist who traveled between 1799 and 1804 throughout the region that would become Latin America. He gathers some 60,000 specimens collected into 55 boxes, which are sent to Europe.

Humboldt traveled with French botanist Aimee Bonpland, writing up his adventures in *Personal Narrative*. Due to it, Humboldt becomes a hero fiction book of day—an "Indiana Jones" type figure who goes to exotic places, facing many dangers and adventures. His stories do make up for good storytelling, and it is not unrealistic to suggest that Humboldt might have actually served as the basis for the fictional character.

At one point, while walking along a river, he runs into a panther, which could have easily killed him. He slowly backs away and retreats without ever placing his back to the panther; as he did not have weapon on hand, he would surely have been killed should the recently-fed panther have chosen to strike. He also has an encounter with an electric eel. He realizes that, contrary to common belief, eels did not kill with their charge, but rather stunned their victims; a horse whose underbelly had been shocked on a water path would likely drown by immobility. Humboldt also related the story that when climbing the peaks in Ecuador, such as el Chimborazo, his body was suddenly covered entirely with bees. It turned out that the bees were stingless and totally harmless; however, being on the ledge of a precipice, this 'bee invasion' could have also resulted in a nefarious outcome. The adventures come one right after the other in the books' narrative, as in an Indiana Jones movie.

Yet Humboldt was much more than a mere adventurer; he is a scientist who plays an important role in European social and intellectual history. He identifies guano (bird drippings) off the coast of Peru, and performs chemical analysis detecting high quantities of phosphorus which could be used for fertilizer. Enormous 'mountains' of guano had formed on small costal islands: dried bird poop that had accumulated through the centuries. This guano quickly turned into a profitable industry, leading to international tensions and ultimately a war between Peru and Chile over it. Humboldt detects an enormous current of cold water that flows to the equator along the western coast of Latin America (Chile). He invents the notion of 'isothermal lines,' or lines of equal temperature which become key instruments in detecting broader climatological patterns.

As Darwin, he also provides a detailed social description of the Americas, which is still avidly used by scholars as Michael Zeuske today to peer into the social institutions of the era. As noted by Zeuske, Humboldt offers a unique glance given that he interacts with the entire social strata of the region, from Simon Bolivar to the peasant farmer. Humboldt had been very fortunate; if he had attempted the journey 50 years earlier, he would have been prohibited by the sheer bulk and weight of instruments, whereas if he had traveled 50 years later, the independence movements of the region would have prohibited his mobility. Incidentally, Humboldt interacts with Francisco de Caldas, one of brightest minds of Latin America whom would eventually be killed during the revolution.

Perhaps typical of his era, Humboldt wanted to 'conquer' the world with his mind, and towards the end of his life he tries to write up synthesis of all human knowledge in *Kosmos*, a multivolume project which goes unfinished due to his death. This was perhaps the last period in human history where a single human could accumulate all existing knowledge on his own. As Derek de Sola Price has demonstrated, the attempt would be futile today.

Humboldt's philosophical influence might be referred to as 'romantic empiricism,' analogous to the *Naturphilosophie* movement but distinct form it. It is the view that there is an underlying unity to nature: order and coherence that is not immediately obvious, but which is counterbalanced by an intense and active exploration of nature. From a scientific point of view, we can point out that Humboldt was aware of the impact of geology on biology. As he climbed the Chimborazo, the Andes's highest peak, he noted that species varied enormously with elevation, and drew maps plotting the relation between the two, an isothermal species map of sorts.

Darwin was fortunate in having such able mentors.

Days of trip

Darwin actually got along very well with the crew, whom referred to him as "Philos." Their appreciation and respect can be gleaned in that, when on land, if they came across something a curious specimen or stone, the crew would often pick it up for Darwin.

By contrast, Fitzroy was a particular fellow. Hired to do 'her majesty's' cartographic work, he kept 21 clocks in his quarters, each set at different times; some were set to Greenwich time to help determine the ship's longitudinal position. England was attempting to lay claim to as much land as possible, and detailed readings helped settle international disputes. Beagle had only been one of many such ships sent for this purpose, which serves as a distinct reminder that this was the period of British colonialism. The nineteenth century would come to be known as 'Britain's' century,' given their ample control of lands, peoples, and resources throughout the globe. British colonialism helped provide the 'path' to Darwin's discoveries.

In spite of their later conflicts, Fitzroy does help Darwin in one crucial aspect: he recommends that Darwin take notes immediately after an event or day trip, when his memory was still fresh so as to allow for a more detailed and richer exposition of the event on paper. If, by contrast, Darwin were to delay his note taking for too long, his memory would fade and lack important details for later use. It would be a tip that Darwin practiced religiously throughout the rest of his trip. Fitzroy aids Darwin in other ways. He began giving preference to the shipment of Darwin discoveries to England, rather than to those of Robert McCormick, the ship's official naturalist, which naturally led to a rupture between the two. Darwin had usurped this position for himself, and McCormick ends up resigning from his post due to the crew's bias.

Yet to suggest a close friendship between Fitzroy and Darwin would be a mischaracterization. There were enormous social and cultural differences between the two men that could not be easily dismissed. Darwin belonged to a family that had historically sought abolition. Although England had passed a law banning the slave trade by 1807, it took much longer to implement it. Hence, law and actual practice did not go hand in hand, particularly the case when the slow pace of travel and the long distances implied a 'large' world, as opposed to today's networked societies. Fitzroy also belonged to the elite of England, and hence believed in inherent racial differences between the races. Fitzroy and Darwin would dine together, as civilized men, but began having increasingly acrimonious conversations. The radical difference in assumptions between the two finally blew up when the Beagle arrives at Brazil on Feb. 1832 where slavery would remain an active institution until the end of the century. Their personal debate became so acrimonious, that Darwin again seriously considered leaving. Both resolved to recognize differences and to not talk on the subject again; too much was at stake for both.

Yet the issue indirectly rises again late in Darwin's narrative. A group of Native-American Indians, the Fuegians, had actually been on the Beagle all along, but are not mentioned until very late in book. The ships' prior captain had been the one to have first picked them up. The intention was to civilize the four indigenous members: Yokcushlu (Fuegia), Orundellico (Jemmy) and El'leparu (Minister York). The fourth member, who had been the smartest of all, died from smallpox in England. Interestingly, Darwin at first believed them to have been British farmers, and upon hearing their origins, Darwin comes to realize the great deal of acculturation that had occurred during their journey. Jeremy had been the friendliest of the group, and had nearly lost all of his original language.

For Darwin, the Fuegians were of a generally low cultural level; upon landing, its tribe members were most concerned with buttons and trinkets rather than the impressive ship which had carried its tripulation across the vast ocean. Darwin comes to realize the important role which culture plays in accounting for many differences between human social groups; for him, most human differences were of a cultural nature rather than of racial origin. He then begins to consider universality of man, regardless of origin and ethnicity. There can be no doubt that trip set the background for Darwinism in various aspects.

Too much emphasis is perhaps given to the Galapagos in popular accounts of Darwin's 'evolutionary theory.' The Galapagos islands were just one of many examples, which while possibly emphasized for the sake of simplicity, fails to give due importance to the diverse sets of data which came to influence Darwin's ideas.

For example, he became aware of the *"Gran Seco"* (1827-30) which had briefly preceded him: a great drought that killed thousands of animals. One farmer alone lost 20,000 cows to the drought. In one particular location, animals had died 'en mass,' the reason of which was easy to understand at the time, but could be so gravely misinterpreted in the future. There was a steep water embankment which the animals would descend to have a drink of the precious liquid; however, they would be too weak to climb back up its ledge, thus resulting in an enormous accumulation of animal remains. It left Darwin wondering what the scene would look like ages hence. Some would come to interpret the scene that a great catastrophe had occurred: a huge flood.

What would the opinion of a geologist viewing such an enormous collection of bones, of all kinds of animals and of all ages, thus embedded in one thick Earthly mass? Would he not attribute it to a

201

flood having swept over the surface of the land, rather than to the common order of things?

Darwin also comes across the fossil of a giant sloth in Brazil. Darwin realized that it likely used its corpulent weight to push trees down, rather than east its uppermost branches as a giraffe. As no other animal of its size was observed by Darwin, it was clear that it had gone extinct. As he explores the fossil record, he also observes a particular pattern: certain animals predominate during certain eras, to then suddenly disappear or become minor players in later epochs. Epochs tended to have particular 'dominant' species, whose type varied greatly over long time spans.

In the Argentinean Pampas, Darwin also witnessed the enormous impact of domesticated species: horse, cattle and sheep, came to substitute local ones as the guanaco (llama), deer and ostrich. The European pig replaced pecarrientire species, and in turn affected the local ecology, which altered the life of other animals in region. While at the Falkland Island (*Islas Malvinas*) of the coast of Argentina, Darwin also comes across smaller horses. He finds that the distribution of horses was rather peculiar; although there were no natural barriers, horses tended to aggregate to one side. If were animals affected by geology as noted by Humboldt, why was this not the case here?

Perhaps the most striking experience was that of an earthquake in Chile where Darwin witnesses land rising 10 feet; what had been a minor tidal pool became a small cliff. Land had been raised, leaving Darwin in a state of 'shock.' "A bad earthquake at once destroys our oldest associations: The Earth, the very emblem of solidity, has moved beneath our feet like a thin crust of fluid." The event might appear to be trivial, but it became a key to Darwinian evolution, revealing the influence of Charles Lyell.

Prior to leaving, Darwin took various books along with him, one of which was Lyell's *Principles of Geology* (1830). In it, Lyell describes his principle of "uniformatism": the notion that geological change is slow and cumulatively, occurring over eons. One of its key presumption was that the same geological changes that can be observed in the present are the same as those which have occurred in the past. We can easily begin to see how these two notions would have merged in Darwin's mind: biological change in the present was the same as that which had occurred in the past, and one needed only to critically analyze the observable evidence it to identify the factors underlying such change.

Lyell's work was a reaction to existing theories of "cataclysm" which, as the name suggests, was a religiously influenced theory claiming that all geological change occurs drastically, radically in huge spurts. Geological processes were characterized by an inverse series of motions akin to Noah's Flood; rare, non-repeatable events in nature whose unique historical conditionality meant they were not amenable to observation and scientific analysis. One can easily detect why Lyell would be opposed to it: as extreme events were unpredictable and non-repeatable, they were not amenable to rational analysis and empirical evaluation. As in the Arab world, one could not do science if God could so quickly change nature from one day to the next, leaving no trace behind.

Upon his return, the two men naturally became good friends as Darwin provided much supporting evidence for Lyell's theories. Darwin's earthquake had only been one of the many that regularly occurred in the *'Tierra del fuego'* region; even Fitzroy noted that after an earthquake, all volcanoes in region seemed to become active. That the earthquake had a powerful impression on Darwin cannot be underestimated; he realized that while a change in 3 or 10 feet might not seem like much, when added together over long periods of time, these would lead to the creation of mountains and mountain ranges. This observation in turned helped to account of the oddity that ocean animal fossils were often discovered high on mountain tops, suggesting that a lake had dried and risen in elevation.

Earthquakes as a geological mechanism also had biological implications. If in geology what had felt like 'eternal' and 'unchanging' could be subject to change, could not the same notion also apply to biological species?

We notice a hint of evolution in book, but Darwin is not there yet. The theory of evolution out of natural selection did not emerge 'fully grown' during a 'eureka' moment of serendipity. It emerges as the final outcome from a gradual accumulation of data and contingent circumstances. While the famous Darwin finches are typically the cased used for 'proof,' it was only after completing his that James Gould informs Darwin that his finches were actually of different species. Darwin initially believed they all belonged to the same species, and as a result has difficulty in identifying their island of origin due to improper labeling.

In fact, while at Galapagos, a cook suggests the species unique distribution by telling Darwin he could identify where each turtle came from simply by looking at its shell.

Our geological puzzle immediately reemerges: if all creatures in Galapagos live under the same conditions, they should in theory all have the same biological forms. But they did not, contrary to Humboldt's close association of geology and biology. Why was this? Darwin eventually realized that he could not tie biological theory too closely to geology, for if the geological science proved to be wrong, it would undermine its biological claims on which it stood. Thus, even though Lyell served as a key paradigm for the emergence of Darwinian 'evolution,' Darwin realized that he had to break intellectually free from geology so as to withstand scrutiny and create a more permanent biological theory.

During the five years of his trip, Darwin accumulated an enormous amount of information: animal specimens, fossils, dissection data, etc. Upon his return, it came time to make sense of it all.

Return to England

During his trip, Darwin picked up a lifelong illness, whose exact nature is hard to determine. It is certainly the case that he never again has the same stamina after to the trip, leading some to suggest he had been infected by Chagas disease. Incidentally, the voyage had not been without its risks, and on various occasions Darwin was a bit too close to danger. While in Patagonia, he leaves the Beagle on a smaller boat to explore the region, only to see an iceberg break up a bit too close for comfort. As the small boat was 100 miles away from The Beagle, it would not have been spotted had it become shipwrecked. On another occasion, he comes close to clashing with the native Americans in Argentina. General Juan Manuel de Rosa's Native American extermination policy had left a violent wake in its path, and had Rosas not sent men for his protection, Darwin would likely have been killed.

Upon returning to England Darwin gave a number of public presentations about his trip, published as *Voyages of the Beagle*. The book helps establish Darwin's early positive scientific reputation. He also began the process of assimilating all of the information he had slowly gathered: speciation in absence of geological barriers, Lyellian uniformatism in earthquake creating mountains over eons of years, fossil records showing epochal variation in species—some of which had previously been dominant during certain periods only to become minor groups later, and so forth.

He also began reading Robert Mathus's *Essay on Population* (1798).

Unlike Alfred Russell Wallace, Darwin had not read the work during his trip but rather after his return home. However, as with Russell Wallace, it played a very important role in the development of his ideas; both thinkers, Lyell and Malthus, constitute a sort of 'idea bank.' Malthus noted that population grows exponentially, whereas resources (food) on which they thrive grow arithmetically. As a result, because populations have a tendency to outpace their resources, there are never enough resources for a given population, which inevitably led to a clashing disparity in which something has to give. Darwin immediately realized what was going on.

In 1838, the same year which he read Malthus, Darwin comes up with the basic notion of biological evolution. Only a few months after returning from his trip, he draws the now famous diagram "tree of speciation."

In spite of his early romantic setback, Darwin recuperates and ends up marrying his cousin Emma in 1839—unusual today but more common at the time. As he did not have to go through the usual dating rituals, his relatives were generally happy about the selection. Emma herself comments that had they even tried to separate, it would have been extremely difficult to do so given the social pressures; money stayed in the family. In a private meeting, his father curiously tells Darwin not to share atheist views with his new wife; in this manner, he would avoid many difficulties suffered by other honest men. The couple settles into Down Street and have eight children. One daughter dies at age of 10, causing a traumatic experience for the entire family—but particularly for her parents.

Darwin is no fool, and was well aware of the implications of his biological notions, and even the impact the suggestion that 'men evolved from brutes' would evoke. It was for this reason his analysis of human evolution, published as *Descent of Man* (1871), would be delayed for publication much later on in life— perhaps for the same reason as Copernicus. In order to bolster his theory, which was discovered at the beginning of his research rather than at the end, Darwin slowly begins building up case studies so as to gather the data necessary for the acceptance of his theory. He was well aware that the religious repercussions of his work would be enormous.

The Bible, as we all know, states that the world was created in seven days, Eve came from Adam's rib, and that Noah's flood created the human world anew.

Notions of biological change

When *Origin of Species* was published in 1859, it was an enormous hit. It is one of the few science books that was accessible to the general public, of a non-mathematical nature. Its simple exposition discussed phenomena that was readily available and observable for all to see: animals, plants. The book addressed some of the most universal questions there can exist: Where do we come from? How did life begin? Every person at some point ponders in these issues, and hence its core themes are not unusual or particular. The book is also one of the most revolutionary ever to have been published, for it proposes a non-religious mechanism of biological change. It is, at its core, an alternative paradigm to that provided in the Bible of Gods' creation in Genesis. It was a direct rival to any and all religious views of nature. Its revolutionary implications were of a much broader social scale, analogous to the work of Newton, if not more so. That both Darwin and Newton are buried in Westminster Abbey (London) is a testament to the fundamental change in world view brought about by the book.

In the context of its broad reach and social implications, it is not then to be surprising that Darwin's *Origin of Species* is one of the most widely translated science books of all time. So many copies have been produced, translated and purchased, that it is actually hard to calculate their total number. The following facts provides at least some idea of their size and scope. During the author's lifetime it underwent six editions; by 1898 it had been translated into nine other languages. By 1977, there existed at least 27 language translations of *Origin of Species*. Whereas everyone could speak of Newton, everyone can in fact read Darwin.

Yet the first book to receive the brunt of the social reaction to the notion of 'biological evolution' had in fact been Robert Chambers' *Vestiges of Creation* (1844). In it, the notion that species change over time is referred to as "transmutation." It was a widely criticized book—by all sectors, both conservative and liberal. The book was a mixture of odd properties. It was not a scientific book properly speaking, bogged down as it was by many religious notions and affinities. Its argument at its core was a teleological one, proposing that all biological change was driven towards a final goal: greater perfection. These core religious aspects of the work forced thinkers favorable to the notion of biological change to attack it, on the 'left.' T.H. Huxley issued a scathing critique as he severely wanted to remove all religious criteria from scientific debates. Such debates had to strictly be based on scientific merits, and could not be led astray by non-scientific criteria. Later in life, Huxley recognizes that perhaps he had gone overboard in his attack.

It's important to distinguish various aspects of 'evolution.' Do species change, and if so, what is the mechanism by which they change? Do they change to certain ends, as Chambers suggested? Readily, the acceptance of change did not necessarily imply the acceptance of any one particular mechanism underlying such change. As Chambers' case showed, one could believe in 'nonscientific' evolution. In other words, 'biological change' did not necessarily imply Darwinian evolution; otherwise, Darwin's grandfather Erasmus of the Lunar Society would have long ago received recognition for the original idea.

The ample public discussion of the topic, however, does suggest that the notion of change was 'in the air.' The fossil record had been more widely explored due to industrialization's need for coal and canal building. The economic incentives of industrialization thus push forth the sciences of stratigraphy and mineralogy, which could be used to detect coal reserves. The financial benefits of geology meant that it ultimately contributed to the study of fossils, as these could also help identify the location of coal reserves. Although sedimentary layers are gradually deposited, 'reading' these layers are not as simple as merely digging a hole given that the processes of uplift distorted the shape and arrangement of such layers. In spite of all of this, fossil deposits did reveal the existence of strange forms of animals that no longer roamed the Earth, yet the full extent of the scientific meaning of fossils was not yet understood.

There were also many deep connections between religion and biology, specifically that of natural theology which used biology to prove the existence of God and 'His' greatness and perfection. John Ray was one of most prominent authors of the era. He argued that life in the world had a natural and harmonious order as the relationships between bees and flowers, which in turn proved the existence of God. Ray sought to show how all species were harmoniously interrelated, but was actually unable to do so. There were too many species to plot out by a single man. In spite of this, his writings were widely read and exerted a powerful influence, nicely correlating with the needs of a Victorian industrial mindset: each worker was a cog in a larger wheel.

Oddly enough, Darwin's own entry into biology was through religion. He had not been interested in the traditional careers. He first went to the University of Edinburgh to study medicine, but would get sick at the dissections table. It was clear he would be unable to follow his father's footsteps. The other alternatives had been either the military or theology. John Henslow had actually been Darwin's professor of natural theology. Darwin's' first scientific work was undertaken in local expeditions headed by Henslow, which is how Darwin came to be aware of his interest and abilities. Paradoxically, it is through natural theology that Darwin actually becomes an atheist, a point he made sure not to comment in public, as he detested public confrontations. It is T. H. Huxley who coins the word "agnostic," arguing that God's existence simply could not be validly proven.

The *Origin of Species* would jolt the religious world in part due to the idea that order out of nothing (disorder) could emerge. Evolution had no divine plan, no purpose, and no end in sight. In contrast to Chamber's views, there were no predetermined biological form inherent to the process of evolution. It is a never ending process with no final design; 'perfection' is in the eye of the beholder, and does not exist. Forms do not necessarily evolve from the simple to the more complex, from the less perfect to more 'perfect' creatures.

The idea of evolution without an external end or guide was a very tough pill to swallow. North Americans (United States) also had hard time accepting strict Darwinian evolution, which helps account for the rise of Neo-Lamarckism during early twentieth century. The school argued there was an 'inner driving principle' that shaped the evolution of "orthogenesis." It was a philosophy typical of a Protestant world view, supposing the existence of an inner will be driving change and history—as well as an example of how values can greatly influence 'impartial' scientific activity.

Lamarckism, however, as a whole is generally misunderstood, typified by 'acquired traits': the giraffe has a long neck because stretched while seeking food, thereby passing the 'new' neck down to its progeny, who in turn stretched it more. However, this aspect was only a minor feature of Lamarckism. Lamarckism's key difference to Darwinian evolution rested in that animals were not historically inter-related; rather they emerged along different times. A more accurate notion of the structure of its historical development would be the idea of 'parallel evolution': simple forms of life were simple because they had spent less time evolving. The more complex, as man, the longer the species had existed on Earth undergoing change.

Origin of Species

Darwin begins *Origin* by first looking at the cases of domesticated animals. In these, the hand of man has acted akin to that of natural selection by selecting those traits deemed valuable. Breeders were able to mold the shape of animals to their desires by picking and choosing breeding partners. It is curious to note that, in this regard, man-made genetic change has been occurring for a much longer amount of time than is typically presumed, although on a different scale and rate. England had a large number of breeders affecting many different types of animals. Dogs were perhaps the most common. All canines emerged from wolves, drastically changing their sizes: from the tiny Chihuahua to the large Rottweiler. The 'variety' of all dog shapes have been imposed by man, sometimes with horrific consequences. He finds a similar process in the case of pigeons, who have long been used in Europe. Various shapes and forms of plumage have been selected by their breeders, again with radically differing outcomes.

While the issue of breeders might appear to be quaint and trivial, it was an important one for Darwin who wanted to establish beyond any doubt his case. In particular, he wanted to: a) unquestionably show that change in biological forms existed, establishing the precedent case of artificial selection (breeding) and b) reproduction was an important component of evolution. Favorable traits that were selected by nature (i.e. natural selection) were passed down to future generations.

Malthusian competition, however, was key engine underlying Darwinian evolution. As mentioned before, populations grew at such rapid pace that they outstripped their available resources, which in turn placed selective pressure on these. Any evolutionary advantage conferred on a given individual that would increase its survival relative to that of its peers would hence be a trait successfully passed down via progeny in an endless cycle. Again, this view of nature is the complete opposite to that of natural theology; nature was not harmonious at all in that each was looking after its own self-interest. Nature was 'red in tooth and claw,' lacking the harmonious order natural theologians presumed it to have.

However, greatest competition was not between predator and prey but rather between animals who were most closely related: those similar in form who depended on the same food source and who were in closest geographical proximity. It is the old story of Cain and Able. The best case were rhea birds, akin to the ostrich. The Beagle party regularly caught and ate these, and Darwin noticed that there were two species in adjoining regions. The area of greatest competition was where their respective turfs crossed; these areas had the greatest selective pressure as the eternally increasing rhea populations competing for the same relatively stable food supply.

The creation of new species had a particular benefit in that it allowed a much greater number of animals to exist over a given area of land as each new variant (species) could occupy a particular niche different from the other. In other words, each animal could dedicate itself to a particular food source and thus not have to directly compete for the same resources. Hypothetically speaking, if one rhea changed food source, then all could actually have increased in population and coexist in same territory as they no longer had 'overlapping boundaries' because of the absence of competition for the same food source.

This is why the case of Galapagos finches was so important: each finch had developed specialized beaks suitable for differing ecological niches and food sources. Long beaks, for example, allowed fiches to pry into wood to get insects out, while short beaks allowed others to crack open tough nuts. Some even became cactus eaters, as their particular beak shapes allowed them to avoid long spine needles; other had overlapping beaks to get at seed.

Yet animals could only use the traits that were already available; nature could only make do with available material. Bodily forms would gradually adapt to existing circumstances in that only the old bodies would be gradually 'molded' to the new circumstances. For example, the Panda's "thumb" is actually made from a wrist bone adapted to help it grasp and eat bamboo; the 'wings' of a bat emerged from its hand. In other words, new animal forms did not emerge 'out of nothing,' but were rather molded only from the ancestral biological material already available—which had very important implications.

This meant that imperfection was best proof of evolution. Given that not all species had time to adapt to existing conditions, imperfect prior structures could be 'caught' undergoing change due to selective pressures. Vestigial organs were hence a definitive proof as to the existence of evolution. Whales and snakes have hip bones, even when they clearly do not need them. This in turn implied one thing: a common ancestor.

But, at what level did natural selection operate? Is natural selection active only on the individual or does it operate at the broader species level? This was a problematic issue, which actually leads Darwin down the wrong path. He argues that natural selection occurred at species level as he was trying to account for species with non-reproductive members as ant and bee colonies. Usually it is only a single queen that reproduces; all other members of the colony are asexual drones carrying out specific tasks required for the colony's survival. Most drones were 'disposable,' and left no future descendants. Darwin argued that one could only make sense of such insect colonies if these were analyzed at the species level. Otherwise, the existence of non-reproductive individuals made no sense. Although later evolutionary theorists would show Darwin to have been wrong on this point, this idea became a central tenant of biology after him.

Another important component to Darwinian evolution was the existence of variation. The issue of variation within species was a troublesome issue. There are a great number of variants in a community: individuals with slightly different but unique and particular traits. How do you distinguish variants from species? Empirically, variants are troublesome, often requiring long years of observation so as to be able to differentiate between them. Speciation occurs only when variants can no longer interbreed amongst themselves, either for physical or behavioral reasons. This sexual wall allowed each species to enter their own respective biological 'paths' as they could no longer 'swap' traits amongst themselves. A clear and distinctive dividing line was drawn between them. Yet how did variations biologically emerge? Darwin had no answer, nor could he have been able to provide one.

Darwin also noted that the larger any given population size, the greater the range of variation, and in turn, the rate of evolution would actually increase relative to the rate found in smaller populations. Darwin realized that animals formed interdependent ecosystems. Long chains of relationships were established in a locale which may not be immediately obvious. For example, he notes that the number of flowers were dependent on the number of cats. That flower pollination depended on bumble bee was a well-known fact. However, the nests of bumble bee were affected by mice, as greater mouse populations tended to increase the rate of bee nest destruction. The greater number of cats would thus keep the total mice population in check, and in turn the rate of flower pollination from bees. Biological change in one species would thus ricochet throughout the network of animal relationships tied to that species; change in form of some animals would, in turn, affect the forms and behaviors of others in the same system.

Darwin is here removing biology from the link to geology. Not that geology was not important but that internal intra-species dynamic was equally as important. During his voyages he became aware that if he tied the animal form too closely to the geological landscape, it might ultimately hinder theory. Biological change had its own internal dynamics, that while affected by geological change were not exclusively influenced by it as perhaps suggested by Humboldt.

The causality of natural evolution is actually left open by Darwin, who was fairly open minded in this respect—unlike other biologists. The factors of change could vary widely, and were not absolutely set in stone.

Darwin, for example, recognized the important role of sexual selection. The female selection of males for breeding had an enormous impact on historical biological change. The cost of pregnancy is generally higher for the female than the male. As a result, this pressure to find the most suitable mate led the mind or 'perception' of a female to act akin to that of the natural selection or the human influence on domesticated animals. The preference for one or other particular trait in a male as evidence of his fitness would lead to the expansion of such trait relative to other traits, and the gradual modification of species.

In this Darwin is very different from Alfred Russell Wallace who believed that natural selection, and only natural selection, was the impinging factor on biological change. Wallace's view is best summarized by the catchphrase 'survival of the fittest.' Weaklings that died prior to reproduction were weeded out from tree of natural history in that they did not leave progeny.

Darwin, to repeat, did not abide by these ideas. As he had actually been studying the issue for 30 years prior to Wallace, he was much more attune to the many nuances of evolutionary biological change

Reaction to *Origin of Species*

The reactions to *Origin of Species* were perhaps as interesting as the book itself, raising many important issues that would ultimately further strengthen Darwin's ideas.

The first to emerge were religious controversies, the most famous of which was the encounter between Bishop Samuel Wilberforce and Thomas Henry Huxley on June 30, 1860. During that long debate at the Linnaean Society, Wilberforce asked Huxley from which ape his line came from, on father's or mother's side? Huxley famously retorted: better from an ape than a preacher. Huxley became an able defender of Darwinism at a much needed moment. Darwin actually had not attended the debate, preferring to avoid public controversy in the spirit of Copernicus. Darwin is more focused on creating viable interpretations than in publicly defending them. In this sense, Huxley resembles Galileo.

Darwinism thus held a brief period of success, until the 1880's when it received its first valid scientific criticisms, which consisted of three principal lines of argument.

The first was the obvious lack of transitional forms in the fossil record. In theory, if Darwin evolutionary gradualism was correct, evolution consisting of intermediary forms (variations) prior to speciation, there should occur an overabundance of transitional forms in the fossil records. But in fact, these were wholly absent.

Darwin tended to downplay this critique by pointing out that the favorable conditions under which animals could be fossilized tended to be extremely rare and unique. Today we know that for fossilization to occur, death needs to be quick and in absence of oxygen as in a mudslide where suddenly killed animals are quickly covered by the non-aerobic material. The pit bogs of England provided ample human samples, for example. Fossilization is a substitution of organic material, or the replacement of bone and muscle by 'clay' minerals. Given that such events are rare, they would therefore leave an imperfect record of historical biological change.

The issue, however, was far more problematic than Darwin himself presumed. What if it revealed something about nature of evolution? What if evolution was not as smooth as suggested by his theory? Yet the critique was not 'crucial' to undermining Darwinism. Fossil discovery itself requires a great deal of luck and patience. Paleontologist can be entire careers without any finding anything, perhaps one or two trinkets, if they are lucky.

A more powerful critique, however, was issued by one of the most well recognized scientists of the day. For evolution to happen, minute gradual changes over thousands of generations, it required that the Earth have a much longer historical time span than that ever previously suggested, of a magnitude and scale that had been unheard of: a billion years rather than the religious argument of a 6,000 or so year old history. Yet William Thompson (Lord Kelvin), one of founders of thermodynamics, pointed out that his calculations revealed that the Earth was roughly 100 million years old, using its cooling rate. While this was a time span that was much longer than those previously suggested, it still was not long enough of a 'chronological space' in which for Darwinian evolution to occur. So powerful did Lord Kelvin's argument appears, that it began to erode the confidence in Darwinism.

His arguments were assisted by those of Fleming Jenkins, a colleague and friend of Kelvin's who had also collaborated in the creation and building of the first transatlantic telecommunications cable.

Jenkins pointed out that any new traits in a creature would be immediately 'averaged out' over the course of interbreeding with members of its own species, in what is referred to a 'blending inheritance.' Using his 'Jenkins distribution curves' (bell distribution curves), he showed that although new traits might push the average to one extreme or another in the distribution of traits, these would not 'break the mold' of the existing curve. In other words, one might get faster horses, but one would not stop getting horses. Even a rare and extreme divergence form the norm would eventually be 'averaged out' and blended back into the originating population, as each descendant obtained traits in equal portions from their respective parents.

Things did not look well for Darwin at the turn of the century.

Kinetic Theory and the Rise of Theoretical Physics

On the Abuse of Institutional Power

THE RISE OF KINETIC THEORY is one of most important achievements of nineteenth century science. Along with the development of electromagnetic theory, to which it is intimately related, kinetic theory was the immediate precursor to the quantum revolution.

In its early form, this theory of heat was the strict application of Newtonian mechanics to gases. Its later statistical formulas came to allow the prediction of chemical reactions. Some of its participants had also taken part in both it and the electromagnetic revolution. James Clerk Maxwell, recognized as the 'Newton' of the period, clarified both fields with rigorous mathematical formulation. As with kinetic phenomena, one could not 'see' electromagnetic phenomena directly. Maxwell wrote of fields that were independent of physical objects that created them, ghostly entities that existed in empty space.

Kinetic theorists pushed Newtonian theory to its limits and helped create a new definition of physics: theoretical physics.

The study of intangible objects also pushed forth the boundaries of mathematics. Physics would be drastically altered in method and definition; from the Newtonian analysis of interacting particles influenced by the gravitational law to the statistical approximation of random colliding particles. Their physical properties and chemical behavior could be reducible to probabilities, and did not require chemistry proper. This type of reductionism helped endow physics with an aura of 'magic'; could all sciences be similarly reducible to physics, and be studied independent of their distinct characteristics?

One may observe that many of the today's classical formulas of physics (or 'laws') were actually produced by others who simplified the work of original authors. Newton's F=MA was by first expressed Euler (1750), and Boyle's gas pressure law by Henry Power, Boyle's contemporary. The simple expression of Boltzmann's law was actually prepared by Max Planck, and Maxwell's complex electrodynamics study was translated into its four lines of 'code' by Oliver Heavyside. Although the 'discoveries' constituted true and original work by their creators, they were so complicated so as to be dysfunctional for efficient and practical purposes. Such simplified formulation thus constitutes an important but unrecognized aspect of their history, allowing engineers and applied scientists to more easily use them.

It is equally important to point out that the entities under study by are not 'common everyday objects.' The reader will please excuse of if we briefly jump into quantum revolution, to which it is interrelated, but it is done to show that the road to discovery was a complicated one.

'X-rays' were not discovered or "seen" from one day to the next, being partly 'constructed' objects even if they did exist in nature. They were hard to discern, very unlike seeing a common object as a 'car' or a 'house' with a recognizable shape and size; metaphorically, the phenomenon under study was more akin to the observation of a camouflaged moth hidden on a tree's bark. These objects were difficult to identify and to analyze. It took Roentgen a year before he made the discovery public.

In his own experiment Rutherford claims that when saw result of gold experiment, the notion of an empty atom appeared to him in a flash. However, as Helge Kraugh notes, this is an outright propagandistic deception

The inherent uncertainty surrounding the objects of theoretical physics meant that authors were often unaware of the implications of their discoveries. Planck in 1900 postulated the 'quantum,' which in German only means a unit of measurement, as a mathematical trick to solve the puzzling blackbody problem. Planck most certainly did not view the quantum as 'entity' per se, and even as late as in 1912 he was still trying to get rid of it. Prior to 1905, Einstein's' *annus mirabilis*, few papers were written on it. Thus Planck's 1900 paper on 'quantum' did not visibly usher a 'revolution.' Einstein became a key figure simply for recognizing its importance. Yet even the most famous scientist in history did not accept its ultimate conclusions. "God does not play dice" he complained.

Kinetic theory also has a close and complicated relationship to thermodynamics. Thermodynamics is the macrocosmic analysis of energy and the interrelationships between its different forms. Both also have some common participants. Rudolph Clausius, who established the first formulas in kinetic theory, is also responsible for the Second law of Thermodynamics, or what today we call entropy. Hermann von Helmholtz, who came up with the first law of thermodynamics (conservation of energy), providing its first math description, had been an associate with the principal founder of kinetic theory, Ludwig Boltzmann. Yet whereas thermodynamics operated at the macro level, kinetic theory adopts its framework at the micro level, and hence why it is 'theoretical': it dealt with atoms, entities that could not be directly seen or experimented upon.

This trait helps account for its wide early rejection, and its revolutionary nature.

Kinetic theory began to alter the fundamental paradigm of physics, from the examination of tangible objects to intangible ones. Many physicists simply did not believe this was a viable model for their discipline. In fact, thermodynamics was often used to attack kinetic theory. One of its main opponents had been Ernst Mach, a physicist who went into philosophy and established logical positivism around his Vienna Circle of adherents. Unfortunately, Mach obtained greater support due to social dynamics rather from than the actual merits of his theory. He was a physicist who spewed complex terms to non-scientists: impressive in a café table talk with a strong European accent which accentuated his public prestige. Secondary traits are not to be confused for philosophical rigor.

Mach believed that the role of physics was only to look for correlations between visible concrete phenomena, and tried to ban all theorization from physics. As a result, atoms could not be studied because they could not be tangibly verified. "I do not believe in atoms," Mach once proclaimed to Boltzmann at a conference; his critique of Boltzmann would be persistent and, ultimately, debilitating. Inversely, he praised thermodynamics, sharply contrasting it to kinetic theory.

Yet Mach's logical positivism was an absurd philosophy which distorted the actual manner in which important physicists undertook their practice. Both Einstein and Planck condemned Mach, as logical positivism is one of the most disingenuous philosophies of science to ever have existed. Had physicists taken it seriously, neither the quantum revolution or kinetic theory would ever have emerged. Oddly enough, logical positivism was so popular, that it can still today be found lingering like a ghost in some halls of modern academia, being common in Latin America.

There is thus an inherently tragic aspect to the history of kinetic theory. Those who worked on it were often rejected throughout their entire careers, no matter how brilliant and original their work had been. Under this never-ending scrutiny and rejection, two important kinetic theorists commit suicide: John James Waterston (1811-1883) and Ludwig Boltzmann (1844-1906).

Waterston is one of the founders of kinetic theory, his 1845 paper being far above those of his contemporaries. In spite of putting forth the first notions of an equipartition theorem, his papers were rejected by leading scientific societies as the Royal Society and the Astronomical Society of London. To make things worse, his articles were not to be returned to him, in a pre-internet era where copies were hard to obtain and originals were often the only versions available. His original work was just shelved in the organization's archives, accumulating dust and undetectable plagiarism.

As there was no knowledge of his work, Waterston would have no direct influence nor receive any recognition during his lifetime for his important achievements—even as others began to make similar discoveries. Rudolph Clausius publishes his first article in 1857, thirty years after Waterston had first submitted his seminal work for publication. Waterston was not 'discovered' until Lord Raleigh accidentally came across his papers the Royal Society's archives in 1891, and immediately recognized their significance. Only posthumously would his body of work be published in 1893 and receive the merit it deserved.

Certainly, Ludwig Boltzmann was a published and well recognized physicist; few doubted his mathematical abilities. However, during a trip to the beach with wife and three children, Boltzmann hangs himself in 1906. At 62 years old, he tragically ends his life. We can get a sense of Boltzmann's state of discouragement by noticing that he had begun writing books for the sole purpose of quietly sitting in libraries until discovered by future generations. Neither colleagues or students follow him; he was under constant scrutiny by the Vienna Circle for his 'ghostly' topics. As Waterston, Boltzmann had simply had given up hope of having any influence during his own lifetime.

To understand their story, we have to turn to the chemistry of the eighteenth century, specifically to the caloric theory of heat with which kinetic theory competed.

Caloric and the origins of kinetics

As we have seen, the study of gases during the eighteenth century had been the key to the rise of chemistry. Lavoisier's discovery of oxygen not only defeated phlogiston theory, but also introduced an important new concept: the caloric.

"Caloric" became a central feature of the chemical revolution, substituting phlogiston. Heat was defined as a 'substance' rather than a 'principle' or a 'process.' The caloric was a type of atmosphere that surrounded atoms; when 'squeezed' under pressure, the atom contracted and released caloric, shrinking in total diameter. Inversely, when an atom gained heat, its amount of caloric increased, and thus its size, hence accounting for the expansion of heated gases. The term was actually coined by Berthollet and Lavoisier in 1787, and was tied to the anti-phlogiston battles of the French Newtonians in Paris. The exchange of caloric was the underlying mechanism of oxygen, and Lavoisier actually identifies 'caloric' as an element in his 1789 chemical treatise.

Yet there were strong debates as to the validity of this interpretation of heat, and, more broadly, of heat as a substance. Its principal opponents, Benjamin Thompson (Count Rumford) and Humphrey Davy, had performed experiments showing the difficulty in accepting its validity. How could one explain that a canon bore could forever generate heat as it drilled a hole? If a 'caloric' were a valid substance, one would naturally assume that it would have been quickly 'used up.' Other experiments led to similar conclusions, as the melting of ice by friction.

The supporters of caloric theory—Laplace, Berthollet, and Lavoisier—countered that if kinetic theory were valid, heat would be quickly diffused and differences in temperature would be immediately smoothed out. The inability of 'kinetic adherents' to address these objections meant that caloric theory would remain the dominant interpretation of heat until approximately the 1840s. Contrary to what one might suppose, the 'caloric' was actually used as a successful mathematical entity by Laplace and Poisson to make formula for speed of sound in a gas at a specific temperature, a formula which is still valid today even though we reject the physical basis (caloric) on which it is based.

As argued by Stephen Brush, the success of the caloric helps explain the century-long delay of kinetic theory, including the rejection of Waterstons' first papers. His notions too drastically conflicted with the broader intellectual background of the period, and it would not be until the 'caloric' was 'defeated' that kinetic theory—a rival interpretation of heat—had any possibility in rising to the fore. Ironically, it is not until the rise of the wave theory of light by Young, Fresnel and others, that we begin to see the debilitation of caloric theory.

Once it had been accepted that light could be accounted for as a wave in an aether medium, as Poisson and Laplace did, a way was then opened for an alternative interpretation of heat that could also be accounted for on a similar basis (the aether). That notion of a wave required the existence of a medium composed of tiny particles (aether) changed the intellectual background on which the kinetic interpretation was perceived. In fact, Waterston was very careful in his early papers to suggest that his ideas did not contradict the wave model.

In effect, the formation and revolution of a scientific discipline in one area, the study of gases, had itself served as an obstacle for the emergence of another scientific discipline in another area, the study of heat. Not until changes in a third discipline, the study light, emerge that kinetic theory begins to have the possibility of intellectual acceptance and, ultimately, success.

Particle Gravity: Herapath and Waterston

The first kinetic analysis of gases was undertaken by Daniel Bernoulli in 1738, but under a restrictive case accounting only for pressure. It would take a century before the publication of another work in kinetic theory emerged, which again raises the issue of its long delay given such early origins. It might appear obvious that heat increases with pressure, a phenomenon that can be widely and regularly observed. However, the available evidence did not unquestionably suggest that heat was necessarily the result of clashing particles. The 'data' was in fact inconsistent, and did not necessarily point towards kinetic theory.

For example, Boyle demonstrated a range of phenomenon requiring air to propagate, and others which did not. Experiments involving vacuum pump evacuation of bell jars showed that sound required air. However, light was not affected by the vacuum. As heat also traveled through a vacuum, as that which comes to the Earth from the sun through empty space, the kinetic interpretation of heat initially held little weight. Common sense dictated that a medium was not necessary to for heat to exist or be imparted, which became another factor in the delayed emergence of kinetic theory.

It is to be noted that 'atomic' explanations also tended to be notoriously arbitrary. Even our praiseworthy and critical Newton fell into this trap, assigning to atoms the properties of the very objects he sought to explain. This problematic feature had already been noticed during the Greek period, and also helps account for the slow rise of kinetic theory. If it is the property of 'coldness' one wants to explain, then the constituting atoms would be imbued with a 'cold' property. Such explanations were regularly used for all manner of traits, and even if internally coherent, as a whole they were mutually contradictory. All atoms could not be both round and sharp at the same time, hot or cold, etc. Atomism thus required a mutually consistent set of properties under diverse conditions before being considered a valid scientific theory.

Oddly enough, the first consistent efforts to develop kinetic theory were not used to account for heat but rather for gravity. It was reasoned that if one could eliminate 'action at a distance' via an atomic mechanism, one would be able to resolve the final 'holes' and 'gaps' in Newtonian gravitational theory. It is through 'gravity particles' that the first exponents of kinetic theory first begin to consider the consequences of particles in collision.

John James Waterston's first scientific papers proposed a kinetic model of gravity using cylindrical shaped atoms. A bright young student 19 years old at the time, his failed hypothesis helped lay the groundwork of his future researches. Waterston realized that rectilinear motion could be translated to rotary motion. Waterston had argued that temperature changes affected gravity; as the Earth revolved around the sun, distance variations produced thermal effects, which in turn influenced gravity. Consequently, he believed that by studying heat through a kinetic analysis, he would be able to better explain the force of gravity. Waterston thus embarks on kinetic study of heat at an early age.

The notion was also shared by a predecessor, John Herapath. In 1820, Herapath sent a paper to the Royal Society who, like Waterston, tries to provide a mechanical view of gravity. Also similar to Waterston, his paper is not accepted by the Royal Society for publication. Shortly thereafter, a new president is elected to the society, Humphrey Davy, whom had been opposed to Berthollet's caloric model. Herapath goes to the Society to argue his case, and is warmly greeted by Davy, who proposes an experimental verification prior to acceptance of the paper. Herapath complies and resubmits his article, but is again rejected.

This leads the author to then submit the article to other journals. One day, he accidentally runs across Davy, who showed an unexpected change of attitude towards him, and Herapath begins a publicity campaign against his prior supporter. Herapath alleged that Davy could not understand his mathematics, and ends up publishing the book *Mathematical Physics*. Worst yet, he provides all letters written between Davy and himself to a newspaper, oddly humiliating Davy when these are published, in spite of the fact both had been on the same 'team.' (While Davy does resign as president from the Royal Society after this, it was due to his poor state of health rather than the unpleasant exchange.) Herapath believed there had been a defamation campaign against his person.

The story is 'tragic' simply because two thinkers who shared similar philosophical assumptions end up needlessly fighting one another. Both in principle agreed with the 'corpuscular' view. Davy had told Herapath that two individuals in the review committee simply did not want to publish his article, likely on the pro-caloric side. One might suggest that Davy was likely fighting an internal institutional battle on Herapath's behalf for kinetic theory. By publishing his article outside the journal and in making public their private correspondence, Herapath lost any sort of support by the powerful elderly figure. Herapath's energetic but ignorant impatience ultimately leads him to unwarranted treason. Herapath's fortunes, however, would be much more successful than those of Waterston's.

In his work, Herapath comes across the 'hard body paradox.' If atoms are truly inelastic hard bodies, he realizes that a perfect direct collision would cease all motion in either particle. With nothing to absorb force, the equal and opposing forces in the collision would perfectly counterbalance each other out, leading to a resulting absence of motion as its odd outcome. Herapath, however, believing that it was impossible to have elastic collisions on a microscopic level, argues that all collisions were instantaneous. He then uses conservation of momentum to calculate the properties of gas. Although he makes the error of assuming that f=mv rather than mv^2, Herapath demonstrates that kinetic theory could be used to derive certain properties of matter, such as changes of physical state or the propagation of sound. Thus Herapath's work, independently published in a self-run journal, marks the true beginnings of kinetic theory

Yet it is John James Waterston who is the most original of the two. Of Scottish origin, Waterston studied at an elite high school in Edinburgh, while also informally attending university lectures. However, he never enters college, which likely plays a role in his future scientific relations. His minority status as a 'Scotsman' in England probably does as well. He moves to London at the age of 21, and works as a surveyor for railroads lines, which were then undergoing a big expansion (1832). He joins an engineering society, whose president recognizes his talents and gets him a teaching position with a high salary in Bombay, India for the British East India Company, a role he would occupy the majority of his adult life (1839-1857). He is also fortunate in having the Grant College library at his disposal during this period.

Waterston's first publication is an anonymous book on the nervous system, trying to use kinetic theory to account for blood pressure. The book is historically interesting because presents a lot of ideas that are suggestive of his future work, such that that it would be ideal if all gases had the same number of atoms at a given pressure and volume. He recognizes the enormous scientific benefit of this relationship.

In 1845 he sends his 'infamous' paper to Royal Society, which is first read by Baden Powell, a geometry professor at Oxford. Powell actually likes the work, as it demonstrated ability and was in concordance with the facts. However, Powell recognizes his lack of expertise in the field, and sends it to a J. Lubbock for further review. Lubbock detests it, claiming the paper was rubbish. Not only that, he also recommends that it not be read to the Society, a wish which is fortunately not complied with. On March 1846, the paper is entered into the Society's records, but when Waterston tries to get his work returned, he is ignored. (It was the institution's policy to retain unpublished papers in its archives.)

During the late 1850s, Waterston returns to England and is able to present articles at various conferences. The *Philosophical Magazine* had accepted some articles but in 1856, its editor dies and no longer does so. When in 1878 Waterston sends articles to the Royal Astronomical Society, these are also rejected. As he had been member since 1852, he does not renew his membership henceforth.

Oddly, Waterston was more well known in Germany than in England. Helmholtz in a journal entry notes his name, suggesting the importance of Waterston's work but also pointed out that his article was nowhere to be found. The issue of cross cultural reception is interesting in that the inverse was also the case, having a greater influence outside his nation than within it.

These painful experience naturally left Waterston very resentful, as testified by his nephew.

> I distinctly remember the Royal Society was characterized in very strong terms useless now to repeat. ...any mention of them [learned societies] generally brought out considerable abuse... The last time I remember him very angry on a scientific subject was in regard to Mr. Crookes and his radiometer, as to which he used some unparliamentary language.

Waterston would rant and rave at mere mention of scientific societies and, worst of all, self-serving scientific men.

It would seem to be impossible to determine whether different gases had the same number of atoms but in fact had been demonstrated by Avogadro's study of vapor. In the creation of water vapor retained volume it was as if oxygen was dividing and uniting with hydrogen at a 2 to 1 ratio—a suggestion Robert Dalton would have found odd as their caloric would lead to repulsion between the same oxygen atoms. Given that like-repulsed-like, whereas opposites attracted, even when formed it would lead to an unstable configuration, according to the caloric theory. While this description might suggest Dalton was a 'conservative reactionary' chemist, it is just the opposite.

Dalton was a meteorologist who wanted to find out why gases in the atmosphere did not settle out into distinct bands according to weight; why do they mix? He argues that different sized molecules existed and were thus kept in an uncertain state of flux. Dalton argued that substances interacted in fixed measures, and Dalton might be said to be the father of atomic theory and first to determine atomic weights. Atoms are defined as the smallest units in chemical reactions, whose components mixed in specific ratios.

By contrast, Berthollet argued that there were an infinite number of combinations between substances, infinite ratios of endless possibility, leading to the same outcome as that by first Greek atomists when compared to Plato's contributions. If all chemical combinations are infinite and endless, it would be impossible to analyze them at all, discouraging the empirical study of chemistry.

Atomism continued to be problematic in the 18th century.

Rudolph Clausis, first approaches

It is wrongly claimed that A. R. Körnig began kinetic theory. As editor of the influential *Fortschritte der Physik*, an annual review of the progress in physics, Körnig had published a review article of kinetic theory which received a great deal of attention. Contrary to what one might suppose, he had not done original work and much of the review's findings can be found in work of Herapath. After his conflict with Davy, Herapath went into the railroad business, becoming editor of a railroad magazine where much of his work appeared.

James Joule had also began working on these issues, and at first considered mainly rotatory motion in his analysis. Once he came across Herapath's article, with focused principally on translational motion, Joule realized that calculations could be greatly simplified, more so than he initially supposed.

He quickly did a 'napkin calculation' in 1847 to determine the average speed of atoms in a gas. He presumed 3 particles of hydrogen in 1 cubic foot space at 60° under a given pressure, each hitting the other. Although only a model, it was good enough for analysis, concluding that particles traveled with estimated velocity of 6,225 ft. per second—a remarkable speed. He also realized that this fast speed would remain true even if the container was divided into 'infinitely' smaller number of particles, allowing him to calculate specific heat of basic gases. Although not 'exact' his rough estimates were a solid beginning.

Gas	Joule (cal/gr)	Experimental 1848 (cal/gr)	Modern (cal/gr)	% variation (joule to modern)
Hydrogen	3.012	2.352	2.402	25.40%
Oxygen	0.188	0.168	0.1554	20.98%
Nitrogen	0.214	0.192	0.1765	21.25%

Yet the man who formally launches the kinetic revolution was Rudolph Clausius (1822-1888). He began his kinetic analysis not by 'particle gravity,' as Herapath and Waterston, but rather via the wave theory of light. The properties of light suggested that heat could be treated in similar manner, and realizes that gases would be a good case study as intermolecular forces would not be at work—or at least their influenced would be greatly reduced when compared to the study of liquids.

As he was a distinguished professor in a German university, Clausius thus held enough of a high reputation to have his analysis seriously considered and viewed in a more favorable light. His papers were the first true mathematical description of kinetics.

Clausius noticed that Herapath and Joule had based their work only on the translational motion of particles, which was inherently problematic. Certainly, Clausius reasoned, atoms had to have more degrees of freedom, specifically a total of six degrees, along three main 'axes': translational, rotational, and vibrational. Herapath had also presumed that particle motion was parallel to walls of container in his calculations, but Clausius realized that the angle of incidence would have no net effect, so he discarded this approach.

He also notices that speeds could not be homogeneous as they had presumed. The best case of this observation was the condition of evaporation. Variation in speeds at the surface meant that one particle would break surface tension and escape; in other words, evaporation occurred becomes some particles were traveling faster than the average of their nearby companions.

These observations, however, meant that calculations would be too complicated, and in order to be able to realistically calculate them, he would have to assume certain properties to simplify his calculation—a common practice undertaken in the first attempts of a field.

Clausius reduced all of these to their average velocity. Once a gas had reached thermal equilibrium, all energies would be evenly distributed between their various degrees of freedom, or what is better known as the equipartition theorem. He publishes his important paper in 1857, which solidly establishes that heat was the result of its kinetic energy, or the average speed of particles in given space. Formal kinetic theory begins with Clausius—but does not end with him.

These ideas were strongly criticized by the Dutch meteorologist C.D. Buys-Ballot who argued that they would imply that particles travelled at near instant speed, which contradicted common experience and repeated similar arguments made a century before. If Clausius were correct, we would smell odors immediately upon the opening a bottle of perfume or a container of chlorine at the other end of a room. It was also known that carbon dioxide could remain at bottom of vessel if undisturbed, violating the apparent consequences of Clausius's work.

As a meteorologist, Buys-Ballot was concerned with the distribution of gases in the atmosphere, and hence very attune to the implications of kinetic theory. However, these solidly grounded critiques have the positive effect of pushing Clausius into a more sophisticated formulation. The following year, he realized that smells are not translated immediately because particles were all colliding with each other, hence 'slowing' down the average translation time. He therefore undertakes a new analysis, evaluating how many layers a particle would need to pass through before hitting a target, or a particle's 'mean free path.'

This notion became more revolutionary than one might suppose. In essence, Clausius begins to incorporate blind statistical analysis to define a particle's mean free path: the first time statistical analysis was used within physics in such a manner. Clausius's analysis was independent of the actual mechanics of the system, and amounted to the application of probability theory as its key scientific epistemology. Maxwell would build on this important insight.

Clausius first coins the term 'entropy,' or the tendency to disorder was the norm in universe. Objects do not heat up normally, but rather their tendency to cool without an external energy source. As a result, objects in a closed system would ultimately result in 'heat death' where no work could be performed in the system as no heat differentials existed—suggesting the inevitable end of the Universe. The tendency to disorder thus becomes the Second law of Thermodynamics, which had already been implicit in work of Sadi Carnot.

These German views were rather depressing for British Victorians, whose implicit view of progress as never ending movement to ultimate perfectibility became a myth in this new context. All systems eventually 'die' without external source of energy; if all British coal were used up, industrialization would come to a sudden halt, pushing Britain back into the Medieval Ages. It would have later implications for biology.

The evolution towards more complex forms was simply due to an external energy source, the sun. The continual pouring of external energy allowed this energy to be redistributed throughout system. Thermodynamics also had implication for ecological macrostructures; ecology, study of distribution of energy in system, owes its origin to it. The arrow of progress of history was only due to existence of the central fire.

Maxwell's brief foray

As a teenager, James Clerk Maxwell (1831-79) read a book review on Quetelet. It was a typical Victorian book review of more than fifty pages whose author puts forth his own view on the subject. What is the probability of error? At the heart of the matter was the identification of deviation from the norm (average). He specifically asks, how do you calculate the probability that a falling object will not hit its target? He uses probabilities along two different axes to arrive at result: a function of the square of its distance. Maxwell eventually uses this work to revolutionize kinetic theory in his 1866 paper "On the Dynamical Theory of Gases."

There can be no doubt that Maxwell was a mathematical genius, and truly the 'Newton' of the nineteenth century. He wins the Adams prize for his study of Saturn's rings in 1859 when only 28 years old. The rings of Saturn had first been identified by Huygens, and it was unclear what exactly they were made up of or what they were. Maxwell's paper showed that the rings could not be solid, but rather had to be composed of particles as the forces of stress were so strong that any solid would be broken up. He also shows that each ring rotated at a different speed. The unexpected results were a truly majestic piece of work. Maxwell showed talent from an early age.

He ends up revolutionizing the study of electromagnetism, specifically via the rigorous mathematical analysis of fields which had been experimentally shown by Michael Faraday to exist—but whom lacked the mathematical skills required for their analysis. Maxwell establishes laws that unify all electromagnetic phenomena. Oddly, Maxwell's impact on kinetic theory is by contrast limited.

One reason for this is that early on he realizes that the Second law of Thermodynamics could not possibly be a law, but rather a probability. Points of increased order are unlikely but not impossible within the random behavior of atoms. He jokes in satirical poems to his friend P.G. Tait that it was funny to see the Germans all worked up about his demon. Although Maxwell would later change his mind, his conclusions lead him to abandon the field; by time he takes it up again, he is unable to long remain in it.

His mathematical style greatly contrasts to that of Ludwig Boltzmann, the principal figure in kinetic theory. Whereas Maxwell's articles are concise and tight, Boltzmann's are eternally longwinded. Maxwell liked to work on a problem until he finished looking at it from all sides and could answer all possible objections. Thus his final papers end up coming out well structured, requiring only a few minor corrections prior to their publication; they are coherent, and 'easy to read' even when consisting of complex mathematics. Boltzmann's style, on the other hand was more akin to Kepler's: a long winded stream of consciousness including any and all trivial points along the way which may or may not have been directly related to the issue. Boltzmann basically bulldozes through, having faith that a solution in the end would be found. After many unsuccessful paths were tried, an answer was eventually found.

Botlzmann's papers are hence not 'elegant' like Maxwell's, and the former is known for sarcastically noting that "elegance is for the shoemaker and the tailor." It is certainly the case, however, that although conciseness is by no means a requisite for originality, it certainly improves the author's chances of gaining adherents. Boltzman's own famous 1872 article was a hundred pages long, which greatly hindered its reading. Only a few dedicated physicists had the patience and fortitude to read through its entirety, one of them being Lord Kelvin. Years later Paul Ehrenfest, whom had been Boltzmann's student, and his wife Tatiana simplified it into a coherent piece when writing a posthumous encyclopedia article.

Had Boltzmann simplified his finalized papers, there is a high likelihood that his impact during his own lifetime would have been greater than it had been. It also would have helped clarify vague ideas, as his H-theorem was conditioned on a contradictory set of presumptions he had not detected due to its complex expression. Towards the end of his life, Maxwell began reevaluating Boltzmann's work as he had not taken the time to read through the long 1872 paper. Had he not died at the early age of 48 from stomach cancer, kinetic theory would have had a much brighter future.

Maxwell realizes applicability of probability to kinetic theory, and revolutionizes the field. Prior to him, the mathematics of probability generally had a low prestige in academia as it was perceived mainly for games of chance and luck, betting being an abhorrent practice under Victorian morality. However, as it is impossible to study individual particles— 1 cubic foot has 10^{24} particles (1,000,000,000,000,000,000,000,000)—the only way to properly study the topic was statistically. Heat was only the 'average' of particles for Clausius. As we have seen, a box of gases has particles moving about randomly, in different directions and speeds. Maxwell realized that even at higher temperature, there was a distribution curve of speeds but whose average point varied according to heat. It is from this 'distribution of velocities curve' that Maxwell makes his contribution to the topic.

Maxwell, who was not 'scared' away at all from the complexity of Clausius's physics, creates a distribution function that is much more precise and accurate than that of his predecessor. It is Clausius's mathematical limitations which prevent him from further pushing forth kinetic theory. So successful is Maxwell in his interpretation, that he arrives at conclusions that were not obvious, a magic trick he repeats on various occasions. He shows that the viscosity of a medium was not related to density, a conclusion which goes against 'common sense' as we tend to presume that denser materials will be more viscous. However, he shows that as density declines, the mean free path of particles increased; in effect, particles move about more because have more space in which to move. At first Maxwell believes this conclusion refuted kinetic theory, but experiments prove him correct. When he repeated the experiment, obtaining the same conclusion, he becomes a later believer in kinetic theory, perhaps as Poisson at the beginning of the century.

Ludwig Boltzmann, tragic hero

In spite of the important contributions of his predecessors, the title of 'father of kinetic theory' can justly go to only one man, Ludwig Boltzmann, who more actively and persistently defended it than anyone else. Although both Clausius and Maxwell had published important works in the field, they had not followed through for their own respective reasons. Boltzmann certainly greatly admired Maxwell, who in turn had written early praise of his 1827 piece. The two, however, could not have had more characteristically different scientific styles. Whereas Maxwell can be defined as a 'fox' by working in many different areas, Boltzmann was a 'hedge hound' sticking only to one. He defends kinetic theory for three decades (1872-1906), and publishes more articles (57) than any of his colleagues or competitors—more than double even than those by Samuel Burbury (25 papers).

Boltzmann alone, and against all the opposition in 'world'—or at least in Vienna—suffered vicious attacks from rivals as Ernst Mach, for a time also a professor at the same university. The Germans generally did not like kinetic theory at all, and sought any opportunity to undermine it. Boltzmann's work, however, received a much warmer reception in England, where Lord Kelvin (William Thompson) recognized its tremendous importance.

His most important work are two articles published respectively in 1872 and 1877, the former when he introduced the "H-theorem." A misreading of German in England of the paper mistook the symbol 'E' for 'H,' influencing its final naming. In 1877 Boltzmann creates the Maxwell-Boltzmann distribution law, later coined by Planck as $S=k\log W$. While his 1872 paper retains traces of Newtonian mechanics, by 1877 these are abandoned exclusively for probabilistic approaches. As mentioned before, Boltzmann would redefine the nature of physics from Newtonian differential equations of particle collision to the statistics of their velocity distributions. His influence is such that his 'physics style' would be imitated by two of the most important physicists of the twentieth century, Max Planck and Albert Einstein, which partly account for their condemnation of Mach. More than that, Boltzmann can also be considered as the grandfather of quantum revolution.

How does an incoherent mass of randomly moving atoms obtain coherent properties as a collective? This was a difficult issue to tackle as it is impossible to track down each and every particle during the entire event. In 1872 Boltzmann develops an ingenious procedure: take a snapshot and 'look' at atoms under a specific window, delimited by velocity range and directions. He estimated how many collisions occurred within a certain range of time, how these affected motions post collision analysis, and then performed a grand average of entire thing. He is, again, bulldozing or 'ploughing through' Newtonian mechanics, but 'getting the job done.'

Unfortunately, Boltzmann incorrectly believed he had actually proven the Second Law of Thermodynamics (entropy) in his work. All collisions tended to move to a minimal state, defined by "H," where h is the velocity distribution of atoms. Even when they exceeded H, they would ultimately drop back down to this limit. It was a more 'naturalistic description,' as the bottom position as default. The tendency towards entropy increase was not as 'intuitive' in that entropy equaled negative H (-H) or H 'inverted.' He also detects problems with Clausius's equipartition theorem, previously shown to have an anomaly in not accounting for the specific heat of diatomics as oxygen. He postulated that that one of degrees of freedom was at unity, n=5, and thereby solving a problem which had perplexed even Maxwell.

In 1877 he continues to refine these ideas, further extending the probabilistic approach. As particles in a gas hit each other, they affect each other's motions and the totality of collisions at any given moment. However, as there will be an infinite energy states per atom, this created apparently insurmountable difficulties in calculation as there would be a consequent infinite number of distributions. Simply put, although there were a finite number of atoms in the formulation, these would reach an infinite number of decimals which were impossible to mathematically determine.

To solve this puzzle, Boltzmann assigns to the atoms finite energy states—a crucially important decision that solves the immediate calculation at hand and is later imitated by Einstein and Planck. Specifically, Boltzmann observed that each atom moved between energy states or 'holes' as each collided with one another, constantly changing states as individual units while the total state of the system remained the same. In other words, Boltzmann postulated a set of fixed energy locations in his calculations, into which the atoms in the gas were constantly moving in and out of—as in a golf course—thus bringing coherence to the system. The statistical distribution of these energy states came to be called Maxwell-Boltzmann law. More importantly, it became a precise mathematical definition of entropy of a given system. As S=klogW is a log function, the curve declined quickly as W, symbolizing the velocity distribution, increased.

Newtonian mechanics had been eliminated by Boltzmann from the calculation, and from here on statistics and probability would take main role and redefine the nature of physics. Physics became the study of the probability distributions rather than of the mechanics of colliding particles. It constituted a monumental shift in the field, as when Darwin felt the Earth 'liquefy' beneath his feet during the Chile earthquake.

Given the enormous importance of work, why did Boltzmann commit suicide?

Boltzmann's death in 1906 is ironic and tragic at the same time. It is ironic because of the theory of probability shows that, whatever the individual personal circumstances or the sufferings that might lead a person to suicide, the rate of suicide remains fairly constant in a given population. While one might think individual decisions to be completely private affairs, the result of chance experiences and random events in life, at the collective level common patterns emerge.

His death is tragic because it nearly coincides with the year (1905) when Einstein had proven validity of kinetic atoms using Brownian motion.

In 1827 Robert Brown had published a famous article, which was ignored completely by physicists during nineteenth century, possibly because it was classified strictly within the field of biology. Prior to his work, 'organic molecules' were seen as the basis of life; the source of 'living' could be reduced to tiny organic granules which were inherently 'animated,' a theory proposed by the Comte de Buffon (Leclerc). Brown decided to put it to the test, at first using pollen. He also cuts up other organics into tiny pieces, which demonstrated the same behavior. What was original and so important in Brown's work, however, was that he also showed the same behavior could be observed in minute particles of inorganic matter. When cut into tiny granules, inorganic matter consistently behaved as if 'animated,' which was very odd.

It is certainly the case Boltzmann had troubled circumstances in life. He was raised in a very cultured family, whose father was a government bureaucrat who valued education. Traditional agricultural opportunities had drastically declined, and the aim of every worker was to obtain a cushy government job. We can gain an example of the cultured atmosphere in which Boltzmann was raised by noting the sophisticated debates between himself and his younger brother Albert as children. Boltzmann wanted to categorize the world, to which Albert responded that he could not, as some items were not directly comparable. Specifically, a translation of Hume to German ruined its meaning. One had to read the original in English, wisely notes the younger brother.

However, both father and brother die of tuberculosis when Ludwig was still young, which has an enormous impact on him. His sister is known to have commented that her brother's attitude became forever serious after the incident. His mother obtains a pension after her husband's death, and places all of the remaining family resources on the education of her son. Both sister and mother dote on Boltzmann their full attention, which helps him surpass the crises and become a successful student.

226

Upon entering the University of Vienna, he adopts Joseph Loschmidt as a father figure. While Loschmidt's scientific contributions were not 'revolutionary' as those of his student in the future, they were not insubstantial either. He calculates the size of atoms at 1/1 M of a meter using Maxwell's work. It was a brilliant solution where unknowns are used to cancel out each other. Loschmidt in turn had been the student of Joseph Stepan, Director of Institute of Physics who had snapped away the directorship from Ernst Mach in the past.

Stepan was certainly the better teacher and creates a nourishing and positive open environment in which Boltzmann thrives and grows as an individual and as a scientist. Boltzmann is known to have said this had been the happiest time of his life, and it is curious to mention that he presumed all scientific practice was similarly carried about in an open ended discussion with few restrictions between student and professor. When he visits Helmholtz in Prussia, Boltzmann is taken aback by the dry and hierarchical tone of the laboratory, leading Boltzmann to allude to 'Vienna' and 'Prussian' scientific styles.

Boltzmann was also fortunate in that both Loschmidt and Stepan were also at the forefront of physics, first introducing him to the kinetic theory of Herapath, Joule and Clausius.

Upon graduation, Boltzmann is very productive, writing 8 papers in the first 5 years; it is a trait that would characterize most of his academic life, solving one problem after another. He goes on tour through Europe to make himself known, when he first meets Hermann Helmholtz and Gustav Kirchhoff, who as an elderly man is startled by his enthusiasm. Boltzmann's work wins him a favorable reputation as a professor at the University of Gratz (Vienna), the same city where Kepler taught high school. He is invited to give colloquiums in England, where he finds a positive environment seeking to hear and understand what he had to say.

The death of his mother later in life, however, profoundly affects him, and his productivity rate briefly drops. There also occur a number of odd incidents with his academic employment, which affects his reputation. Boltzmann applies for a job in Berlin—a job application which turns into a fiasco. For some reason, it became hard for Boltzmann to make a decision, and during the negotiations he routinely waivers back and forth between staying or leaving. He was certainly caught in a bind, as he had privately accepted the job offer while publicly declaring his loyalty to Vienna. Berlin gets tired of the ordeal, and decides to let him go, while making sure the indecisive candidate had not forfeited his original employment. Berlin instead hires a young and promising physicist, Max Planck. The infuriated Helmholtz gives Boltzmann a piece of his mind.

More oddly still, his employment indecisiveness would again occur on a number of other occasions, Munich and Leipzig. After various stays at different universities, Boltzmann ends up back as Director of his former alma matter, the Institute of Physics. However, over the years, it had fallen into such a dilapidated state, that Lise Meitner, a student at the time, noted that if a fire were to break out in its facilities, it could easily have killed everyone in the premises.

Over the years, Boltzmann had certainly accumulated many health problems. When he had first gone on tour through Europe, the few photographs show him as a very skinny fellow. Over the years, he grew corpulently rotund, inducing the many health issues that accompany obesity. He gradually loses his eyesight, a phenomenon typical of diabetes, to the point that students had to read papers that were sent to him. This might account as for his lack of awareness of Einstein's recent work, which had been sent to him. Boltzmann had also lost his manual dexterity. As a young physicist, he had been known for skillful ability in experimentation and dexterous manipulation of materials; he also played the piano very well, but no longer does so in later years. His doctors had been telling him to lose weight—advice that is completely ignored.

The decline of his faculties and overall professorial 'quality' can be detected during a 1905 trip to the United States, where he had been invited to offer a series of lectures at the University of California at Berkeley. In his autobiographical notes, Boltzmann believed himself to have made a good impression, but in fact Boltzmann no longer worked at the leading edge of physics and left many of the attendants disappointed. (Boltzmann also wrongly thought his English to have been better than it actually was, worsening the difficult exchanges during the visit.)

During the summer of 1906, Boltzmann and his family go beach for a brief vacation prior to beginning of school year. It is clear that his mental state had greatly deteriorated by then. So concerned was his wife, that she asks her son, who had recently become a military officer, to come visit as the two had a close relationship. When she then goes to beach with her two daughters late in the afternoon, she notices her husband taking an excessively long amount of time in joining them. She then sends the youngest daughter, then 13 years old, to fetch her father.

Henrietta finds her father's body hanging from a window frame, a traumatic experience of which she never speaks or comments the rest of her life.

Part V: Rebirth

Einstein and Relativity

Using Contradictions in the Right Way

FOR 200 YEARS NEWTON'S MECHANICAL PHILOSOPHY reigned 'supreme,' generally speaking. This period is now referred to as that of 'classical physics,' or the physics of the very large where the 'macrocosm' was studied. The *Principia* of 1687 was the culmination of Cartesian rationality, as well as its complete overhaul.

As a whole, the Newtonian world view presented a tidy picture of the universe. The universe was the result of particles influencing each other, in a manner which could be exactly determined by clear and distinct laws—the most fundamental of which was the law of gravity. As Laplace stated, mathematical prediction reigned supreme; one could theoretically determine beginning and end of universe if the location of all particles could be known. The universe was ultimately rational and understandable. There was only a single universe, the Milky Way: a concentrated core of stars with scattered ones at periphery. The universe was stable, fixed, and unchanging, as were the metaphysical structures on which it was built. Time and space were absolute, fixed, and static. Although the nature of light had not been determined, the aether in which it existed pervaded the entire universe. There was a sense of unity tinged with implicit harmony in the Newtonian world view.

Yet by 1932, give or take a few years, this world view had been overturned. In contrast to Newtonian physics, quantum physics as a science emerged from the study of the very small. Chemists who had previously rejected the atoms were now forced to accept them. But the modern atom was very different from its Greek predecessor. Atoms were not undivided, in that scientists were now able to peer into its structure, itself composed of other particles. The new physics also discovered many new types of energies and rays previously unknown. The notion of a microcosm/macrocosm parallelism so typical of the Medieval world view would turn out to be a gross myth. Quantum phenomena were unique and irreducible. Atoms were not planetary systems, as had once been romantically imagined. Matter and energy were interchangeable; space and time were relative; the aether did not exist, and light was both a wave and a particle. Surprising to all, the universe expanded at an accelerating rate, and the Milky Way was only one of many galaxies.

In a very real sense, the entire picture of the universe was turned upside down during the quantum revolution. Instead of perfect rational harmony, there was irrationality and chaos; light could predict the future and be in two places at once. At its core, no model could account for the phenomena under study; there were no underlying 'gears and shifts' to explain its actions. If Giambattista Vico and others had previously claimed that making and construction were the ways to truth, humanity was left with no device to construct. We were reaching out into the void.

When compared to periods in the 'birth' of other new sciences—Darwinism in biology or the tediously slow revolution in chemistry—the quantum revolution occurred at a rapid pace. Within a span of less than 40 years, it had been mostly completed. In this sense, the use of the term "boys' physics" to describe quantum mechanics is rather apt. Its leaders were relatively young, in their early 20s. At its peak, roughly September 1925, Werner Heisenberg was 23, Wolfgang Pauli 25, Pascual Jordan 22, and Paul Dirac 22. 65% of its contributors were born after 1895.

Changes in policy of the renown *Annalen der Physik* created a situation analogous to today's internet and ArchivX; if a reputable physicist submitted or supported an article, no peer review would be needed. Submitted articles tended to be published within a month, and the pace of intellectual change in the field grew so rapidly, that physicists living outside of Europe, specifically the United States, found it very difficult to make contributions in the field. Articles in the United States were received far too late, and by the time a theoretically 'new' contribution had been submitted, the game in Europe had already changed. Even its 'grandfather' Neils Bohr, whose Copenhagen Institute served as the nucleus for its practice, complained to Ernest Rutherford that it was sheer madness trying to keep up.

Scientific styles would ultimately also be swept under the rug.

Schrödinger's cat

We have to distinguish relativity from quantum mechanics in the physics of the period. The two had very different, actors, traits, and chronologies. Relativity arises with Albert Einstein. His special theory in 1905 and the more general theory 1915 were the result of one man's effort—even if tremendously aided by many individuals along the way. "Relativity" was his "baby," so to speak, emerging out of the blue. His paper on special relativity contains no citations whatsoever.

Quantum mechanics, on the other hand occurs in two stages (1925; 1931), was influenced by many different individuals, and was intrinsically difficult to understand. Whereas relativity was mainly discovered in Einstein's Bern Patent Office, quantum mechanism was developed in different centers and institutions: Copenhagen (Denmark), Göttingen and Munich (Germany), amongst others.

Their diverse outlooks and philosophies are perhaps best illustrated in the famous, but actually brief, Bohr-Einstein debates; Einstein believed it was a sin to abandon causality as Neils Bohr proposed. The debates between the two men occurred mainly in two Slovay Conferences (1927, 1930), and their public argument display their mutual brilliance. For every profound critique offered by Bohr, Einstein had an incredible ability to quickly present counter-examples, as if "pulling rabbits from his hat." Although at the end of the day Bohr "wins" the encounter, it is actually continued at Princeton University's Institute of Advanced Study where Einstein finds exile from Nazi Germany. While his ideas were then allegedly proven by John von Neumann, whose mathematical genius was such that his use of Hilbert spaces in *Mathematical Principles of Quantum Dynamics* (1932) went unquestioned. In the 1960s his argument was ultimately shown to be incorrect by John Bell.

The debate was also explored by Erwin Schrödinger, responsible for developing wave quantum mechanics. Older than the rest of the 'boys,' being in his 30s, Schrödinger provides a grotesque hypothetical cat experiment, in which the poor cat is killed under a given set of conditions. While the hypothetical experiment grew in popularity during the 1970s in the United States, possibly due to the rise in popularity of murder movies, the fact of the matter is that it was never taken seriously. Bohr himself never bothers to answer the thought experiment, possibly due to its morally problematic nature. From a scientific standpoint, the case in itself was invalid as the microscopic world is not subject to macroscopic dynamics. The context of Nazi and World War II, where Denmark is itself invaded by Hitler and Bohr is also forced to flee for life, must have not helped.

One has to emphasize, as Helge Krauge does, that this period in the history of science is particularly 'blurry'; it is not a 'clear cut intellectual development' but is rather characterized by many false twists and turns, both experimentally and theoretically. Energetics theory was in its heyday at the turn of the century, and considered energy as the underlying substructure of matter—the inverse of our contemporary view. Atoms were believed to be the result of electromagnetic phenomena. Its proponents felt that they were close to a revolutionary breakthrough in the understanding of the universe. That the contemporary worldview would be upturned is a surprisingly common theme during period.

It had many important proponents as Wilhelm Ostwald, the father of physical chemistry, and Pierre Duhem, who later became a philosopher of science, believed atoms were merely metaphysical constructs. Pierre Curie, husband of Marie Curie, was another follower. In spite of its early enthusiasm, energetics theory silently died away as new phenomena was discovered. New theoretical

232

developments made their way into the community also tended to reveal a certain irrelevancy to the theory as the objects which it claimed to explain were undergoing radical change.

More curiously still, many objects were 'discovered' that turned out to be complete figments of the imagination. Gustav LeBon argued for the existence of black light—rays akin to x-rays. They repulsed matter but mutually attracted. There were religious undertones to the notion, and its sale of 44,000 copies reveals its public appeal. While LeBon was not a physicist, the incident but reveals something about the cultural atmosphere the time. Nrays, on the other hand, were proposed by Rene Bleondo, a physicist at Nancy University, in 1903. Bleondo, who claims to have made experimental observations, noted the nrays could pass through metal and wood, but could not be refracted or reflected; 40 other scientists also claimed to have observed them. Agusto Righi in Italy also proposes the existence of 'magnetic rays' in 1908. Nearly 70 papers are published on the phenomenon between 1908-1918, two thirds of which had been written by Righi himself. It remained an Italian phenomenon, as most supporters were Italian. There was also the alleged existence of 'red' and 'green' electrons, first proposed to account for cosmic rays. Although taken seriously for a while, these red and green electrons turned out to be a new particle, the 'muon'; all others had been figments of the imagination as well.

Another common theme of the period was the prevalence of a 'universal theory of everything': the belief that one could explain the entire universe on a single particle or formula—harkening back to the universals of Presocratic Greek philosophy. Many wanted to become the new "Newton." J. J. Thompson, discoverer of electron, proposed a theory of everything based on his particle. Though obviously wrong, his electron model had only been based on the analysis of a two dimensional surface, as three dimensional entities would have been too difficult to mathematically handle. Paul Dirac briefly believed he had 'repaired' the theory during the 1930s via his use of the anti-electron.

Finally, Sir Arthur Eddington, who experimentally demonstrated the effects of general relativity, later in life sought magic in universal constants. He believed in the macrocosm-microcosm Renaissance parallel: the existence of an underlying unity between the quantum world and cosmology. He spent many years searching for a 'final theory' that would unify quantum theory, relativity, cosmology, and gravitation. Although a fascinating philosophical topic, he had few followers. The phenomena were well captured by Dirac, "It has always been the dream of philosophers to have all matter built up from one fundamental kind of particle..."

The grueling two world wars of the period also shaped the receptivity of new scientific ideas, and helps account for popularity of general relativity. The wars had psychologically drained individuals; tired of conflict and destruction, many sought emotional refuge in relativity, looking for hope and renewal. As again observed by Paul Dirac,

It is easy to see the reason for this tremendous impact. We had just been living through a terrible and very serious war…. Everyone wanted to forget it. And then relativity came along as a wonderful idea leading to a new domain of thought. It was an escape from the war….

233

It is often forgotten that Einstein lived in Germany throughout World War I. During the war in 1916, when paper scarce, the German government allowed the publication of a 70-page description. Meant strictly for public consumption, it surprisingly went through three consecutive editions during the war. Einstein himself was taken aback by his popularity, which somewhat rather annoyed him. While taking the train one day, he was queried as to his profession by someone who had not recognize him; Einstein's replied in a sardonic one: he was a photographer's model. When assisting a movie premier with Charlie Chaplin, their car was hounded by photographers. Einstein turns to his friend asks what it meant, to which Chaplin replied "absolutely nothing."

The rupture of French and German relations during World War I led to a brief period where the French attacked relativity as an example of German ideology. The general rupture ultimately meant that, with the exception of Louis de Broglie, the French would not participate in the discoveries of the quantum world.

The conflict, however, often drew the worst of an individual's traits. Max Planck signs a public letter declaring Germany as the climax of civilization, an ethnocentrist expression to say the least. He later came to regret it, and we should not judge Planck too harshly. He had actually warned Hitler that anti-Semitism would ultimately hurt German science, making the Führer furious. Planck's home in Berlin was destroyed during World War II, burning all of his personal papers and correspondence—one reason why he is seldom cited as a historical source. All of his children die during the war; one son was even accused for being involved in a plot to kill Hitler, and is consequently captured and executed.

Of all German scientists, Planck was the first to have recognized Einstein's genius, and becomes one of the first 'relativists' in physics. Philippe Lenard, discoverer of the photoelectric effect on which all cell phone cameras are based, would be a notorious counterexample as a Nazi ideologue and apologist.

Max Planck's 'quantum'

It is a common historical misconception to assume that chronological changes coincide with revolutionary ones; we tend to divide time by centuries and oddly presume that new centuries have also have consequently new properties and characteristics. The notion is patently false as there is no necessary correlation between human events and our particular method of measuring time. A good example is the case of the turn of the century 'quantum.'

In December 1900 at a conference in Germany, Max Planck presented a small paper containing the notion of a 'quantum.' As previously mentioned, "quantum" in German merely alludes to a quantity and has no special meaning for Planck, who viewed it merely a mathematical stratagem akin to that used by Boltzmann. Planck, a physicist at the University of Berlin who received the position after the Boltzmann's uncertain wavering, had been working on the 'blackbody problem.' Often referred to as the ultraviolet puzzle that veers to infinity at high wavelengths, the principal problem consisted in the incongruity between data and theory. Max Wien's formula accurately traced its distribution curve at lower temperatures, but failed for curves at higher temperatures. While one could 'tweak' the formula to fit the data, Planck wanted to provide a sounder theoretical resolution to the disparity.

Planck's approach was to rigorously derive a formula from known theory, specifically Maxwell-Boltzmann's distribution law. As Boltzmann, his calculations tended to go off into infinity. Consequently, Planck redefined energy was defined as being smeared over a region rather than occurring in a continuous flow. It is here where the quanta comes into existence. Planck assumes that energy is distributed in discrete packets, or a 'quanta' of energy, and thus able to successfully provide a new formula which accurately predicted curve at higher temperatures.

Again, it is important to note that Planck did not believe in 'quanta' in 1900, who only viewed it as a mathematical operation which simply allowed formula to work. Akin perhaps to an early interpretation of the Copernican universe, it was undertaken for practical purposes in an ad hoc manner, and Planck did not pretend to suggest that it represented 'reality' as is. As late as 1912, we find him still trying to eliminate the stratagem. He was eventually convinced of its validity in a Slovay Conference by the renown Dutch physicist H.A. Lorentz. Lorentz was the first director of the Slovay and very well respected by Einstein.

Maxwell's Electromagnetic Aether

The nineteenth century also saw the rise of active experimentation with electricity. Rubbing a glass rod with a piece of wool, as had been commonly done during the previous century, allowed for brief if funny experiments by its creation of static electricity: a boy held by strings in the air while being sparked, for example. Benjamin Franklin's own testing led to the characterization of electricity as consisting of two different types of 'fluids': positive and negative. One beneficial result was that Franklin realized that he could protect buildings simply by putting a metal rod at top through to the ground; churches typically tended to be destroyed when struck by 'incendiary' lighting. The sources of electricity, however, were not persistent—a situation which changed with the creation of the Voltaic pile, an early battery, in 1800. The chemical generation of electricity allowed for the consistent generation of electrical currents and, in turn, to more rigorous experimentation.

H. C. Oersted in 1820 noticed a relationship between electricity and magnetism. A romantic *'naturphilosophie'* as Humboldt, Oersted believed in the inherent interconnectedness of nature, and comes across the phenomenon accidentally one day while teaching a class. He noticed that the compass needle would change direction whenever it was near a wire containing an electric current. Oersted was not a brilliant experimentalist, but was certainly smart enough to recognize the significance of the event. This awareness of its implications for physical theory is often what counts the most in scientific discovery, requiring a deep understanding of concepts. His discovery began to modify the definition of nature of electricity away from 'fluids.'

The premier experimentalist of electromagnetism of the era, however, was Michael Faraday (1791-1867). Systematically studying electricity and magnets, Faraday discovers magnetic fields that were produced by both magnets (north/south pole) and electricity (positive/negative poles), demonstrated by iron fillings. His dynamo of 1867 is the result of his discovery of electromagnetic induction, based specifically on the dynamics of the field. Faraday was able to tie the force of the field to the current produced by it, creating a motor which increased in power as the two mutually interacted—allowing for the conversion of strong mechanical energy into a strong electric current as that in hydroelectric dams. (The greater the power generated, the greater the resistance in the moving field and the electricity thereby generated.)

His work on polarized light of 1845 also further verified Oersted's assertion, by showing that polarization was influenced by electricity. There appeared something inherently valid about Oersted's faith in the connectedness of nature.

One problem, however, was that Faraday was not a mathematician, and hence unable to more theoretically analyze his results or push the limits of his understanding. By contrast, James Clerk Maxwell's unquestionable intellectual talents allowed him to express Faraday's notion of a field into mathematical language, and to a more abstract analysis of the phenomenon. By 1860, all electromagnetic phenomena could be described in equations, allowing Maxwell to reach conclusions that had not been immediately obvious to Faraday, as Maxwell previously had done with a substance's viscosity. Maxwell discovers that there could be electromagnetic standing waves, or what we refer to as radio waves—a phenomenon which would be portentous not only for the world of communications but for Einstein as well.

The person to first detect the presence of Maxwell's standing (radio) waves was Heinrich Hertz (1857-94), whose life, as Maxwell's, can be characterized as tragic for the talent demonstrated in light of its brevity. Producing an oscillating electric current, Hertz generated a spark in the gap of a ring which could be seen 'with the naked eye.' The standing wave generated by the oscillating electric current was then detected on a copper ring with a gap on it some distance away from the source. Further experiment with these waves also verified Maxwell's assertion of their common properties similar to those of light. Electricity had obtained a life of its own.

There is an irony to our story. Maxwell began work on his theory using a mechanical aether model: a convoluted box of twisting aether 'particles' using rods and rubber bands. Physical models were actually produced by his followers, and did believe them to have been valid mechanical representations of generated currents. The irony of his work resides in the fact that, his formulation ultimately made the aether mechanical models obsolete. They were no longer needed to consistently represent electromagnetic phenomenon, as the scientific focus shifted to fields and formulas. As in the case of Newtonian gravity, one could obtain precise mathematically predicted behavior in spite of the ignorance of the

236

underlying mechanism. QED would show the same dynamic, but with even more puzzling consequences.

At a broader level, Maxwell's work was a profound conceptual discovery in that various phenomena were now defined by the broader and more general concept of electromagnetic waves; Maxwell had provided a 'unified theory of electromagnetism.' As such, the same traits that could be identified in light, reflection and refraction, could also be identified in other electromagnetic phenomena. Maxwell's work, however, was very complicated, and his complex formulation simply did not lead itself to easy handling. A group known as the "Maxwellians," consisting of Oliver Heaviside, Oliver Lodge, and G. F. Fitzgerald, simplified these into their present form. The main figure of the group had been Heaviside.

Oliver Heaviside was a rather unusual character. He had first worked as a cable engineer but retires at an early age to move back into his parent's home, where he spends the rest of life simplifying the mathematics of Maxwell. The quirky aspects of his character and life, however, should not be used to underestimate the magnitude of his achievement. Heaviside also suggests presence of an ionosphere above Earth, which allowed the reflection of radio waves over long distances of its curvature. In other words, rather than traveling directly into outer space in a straight line, radio waves are reflected along the curvature of the ionosphere, allowing these to travel enormous distances.

These were but two of his many substantial achievements, out of which entire new industries were created—all the result of a hermetic working at his parent's garage, alone and isolated.

1905, *Annus mirabilus*

Whereas Planck had rejected the validity of 'quanta,' Albert Einstein viewed it as a valid entity, a testament to his revolutionary genius. In contrast to many of his peers, Einstein is able to get at the essence of topic, discarding all that was irrelevant. He actually viewed his work on quanta as his most important, even more so than relativity. It is from the direct application of Maxwellian electromagnetic theory that relativity theory is born. In fact, his special relativity theory of 1905 is so closely tied to it, that initially his 1905 paper was considered by the greater physics community only within that framework.

The year 1905 is considered his *"annus mirabilis,"* a miracle year where he publishes four original groundbreaking papers placing physics on new basis. But we can only say this in hindsight. In that year, Einstein was an unknown stranger to the greater physics community. As he had been prevented from teaching or obtaining any position at a university, and not without good reason, nobody of importance knew who he was. The preceding years, specifically 1902-3, had been a very difficult time. The death of his father and his uncertain income from odd jobs as tutoring led to uncertain finances, barely making ends meet. He and Mileva had a baby out of wedlock, a shame which is hidden away by placing the unwanted child up for adoption, which must have certainly affected their sentimental relationship.

Einstein's future was by no means certain in 1904, and he could easily have ended up as unknown a figure as Waterston was in his day. His dangerously precarious position had been his own foolish undoing. "Einstein who?," we might be asking ourselves today. Hence, the four papers he wrote and published in the *Annalen der Physik* were truly miraculous from a personal point of view, irrevocably altering the course of his life as well, as that of physics more broadly.

His first paper dealt with the photoelectric effect, a phenomenon whereby electricity will flow through a metal when light is shined on it. The higher the frequency, the more electricity generated in the process, which now allows all cellphone users take millions of selfies every day. Prior theory could not account for it. If light were a wave, the resulting phenomenon made no sense. Einstein shows that by assuming light to be traveling in discrete packets or quanta, the effect could be accounted for.

His second paper dealt with Brownian motion, previously discussed. The wiggly motion of tiny inorganic particulate, as if alive, could be accounted for by considering the velocity distribution of atoms in the surrounding area. Pollen particulate was small enough to be influenced by these fluctuations in velocity, hence proving the existence of Boltzmann's hypothetical atoms in kinetic theory. As icing on the cake, Einstein recalculates atomic sizes—for which he finally receives a doctorate. (Incidentally, when the article was first submitted to the university, it was criticized for being too short. Einstein adds a single sentence, and resubmits the final draft.)

His third paper presents an early formulation of mass-energy equivalence, but would not be finalized until 1907 when he formally presents his famous equation on the equivalence of light and matter: $E=MC^2$.

Finally, his special theory of relativity states that because velocity of light is same for all observers, regardless of frame of reference—for which he called it 'invariant theory'—our notions of time and space had to be drastically modified. Time contracts with motion. His ideas can be better understood when we consider them to be strictly logical conclusions. Consider that velocity is distance divided by time, $v=d/t$. By definition, if velocity remains the same for all observers, distortions will inevitably be wrought on distance and time. T cannot remain the same while v is changing; nature must bend under her own weight.

The key to understanding special relativity, however, is the issue as to why light speed should be independent of all observers, when the velocity of all other objects in the universe is not. Many other thinkers as Henry Poincare had come very close to this realization, but had not been willing to step through its conclusions to question fundamental Euclidean notions. Poincare must have noticed how tantalizingly close he was to relativity; after 1905, he never mentions it again.

Physicists living in 1905 who had read his groundbreaking articles must have asked to themselves, "Just who is this Albert Einstein fellow? I've never heard of him before."

Olympia Academy

Einstein came from Jewish family that had settled in Munich. His father sets up a manufacturing plant producing electromagnetic goods with his brother but they are unsuccessful, and even fail at other similar attempts. His mother was a music teacher who played the piano, and teaches Einstein as young boy to play the violin. Einstein was such a good violinist, quasi-professional, that had all else failed he could have earned a meager living from it. It is to be noted that both parents, although Jewish, were not adherently so, being more concerned with the problems of daily life. He had a younger sister, Maja.

As a young child some 7 years old, Albert got very sick, and when his father brought a magnetic compass, it left a profound impression on him. The compass seemed to violate all natural behavior, apparently not subject to gravity. So magical it had appeared to be that Einstein even later in life could still recall the incident.

Contrary to common presumption, Albert was actually a very good student. His uncle tells Einstein about Pythagorean theorem as a boy, and Einstein is able to write up a proof on his own. He then receives a copy of Euclid as a reward, which becomes a sort of a bible to him. The irrefutable proofs exposed in crystal clear logic become a model to be thoroughly imbibed and imitated, an experience many other outstanding scholars as Bertrand Russell are known to have also experienced. It is certainly the case the boy was very smart—too much so perhaps. A teacher wanted to have him removed from the school as the sheer presence of Einstein in the classroom would make other students turn to mockery. One may well imagine Einstein revealing the many errors in the teacher's expositions.

We may also obtain a notion of his talents and acquired scientific knowledge by the following anecdote. His uncle and assistant had been working for weeks on a business related problem. On a whim, they decide to give it to Albert to work on. Einstein as a teenager solves it in 15 minutes, astounding both uncle and assistant.

His parents move to Milan for a fresh financial start when Einstein was just 16 years of age, leaving him behind in Munich to finish his studies. Yet Einstein was miserable living alone in the big city. To make things worse, his family sent letters describing their wonderful Italian experiences. Albert 'mutinies,' obtains a letter from a teacher testifying that his mathematical knowledge was of a university level, and leaves for Italy. The temptation was too great. His year of merriment finally ends when his father instructs him to return to school; the father's business was again failing.

Although Einstein applies to the Federal Institute of Technology in Zurich, he is rejected. Its director Heinrich Weber, however, personally assists Einstein, and Albert completes his secondary education at the Swiss Cantonal School of Aarau. The years at Aarau became a very special time for him, making a good friend Michel Besso, and he begins to ask himself what a light wave would look like if seen from the side. Would he be able to see his own reflection in mirror while traveling at light speed? He would think on that question for many years.

It is only upon his entrance into the Zurich Polytechnic where he begins to behave like an irresponsible student. More openly demonstrating a strong antiauthoritarian attitude; he attends classes irregularly, choosing instead to dedicate his time to learning on own. It is at this stage when he comes to dominate Maxwell's work on electromagnetism, of which he becomes an expert. In this endeavor, he routinely studies with another student, Mileva Maric, who later becomes his wife. Einstein would also makes another important friend: Marcel Grossman.

The importance of Grossman's help in Einstein life cannot be underestimated, helping our figure at three critical junctions. There can be no doubt whatsoever that without Grossman's critical assistance, Einstein would not have succeeded in his efforts. Grossman, in effect, becomes Albert's personal *deux ex machina*—a point that cannot be emphasized enough.

As one might expect, Grossman was opposite in character to our figure. Grossmann is a very meticulous note taker, eventually studying non-Euclidian geometry. Because Einstein had so routinely failed to assist his classes, he would certainly have failed the graduation exam. Grossman, however, comes to the rescue and lends Albert his detailed class notes, which allows the latter to barely scrape by and obtain his university diploma.

Upon graduation in 1901, however, Einstein fails miserably in obtaining a science related job as a laboratory assistant. Weber, the Zurich University director who had helped him get into university, felt betrayed by Einstein's university behavior, and does not write any letters of recommendation for him. It would not be unreasonable to suppose that Weber likely misspeaks of Einstein to his colleagues, thus hindering his chances of ever obtaining a university position. Albert writes to various places soliciting positions, but none come through. In one particular case, when Einstein writes to Wilhelm Ostwald at Leipzig, Ostwald never replies. Albert's father was so concerned for his son, that four months later, he personally writes to Ostwald pleading in his son's behalf. But again, there was no answer. Einstein's mockery of academic authority had come back to haunt him, and he grew increasingly despondent as a result.

But this was not the only trouble afflicting him. While he had been a university student, Einstein's uncles in Genoa had provided a stipend to the promising student, which was removed upon graduation. The death of his father in 1902 also deteriorates his mood. There was, however, one bright spot to this period (1901-3): The Olympia Academy.

One of Albert's tutoring advertisements is answered by Maurice Slovine, a humanist graduate student who wanted to improve his knowledge of science. Slovine thus 'hires' Einstein as a tutor, or so was the initial intention. Their first meeting was a 'hit,' spending more than two hours debating issues in the philosophy of science. Half an hour later, they continue their intense discussion at a nearby café. During the second tutoring session no instruction is given by Einstein but rather both men continue their intense debate. This meeting of the minds lead to the formation of what came to be referred to as the "Olympia

Academy": a group of three peers whom, along with Konrach Habitch (a friend of Einstein's from Aarau), read the latest books and intensely debate their implications. It became a period of a great deal of intellectual stimulation for Einstein.

Their academy, however, dissolves when both Slovine and Habitch graduate from university and obtain respective jobs in different cities. Einstein once again finds himself adrift.

In light of Albert's worsening financial circumstances, Marcel Grossman again comes to the rescue. His father knew the general director of the Patent Office for all of Switzerland, and puts in a good word for his son's friend. When the Director finally interviews Einstein for 2 hours, it becomes clear to him that Einstein was wholly unqualified for the position. However, his grasp of Maxwell's work was so impressive, that the Director allows for his entry into the system. When an opening for a position in Bern appears a few months later, the charge is given to Einstein in 1903.

Although Einstein becomes a lower patent official, class C, his office quarters would become a 'holy space' where his most important ideas in physics were hatched. Albert somehow manages to gets Michel Besso, now an engineer, to join him at the same Bern patent office, allowing for their continued fruitful conversations. Albert is able to become so proficient in his tasks, that he manages to steal hours here and there for his own research. Einstein formally marries Mileva, and they have two children.

The seven years at Bern would be the happiest in his life.

Invariant Theory

In spite of their diverse scopes, both the Special and the General Theory of Relativity were conceived in a similar fashion. Einstein takes two unquestionable principles and builds an edifice from there on, exploring the full consequences and contradictions of these principles. His bravery consists in accepting whatever odd conclusions that emerge from his reasoning; there indeed exists such a thing as intellectual bravery. In both cases, he also explores the contradictions inherent in the existing interpretations, building up on these; he uses one to feed off the other.

The special theory is related to light, and it is 'special' in the sense that it constitutes a delimited case. The general theory expands the special to a full on treatment of gravity. It is important to note that 'space-time' cannot be imagined, an error commonly seen. The notion appears early, being developed by Hermann Minkowski in 1907. Space-time, however, could not be visualized by him, and is rather a mathematical tool of analysis. It is also often used to iconographically represent general relativity, but is actually only one aspect of the theory.

Special Relativity is built on the work of James Clerk Maxwell. Einstein begins with two principal observations. The first is that the treatment of electromagnetism was inconsistent. Although the same event was occurring, regardless of whether a wire was pushed through a magnetic field or if the magnetic field was moving past the wire, Maxwell's formulation provided separate treatments for each case, as if they were wholly distinct frames of reference. The issue is akin to that Galileo in studying objects in moving ship, and to account for the absence of a sense Earth's rotation and revolution.

He also adopts the work of his predecessor in a different way. Maxwell showed that a light wave has own unique properties; an electromagnetic wave was its own entity, regardless of how it had been formed. Einstein hence concludes that speed of wave was irrelevant to object which caused it—his second principle.

These two foundation principles lead directly to the special theory of relativity, showing that the universal speed of light inevitably leads to distortions of time and space. Again, it took the courage to take that fundamental step to question time and space. Einstein was curiously ignorant of the Lorentz-Fitzgerald contraction formula, but is able to derive it directly from his work. This in itself was an impressive achievement, validating the soundness of his work.

He writes that when a body gives off energy it lost mass, diminishing in proportion to E/c^2. However, the amount of mass lost was miniscule as a result of the high speed of light. A 100-watt bulb turned on for an entire century would lose only $1/1,000,000$ of an ounce. This work would not be fully developed until 1907, and so the years between 1905-7 were when he finally derives his famous formula. In turn, it implied that there exists an enormous amount of energy in matter. Einstein did not realize at the time its full implications.

When the special theory was published, Planck immediately recognized its merit and becomes one of Einstein's early 'groupies.' Planck was impressed by the simplicity and symmetry of the work, and began contributing to it as early as 1906—as well as beginning to assist Einstein's professional career. He writes a letter of recommendation for the young man, attesting to his brilliance, even when it seemed to go overboard. Planck rejected Einstein's work on the photoelectric effect, and requests the evaluating committee to dismiss such flights of fancy from such an able young mind.

Yet the special relativity theory paper was somewhat odd. It was published without any citations, as a rabbit magically taken out of a hat. What did Einstein actually know at the time? He had also argued that the aether did not exist. How did he know and how did he reach conclusion?

The speed of light was first empirically studied by Galileo in an ingenious but naturally unsuccessful attempt to measure it. He gives candles to two respective individuals who walk away from each other, increasing the distance between them. By turning off their respective candles, any consequent delay could be detected. As one would expect, there was no appreciable difference between the two could be found. Was light instantaneous?

Ole Roemer in 1672 took up the issue, and came up with a more 'modern' figure. Previously working at Tycho Brahe's castle, Roemer obtains a position with Dominico Cassini, who suggested the experiment. Given the enormous distances between the Earth and Jupiter's moon Io, Roemer discovers a light speed of 190,000 km/s, remarkably close to today's estimate of 300,000 km/s.

242

Whatever the disparity, what was important to note about Roemer's work, is the 'proof' that light speed was not instantaneous but only appeared to be so due to its very high speeds. However, Roemer's publication made no mention of Cassini, who is naturally offended and becomes his greatest critic.

The first accepted modern measurement of the speed of light, however, was established in the famous Michelson-Morley experiment of 1887. It was a very subtle experiment made in Ohio which sought to produce interference waves of light. Since it was hypothesized that light traveled through the aether, a medium that was presumed to be required for its translation as just as air is required for sound waves, Albert Michelson and Edward Morley deduced that one of two perpendicular beams of light would be alternatively slowed down given the Earth's rapid motion through space and the all-pervasive aether. If the two beams were then crossed, their varying speeds would inevitably produce interference patterns. No such patterns were ever detected.

Contrary to the image he wished to portray, Einstein was indeed cognizant of their results.

It is ironic that Einstein was also aware of Ernst Mach's work, pointed out to him by Besso during their school days. Yet the influence of Mach's writing was complex, and the principal assumptions of logical positivism were obviously never internalized by Einstein. Mach's indirect role was in helping Einstein get rid of any concerns and fears with regard to his questioning of fundamental Euclidian notions. In his work, Mach had criticized Newton's presumption of absolute space and time, arguing that these were metaphysical beliefs that had no grounding in data and hence irrelevant to science. Mach's work thus unconsciously helps Einstein 'jump' the intellectual canyon of classical physics into the abyss—something Poincare was hesitant to do.

Naturally, Einstein's revolutionary 1905 papers eventually began to be diffused, and ultimately the physics community became interested in meeting the man behind them. As Minkowski's work on space-time was based on Einstein's work, his conference participations inevitably resulted in Einstein's own invitation. In this, the Slovay Conferences after 1911 were of an enormous personal significance to him, as well as to the broader physics community in general. The conferences gave Einstein the opportunity to be seen and heard. The community came into direct contact with his clarity of his mind, and Einstein came to greater appreciate his abilities by seeing that he could 'hold his own' when interacting with the rest of the physics community.

The Slovay Conferences had been organized by Walter Nernst, a physical chemist who had been a student of Boltzmann. There was a need in the field of physics to resolve many of its problems, as it was widely recognized to be in a state of 'chaos.' Ernest Slovay, a rich industrialist whose fortunes emerged from the caustic soda industry, assisted Nernst in this venture. Its main purpose was a simple one: to talk about the current problems and puzzles in physics. Only eminent theorists were invited, and the series of conferences became a tour de force, a "Who's who" of physics which can be used to trace the history of quantum revolution.

Although, as we all know, Einstein eventually becomes a sensation, his success was not instantaneous as it took time for work to gradually become known and digested within the physics community and outside of it. He does not leave the Bern Patent Office immediately, but obtains a concurrent position at the University of Zurich, his alma matter. However, this change of status is not as positive as it might first appear, as the position of *privadozent* implied no salary other than that directly paid by students—as well as a greater time burden for lecture preparation. Only three students enroll in his class, one of which was his ever faithful friend Michel Besso.

Einstein is finally able to improve his condition when he is finally offered a position in Berlin due to Planck's enormous efforts. Planck moves 'heaven and Earth' to bring Einstein to Berlin, and is able to get extremely favorable conditions for his colleague. For example, Einstein would be immediately made a member of the Prussian Academy of Sciences, and would not have to abandon his Swiss citizenship, which meant a lot to him. He would also have a guaranteed income without a teaching requirement. Although Einstein finally has the opportunity to dedicate all of his time to research, he becomes hesitant in accepting the generous offer.

Being offered positions in different universities, Einstein felt like a chicken that was expected to lay golden eggs but was uncertain if he had more 'eggs to lay.' Einstein was also wary about the German autocratic mentality that so adored its military. In sharp contrast to his intellectual independence and autonomy, German culture was rigid, structured, disdainful of individuality.

In the end, however, the offer was too good to pass up, and Einstein joins the Berlin faculty in 1914.

Of trains and elevators

The aim of General Relativity was to extend the special case to the understanding of gravity. In a relative short amount of time, Einstein succeeds and publishes the substantial *Die Feldgleichungen der Gravitation* (1915), containing page after page of excruciating field equation formulas that would so typify the modern view of the absent minded scientist.

Einstein first comes upon it while working at Bern Patent Office. The image of a man that falls but that does not feel his weight forms the starting point of his revolutionary theory. The scenario seems trivial but it again helps Einstein realize that there were gross incongruities in the existing explanations, akin to the incongruities seen with Descartes and Maxwell. Einstein's main observation was that Newton was contradicting himself in his varying treatments of inertia and gravity. The work does not deal with uniform motion, which had already received apt treatment by both Galileo and Newton, and in which the 'principle of relativity' can be found. As observed on Earth or in the cabin of a ship, all objects move freely with respect one other in a uniformly moving body.

The new technologies of the era as trains and elevators were creating new experiences, which are then further explored by original thinkers as Einstein. It goes without saying that we have all, at some point or other, tried to jump up in a downward moving elevator, or puzzled at having the sensation of movement when nearby train cars (and automobile) change position while ours is perfectly still. Both these cases are actively used by Einstein in the exploration of physics, and one has to wonder whether they were also in their formulation as well.

Einstein considers two similar cases: a) an elevator being pulled up in empty space with an acceleration of 32 ft./sec and b) an elevator being pushed down by gravity at its usual rate of 32 ft./sec. He realizes that the treatment of both were identical; the varying frames of reference used by Newton to account for inertia and gravity was arbitrary. Inertia could be seen as the force needed to move an object from its current path; the greater the object, the greater the force. This observation this was a particularly useful theoretical tool in that both cases, inertia and gravity, had been characterized by different set of formulaic baggage. It meant that, from point of view of method, any changes successfully applied to one case (inertia) could be applied in the second (gravity), and vise versa. It was a stroke of mathematical / physical brilliance. Light would bend under the weight of gravity.

However, Einstein is aware that he quickly reaches his limit of abilities. He required knowledge of a more sophisticated mathematics to treat the problems at hand, but lacked it. Again, Marcel Grossman comes to the rescue.

Incidentally, one of the reason Einstein chose physics over mathematics as a student was because, whereas in physics he could easily identify key problems, mathematics was so well advanced and developed, that he would have to spend a lifetime of study merely to master one of its subdomains. The cost-benefit ratio did not look positive for the latter.

As it should so fortunately happen, Grossman's area of expertise was precisely the area of mathematics Einstein needed to overcome his existing obstacles: non-Euclidian mathematics. Both enter an intense period of work from 1913 to 1914, applying tensor analysis to the topic, specifically the work of Carl Friedrich Gauss and Bernhard Reinmann. Grossman and Einstein thereafter publish joint papers, where Grossman approached the mathematics and Einstein took care of the physics involved. So effective was their collaboration, that at one point, the work between the two men almost ventures down a path which would have led them to the field equations (general relativity), but desist from doing so. This intense partnership, done while Einstein was still working at the Bern Patent Office, ends upon his acceptance of the Berlin position. Einstein would be on his own again.

Einstein's period of glory was preceded by a stay of pain. His move to Berlin coincides with the separation and eventual divorce from Mileva, who makes him promise the remuneration from any Nobel Prize moneys awarded. In his empty Berlin apartment, he consoles himself by playing the violin. When a bout of illness weakens him, he is taken care by his cousin Elsa, who nurses him back to health. The two eventually marry.

Feeling the competition from other physicists closing in on the prize, Einstein locks himself in and succeeds in establishing the general theory : $ds^2 = \sum g_{mn} dx^m dx^n$ whose solution implied ten lengthy quadratic equations solved over the huge tome, where any deviation or small error in the long calculations would invalidate its entirety. Upon completing it, Einstein enters a period of quasi-religious elation that lasted various days. By the end, he is completely exhausted, and with good reason.

In spite of the surrounding circumstances of World War I, scientists quickly hear of his revolutionary work, and set about testing its implications. Einstein already knew, however, that the disparity in the speed of the planet Mercury, if solved, constituted an immediate proof of the validity of his work. Newton's' predicted speed of the small planet at it perihelion varied from 40 to 50" of arc; Einstein's work set it exactly at 43" of arc. The most important experimental verification of his work, however was done late in the war by Sir Arthur Eddington, as mentioned before. Eddington measured the bending of light by gravity, and the micrometer results fell in line with Einstein's predictions.

The verification led to a momentous recognition in the world of physics and of the general public. Einstein is hailed as the new "Newton," and J.J. Thompson, now President of the Royal Society of London, holds a special conference to commemorate the achievement by his German colleague.

Globally, everyone celebrates the achievement, leading to the popular use of the notion of 'relativity' in areas where it did not apply at all: art, psychology, and so forth. The cases are far too numerous. The term incorrectly suggested a notion inappropriate to its actual achievement, and suddenly 'everything was relative.' Even the renown Spanish scholar Jose Ortega y Gasset is led to formulate the 'new' philosophy of "perspectivism". In essence, he argued that all cultures were 'relative,' each with its distinct and valid normatives and values.

It is not with undue concern that Einstein had wished to coin his ideas by "invariant theory." However, Max Planck's early entry and enthusiasm for his work led to its social baptism with the term 'relativity theory."

It was a name that so contingently and ominously stuck.

Quantum Mechanics

The Irrationality of the Very Small

THE HISTORY OF QUANTUM THEORY could not have been more different from that of relativity. As we have seen, relativity was mainly the result of one man—with help of one or two friends, either as sounding board (Besso or Slovine) or as direct partners (Grossman). Although Einstein gives Grossman due credit, a mishap in the printed article removes the important recognition. After World War II, Einstein obtains worldwide popularity with the mantra of the wise sage given his ruffled white hair and thick mustache. Every public appearance was widely photographed, everyone wanted his signature—the first cases of the infamous paparazzi.

Quantum mechanics, on the other hand, was the work of many silent physicists working privately in closed quarters. Most would cluster in particular cities, the Copenhagen Institute being a key node in the field ably led by Neils Bohr. The work proceeded at such an abstract level that it was mathematically and conceptually difficult to follow, and hence initially drew little public attention onto itself or its philosophical implications.

Its fame and stardom would only occur decades later, after the ultimate 'practical' result of their work became known. The ever powerful atomic bombs in Nagasaki and Hiroshima provided a hint of the enormous significance of their achievement. While both relativity and quantum theory had a distinct German character, the latter became more closely imbued with an American character due to Nazi politics.

After the war, however, physicists suddenly became famous public figures. The period following World War II is marked by a drastic alteration in the social character of physics, becoming a "big science" that required big money, big technology, big journalism, and big politics. So great was the change, that older figures began to reminisce nostalgically on what had been lost: a "little science" that was personal, intimate, and the result of theoretical genius or experimental acuity. The world of paper and pencil was transformed into that of colossal machines and global geopolitical transformations.

Quantum mechanics, however, is born in absentia; the result of the attempt to reconcile classical mechanics with experimental results. It was the physics of the very small: invisible atoms that no longer were hypothetical indivisible constructs, as the Greeks had reasoned, but worlds onto themselves.

Unusual suspects

The turn of the century was marked by many unusual discoveries which, one after the other, did not fit classical mechanics. It is these public results which grab the physics community's attention, rather than Planck's obscure 'quantum' which remained quietly hidden in a closet. Whereas few were aware of the meaning of the quantum, the irregularity and unusual character of the new phenomena immediately drew attention onto itself.

John Joseph Thomson at the Cavendish Laboratory in England had formerly been a student of James Clerk Maxwell. It is Maxwell who first establishes the now famous Cavendish in the basement of a castle at Cambridge. In this cold, dark and dank space, Thomson began his study of the cathode ray tube, first invented by Faraday, an old 'CRT' box television set. Its internals were composed of a vacuum tube with negative cathode node at the other extreme from the screen. When turned on, the cathode shot particles onto the screen, which could be distinctly seen to light up, one by one. Thomson placed a magnet around the discharged particles in 1897, discovering that they would veer away from direction of negative charge, showing that it had a negative charge. Thomson referred to the particle as a 'corpuscle,' perhaps in imitation of Newton. As with Einstein's relativity theory, the name that stuck was not the discoverer's but rather that of "electron" during the 1890s.

Others had also begun important theoretical analyses, as H.A. Lorentz in 1895. A deep theoretical physicist, Lorentz wonders what would happen to Maxwell's equations if electromagnetism was treated as a particle rather than a field. He shows that the two coincided, demonstrating no inherent contradictions. In other words, the properties of matter could be explained as if full of electrons; metals are networks of electrons free to move around, and hence why they are such good conductors. When some began to suggest that other particles aside from atoms existed, Thomson oddly opposed this view. His faith in the Greeks was insurmountable: atom were simply indivisible and that would be the end of it—even though his most important discovery, the electron, disproved such a view.

X-rays were discovered by Wilhelm Roentgen in 1895 also using a cathode ray tube, the same instrument forming the core of J. J. Thomson's experiments. In fact, Thomson had actually witnessed the phenomenon—a sheet that fluoresced when the machine was turned on—but was so focused on the ray itself, that he that did not pay much attention to it. There was an ongoing debate between the respective German and British communities as to what the cathode ray actually was and what it consisted of. The Germans alluded to a field; the British to particles. Hertz had wrongly argued that it was not affected by magnetism.

As Thomson, Roentgen noticed that a fluorescent sheet at the other end of the room radiated when the cathode ray tube was turned on. Even when he covered the cathode ray tube with black paper, he still witnessed same effect. He could even detect it when covered with his hand, although the effect was reduced. Roentgen, in contrast to others as LeBon or Righi, studied the phenomenon for a year before he making the announcement. He needed to determine if the phenomenon was new or if same as other.

It certainly had very odd properties. X-rays traveled in straight lines, but were uninfluenced by magnetic fields and did not seem to show refraction or reflection. It had a very small wavelength, of 10^{-13}m. Max von Laue would later use these in 1912 to study distance between atoms in crystal would, creating the field of crystallography. The x-ray diffraction patterns by crystal revealed their inner structures. (It is Rosalind Franklin's DNA crystallographic data evidencing its heliacal patter that was stolen by Watson.) Roentgen then took an x-ray of his wife's hand, which clearly revealed its inner bones; it suggested medical application. X-rays, in short, became a worldwide sensation.

Another odd experimental discovery was that of radioactivity by Henri Becquerel and the Pierre/Marie Curie team in Paris. X-rays immediately interested Henri Becquerel, a scientist at the Paris Museum of Natural History. Both his grandfather and father had held positions there. It is actually Henri's father who first notices that uranium had a surprising spectral regularity, emphasizing the phenomenon to his son. Henri was an expert on fluorescence, which typically occurs on objects which absorb and emit sunlight. Henri decided to verify if there was any relation between the observed x-rays and fluorescence, so performs a series of experiments on uranium for various days, without getting any results. When he shifts to the use uranium salts, these reveal 'fluorescence.'

The role played by serendipity in his experiment is well known. Seeing that uranium salts created fluorescence on photographic sheet wrapped in black paper, Becquerel apparently believed that effect was being caused by sun, as he would expose these to sunlight. One particular day had been markedly cloudy, but oddly he observes the same effect not expecting to have seen none at all. Eventually, Becquerel realized that it was not the sun causing effect but rather the uranium itself, as Roentgen had noted all along.

Pierre and Marie Curie also begin to investigate the phenomenon. A husband and wife team, Marie had been Pierre's student, they analyze different substances with similar properties. It is curious to note that uranium in its natural form is rather abundant, and is not harmful to human as it was part of our evolutionary natural environment—in sharp contrast to the case of plutonium which is extremely harmful. The Curies identify various radioactive materials as thorium and discover 'polonium,' named after Marie Curie's homeland. Of all the substances tested, radium gave off they most energy.

The team believed these objects were emitting 'uranium rays', but ultimately label the phenomenon as 'radioactivity' in 1908. Marie Curie, surrounded by such harmful materials, horrifically dies from tongue cancer at the relatively early age of 66.

Discovering the nucleus

Ernest Rutherford was another important participant in these early years. A physicist from a humble background in New Zealand, Rutherford ends up studying with Thomson at Cavendish Laboratory and actually invents a radio receiver. Concerned about the wealth it would generate, his mentor consults Lord Kelvin who had been involved in commercial activities as the Transatlantic telegraph cable. Kelvin incorrectly calculates that would not make much money, requiring 10,000£ alone for marketing. Thomson thus confronts Rutherford, dissuading him from the project. Rutherford would have to pick between science or mammon. Although offended, Rutherford prematurely gives up on the project.

He identifies various particles that were being emitted by the strange phenomenon: alpha particles (protons), beta particles (electrons) and gamma particles. Rutherford begins to systematically bombard elements with alpha particles. He has two assistants, Ernest Marsden and Frederic Soddy—the latter a chemist who is hired to help Rutherford identify the various elements created in his experiments. He comes upon the phenomenon 'transmutation' (one element turning into another), which had been the dream of alchemists for ages.

Rutherford realizes that radium was giving off rays, decaying into isotopes until finally becoming lead, and calculates that its half-life was immense. Soddy in 1907 writes book on radium, which influences H. G. Wells, who turns it into an influential utopic vision of the future. Wells predicts that that by 1952 Britain would have 'eternal energy' power plants.

Rutherford continues experimenting, alternating between his small laboratories at McGill University and Manchester. His famous gold film experiment shot alpha particles at 45 degrees onto a gold film, with a lead plate situated directly between the alpha emitter and the detector so as to block unwarranted alpha particles. Strangely enough, the gold film began reflecting particles, astonishing Rutherford. To him, it was if shooting a cannon ball onto a sheet of paper, only to have it bounce back. He eventually realized that atoms consisted of a tiny nucleus surrounded mostly by empty space.

He then begins to build a 'satellite' model of the atom in contrast to Thomson's 'plum pudding model' where negative electrons were softly held in a positive field. Philippe Lenard had already been bombarding substances with cathode rays, and is actually the first to argue that much of the atom was empty space. These discoveries lent themselves to romantic images of the atom, as if it were a solar system onto itself. The greatest exponent of this view was the Japanese physicist Hantaro Nagaoka who named it his "Saturn model," influenced by Maxwell's study of Saturn rings. However, the model was not held to be rigorous or credible, but rather reflected the religious biases of its creator.

The Saturnian atomic model was another case of the microcosm-macrocosm romantic world view.

Yet Rutherford's 1911 model was very similar to it, forming the basis of the atomic iconography commonly seen in old NASA logos. In essence, the core was formed of protons, surrounded by circling electrons on its outer layers. James Chadwick, who had been influenced to enter physics by seeing a public lecture by Rutherford, adds the neutron to the picture in 1914.

However, this 'planetary model,' was inherently flawed as it was unstable. Another interpretation of the structure of the atom was needed.

Neils Bohr

After working with Rutherford, Neils Bohr returns to Copenhagen, where he continues his mentor's work. Bohr would become one of the most important but unrecognized figures in the history of physics. In 1913 he publishes a revolutionary new theory about the structure of atom, paving the way for quantum mechanics.

He realized that the atomic planetary model was incorrect, as the circling electrons would quickly lose energy and collapse into nucleus. Bohr takes a simple but crucial decision and decides to study the hydrogen atom, the simplest of all elements with only one proton and one electron. Bohr then incorporates Planck's 'quanta' into his model, given the theoretical similarities of atomic modeling to the blackbody issue the quanta had first resolved. Bohr obtains excellent results that were verifiable.

In the new Bohr model of the atom, electrons existed at stable levels outside of the nucleus, which reflected the energy of the system. Every time an atom absorbed energy or lost it (quanta), electrons shifted levels. The input of higher energy, meant higher frequency packets of blue light, whereas lower energy meant lower frequency packets as red light. Its mathematics showed discrete quanta being emitted, which could experimentally be detected with spectroscopy. Rowland's diffraction grating, a metal sheet with thousands of tiny engraved lines on it, acted as a sophisticated prism. Such orbital changes left distinct spectral lines that could be readily measured, providing unquestionable evidence of the validity of Bohr's approach. Bohr had demonstrated why spectral lines were what they were for hydrogen.

Given the difficulty of the topic, there were gaps in this model. Many questions remained unanswered, problems which Bohr was well aware of. His early model is now referred to as "Old Quantum Theory," and although it is soon thereafter discarded, it soundly establishes Bohr's career. His successes and recognition lead to the creation of the Neils Bohr Institute in Copenhagen, later affected by World War II. Between 1921-3, Bohr develops a second atomic model. However, there were still many problems, specifically incongruities to experimental results of Hans Geiger and Walter Bothe. Even so, the quality of the discussions at the Copenhagen institute were very influential, imparting a lifelong intellectual style to its students. As a thinker, Bohr tended to be systematic and rigorous in his approach and questioning.

The incompatibilities of the second Bohr model were eventually resolved by Werner Heisenberg, Paul Jordan, and Max Born. Heisenberg's seminal paper of September 18, 1925 in the renown *Zeitschrift fur Physik* formulated a quantum mechanics that was logically consistent. During the spring of 1925, Max Born and Pascual Jordan realize that matrix equations could be used to express Heisenberg's work, becoming a feasible a way to describe an atom's internal dynamics. However, its extremely complicated physics made even the most capable physicist twinge a bit. Matrices were not easily understood, even by Heisenberg.

Fortunately, Erwin Schrödinger, who greatly disliked both Bohr's atomic model and matrix theory, postulated an alternative interpretation and a simpler method of calculation. Even if one disagreed with the wave model, the simplicity of its calculation made it the go-to approach in quantum mechanics. In poor taste, Schrödinger characterizes the matrix atom as a fleeing refugee. His work was influenced by Louis de Broglie whose 1924 Ph.D. thesis showing a strong relation between matter and waves. His effort to bridge quantum theory and relativity led him to combine two of Einstein's formulas. Schrödinger describes his wave model in a 14 page paper appearing in the *Annalen der Physik* of 1926: $\Delta[y] = 8\pi^2 m/h^2 (E - V) = 0$. His work shows atoms in phase changes, and his method becomes the more widely used than the matrix model. In actual practice, the calculations would be preformed in Schrödinger's style, only to be transferred to the matrix model prior to publication.

In 1927 Heisenberg tries to write up an interpretation while at the Copenhagen Institute, which is at first rejected by Bohr. The previous year in September, Heisenberg had been discussing with Wolfgang Pauli the strange phenomena. Pauli retorts that it was as if one could see with one eye (p position), close that eye and look with the other (q momentum) but oddly be blinded when both were opened at the same time. Bohr ruthlessly rejects Heisenberg paper with unclear arguments; however, a third scientist points out the similarities between Heisenberg's "uncertainty principle" to that of Bohr's, showing that it was but a subset of the "complementarity principle."

Heisenberg's ideas would become a cornerstone of quantum mechanics, striking at the heart of the Newtonian paradigm. The absolute certainty implied in Newtonian physics would be a thing of the past.

The world of the small operated by its own set of rules.

Steps to an atomic bomb

The years of 1931-3 are important in retrospect. Chadwick proves the existence of neutron, helping to account for atomic weight and the stability of electrons. Paul Dirac further combines relativity and quantum theory to proposes the wild existence of anti-particles and antimatter. The 'positron' for example was an anti-electron. While in hindsight this work is revolutionary, it received a strange reaction in that matter would only last 10^{-9} seconds. In the end, the neutron becomes the key to atomic bomb as it would pass without impediment directly to the nucleus due to its absence of charge.

Enrico Fermi (1901-54), one of last great physicists who was both a brilliant theorist and experimentalist, started bombarding elements with neutrons, in particular uranium. His result was not what had been expected. Upon neutron bombardment, elements usually became 'heavier' as new neutrons are usually added to their cores. However, with uranium, it actually lost atomic mass. Typical uranium has an atomic weight of 238, made up of 92 protons and 146 neutrons.

Fermi's work was followed up by Lise Meitner and Otto Hahn, the latter an experimentalist, the former a theorist who had worked with Boltzmann. When in December 1938 they also bombarded uranium with neutrons, their chemical testing showed that it had turned into barium, which had half the atomic weight of uranium. Meitner worked on the theoretical physics behind the process, and concluded that the neutrons were splitting uranium atom in half. She also realizes that Einstein's $E=MC^2$ formula could be used to calculate exactly the total energy released in the process called fission. Their work caused a sensation in the world of physics due to its enormous implications.

By January 1939, nearly all active physicists knew of Meitner's and Hahn's discovery, suggesting huge amount of untapped energy was stored within the atom. Fission turned out to be a set of cascading neutrons; each break would lead to neutron release, which in turn caused more neutron releases at an exponential rate—what is known as a chain reaction. However, not all isotopes of uranium were effective fuels. In order for a chain reaction to occur, the set of cascading neutrons needed to occur at a certain rate and speed. The two best candidates were U235 and plutonium (U239). The inevitable problem, of course, was that both were extremely scarce. Ninety-nine percent of the uranium naturally found is U238; only 0.7% was U235. Ernest O. Lawrence's cyclotron becomes the first to make U235, one atom at a time.

To build any sort of bomb would require costly man made materials. It could not be constructed in a home garage or in a small apartment, but would require vast industrial complexes operating continuously for years on end to produce only a few ounces of the deadliest and most expensive material in history.

The Manhattan Project ushers the era of Big Science.

Science and World War II

The Illusion of Scientific Autonomy

THE MODERN ERA REPRESENTS A DEFINITE SHIFT in public support for science. There is an increase in professionalization. Scientists obtain well paid positions as professors in academia as well as researchers in industry. AT&T establishes Bell Labs in 1925 after its merger with Western Union. Other big industrial R&D labs prominently hire scientists include Corning labs, who invent the first optical fiber and becomes the *de facto* standard. Xerox Research Park and Donald Engelbert's team develop a new view of computers, showing their inventions to Steve Jobs, including the mouse, the graphical user interface (GUI) and the Ethernet.

Scientists also prominently obtain positions in government. During the nineteenth century, the United States Standards Bureau had already become a place of employment for physicists, who help establish international standards, units of time, and units of measurement. Broadly speaking global transactions required standardization so as to allow for the coordination of parties; the increase in global capitalism during the nineteenth and twentieth centuries hence leads to an increase in the number stable and well remunerated scientific positions across various economic sectors.

The situation contrasts to the role of science in early modern period where patronage systems of unstable and uncertain employment prevailed, as in the case of Galileo. Universities did not have that many positions for scientists generally speaking; Newton is able to obtain one only because of the position unoccupied by Barrow.

We should not assume, however, that professors in modern university settings have full autonomy as all social settings, by definition, imply social interactions subject to change and compromise. There will exist diverging values and goals at variance, or inter-institutional competition for financial resources. Ultimately, scientists are subject to the state in this new setting. When its leaders and the general population do not recognize the importance of science, dark clouds menace its future. As all universities are state funded, control of the purse is tantamount to holding the fate of science in one's pocket.

These dynamics can be very clearly seen during World War II and the Cold War, in the nations of Germany, United States, and the Soviet Union.

Nazism and Science in Germany

One of the most common images used to describe the rise of Nazism in Germany is that of a frog in a slowly heated kettle, which will allegedly stay in the kettle until boiled to death. The allegation is false in that a frog will jump out when heated. However, the story very aptly captures the growth of German Nazism. At first the party pretended to abide by the law. Its encroachments on civil rights do not scale immediately but are rather gradually increased until the point which later became well known to the world after the end of the war: the mass extermination of Jews and other undesirables in gas chambers.

Upon Hitler's acquisition of power, Albert Einstein became one of the first to be dismissed from German academia. By 1933 he had become a world renown scholar. His relativity theory was labeled by the Nazi party as 'Jewish science,' and hence tended to be publicly dismissed as being of little significance. It was too theoretical and speculative; its strong mathematical component had been biased by Jewish commercial interest always seeking to usuriously calculate interest rates.

An alternative model of science is presented: that of "German science" as defined by Philippe Lenard, by then a much older man. Lenard becomes the ideological leader of Nazi science, and dresses scientific virtues in rather conservative traits. These included: classical Newtonian physics, absolute space, a knowable universe, predictable experimentalist, empirically grounded in reality, and finally entities which could be modeled and visualized. All Jewish science was an inferior science; all Aryan science, superior. Lenard's call was, in essence, a call for a return to Newtonianism masquerading itself under the horrors of ethnic cleansing.

Labelled as rebellious, Einstein criticized Nazi policies and actions. Einstein's God was the god of Spinoza—non human centered, subtle, and complex—leaving him in awe of the sublime. His god was most certainly not the God of the Old Testament, whose focus of attention on man was framed within a relationship of fear. Einstein points out that only rebels advanced science, citing examples such as Giordano Bruno, Alexander von Humboldt, and Voltaire. Only unconscionable self-restraint is what had allowed Hitler to obtain power. He observes, "You can see just where such self-restrain leads… Does not a lack of sense of responsibility lie behind this?"

In contrast to World War I, when scientists still had some degree of interaction with the rest of the world, the Nazi restricted the participation of all

255

scientists from international conferences organized or related to the League of Nations. The League had imposed harsh economic sanctions, whose dire consequences were first warned about by the economist John Maynard Keynes. Sanctions could force Germany into a second war—a warning that went unheeded. Scientists were eventually even forbidden from attending the Nobel Prize ceremonies in Sweden. Their many awards would be given in absentia.

The initial scientific impact of the rise of Nazi was the dismissal of Jewish scientists. Twenty-five percent of the physics community were either fired or forced to resigned—a figure much more significant than might appear at first glance. The total number of German physicists declined from 175 (1931) to 157 (1938). When in 1933 John von Neumann visits Germany, he is shocked at the decline of physics in his country of origin. Typically, the most original and talented were exiled, resulting in irreplaceable losses. These had not just been technicians repeating experiments by rote instruction, but rather original thinkers whose work revolutionized many fields. Many of the founders or important contributors to quantum theory were lost in this way: Max Born, James Franck, Walter Heitler, Heinrich Kuhn, Kothar Nordeim, Eugene Rabinowich, Hertha Sponer, and Edward Teller.

The Nazi policy specifically affected theoretical physics given the particular cultural phenomenon of Jewish prominence in that field. Of the 60 theoretical physics in Germany, 26 went into exile: Einstein, Franck, Hertz, Schrödinger, Hess, and Peter Debye; eight of these would be future Nobel laureates. Forced exile took German physicists all over the world, but at first these tended to migrate to nearby countries, believing that Nazism would have no enduring impact. Many were well received in England, but England was weary about the flood of refugees and clearly defined itself as temporary stepping point between Germany and the United States.

Even physicists who had been global scientific leaders were unsure of their employment. Max Born applied to universities all over globe, until finally receiving tenure at the University of Edinburgh. Others were less fortunate, ending up in Istanbul where they had to teach in the Arabic language. The reason for the half-hearted British policy was to be expected: they did not want to hurt new graduates. The flooding of German refugees to the British academia would have closed all employment opportunities for its new scientists, leading to their formal displacement by the home state. This, England could not do.

The refugees were fortunate in that physics in the United States at the time was undergoing a period of rapid expansion, and its departments more than happy to receive renown German physicists. Hans Bethe, for example, had applied quantum mechanics to analyze star formation, whereby new elements are created. He arrives at Cornell University, and is fascinated by his new circumstances. German academia had been too formally hierarchical, and its student merely assisted lecture and were not expected to raise questions. By contrast, Bethe was amazed that he was often interrupted in middle of lecture at Cornell —leading to cherished exchanges. The massive exile truly became a meeting of cultures.

256

It was noted by German scientists, for example, that United States academia tended not to be weary of commercializing their innovations, became public figures leading the popularization of science (something which was not seen in Europe), held close relationships to the press, and tended to believe in 'the bigger, the better' technologies. They were instrumentally focused, tended to work long hours and weekends, and finally were 'a-philosophical'—which shocked Fritz London who ends up at Duke University. For him, United States culture lacked enthusiasm for any endeavor other than bridge and football.

In spite of the exodus of German physicists, there was oddly no net loss in German's physics community. Many new institutional positions were created that made up for academic losses. The case of fascist physics in Italy was even more In spite of its promising start with the work of Galileo, by the nineteenth century Italy had become very backward scientifically; at the turn of the twentieth century, it was in bad shape. Under these circumstances, one might naturally assume that fascism would have completely destroyed physics in that nation. Strangely enough, the opposite occurred. Italy physics briefly benefits from totalitarism, as the case of Enrico Fermi shows.

Although apolitical, Fermi had a political godfather who helps him excel in his home country. However, Hitler's increasing co-optation of Mussolini leads Fermi to begin publishing in English, hence widening the audience that would have otherwise not read his work or become aware of his significant contributions. The situation worsens for Fermi in 1939 as Italy becomes junior partner in its relationship with Germany; Fermi's wife was Jewish. A plan is arranged and when he goes to receive Nobel Prize in 1932, he is able to obtain asylum in the United States, followed by Bruno Rossi and Emilio Ségre using alternative routes. Ségre migrates via Copenhagen.

For these and many other cases, Bohr truly becomes the grandfather of theoretical physics by saving many young European scientist, in the process putting his own life in danger.

Psychiatry and human experimentation

It would be a mistake to presume too great a differentiation between 'science' and the 'state' in Nazi Germany, however, as there was ample participation by many 'scientists' in the rise of Hitler. This is particularly the case with regard to the physicians and psychiatrists of the German scientific community. Their influence was much more noxious and horrific than one might imagine.

German medicine during the interwar period was typified by the "racial hygiene" movement. The notion is somewhat akin to medical sanitation, but applied to the 'cleansing' of race: the removal of invalids suffering from mental or physical handicaps who had become undue burdens to the state. Tragically, it was a grotesque philosophy internalized even by family members who felt ashamed of having an invalid in their homes. The policy led to the creation of 20 university institutes for the study racial hygiene with ample funding before the Nazi ever came to power. Fifteen journals in the field were established, one which ran a

257

long standing column proposing solutions to the 'Jewish problem.' Strangely enough, its focus on health and salubrity meant that Adolf Hitler himself neither smoked tobacco or drank alcohol, and would not allow anyone else to smoke or drink in his presence—even if he did imbibe much harsher drugs in private.

The movement contributed greatly to the acceptance of Hitler and to his grotesque policies. A prominent racial hygienist, Fritz Lenz, early on praised Hitler in 1930 for his socially beneficial declarations on 'racial hygiene.' German medical practice had been going through an economic crisis at the time; the Great Depression in Germany was ten times worse than that in the United States due to its war reparations. Physicians graduating from medical school were unable to find any work, and quickly began joining the National Socialist party, specifically its "Physician's League." By 1933, some 3,000 doctors had joined—a figure which increased to 38,000 by 1942. Nazism in fact greatly improved the financial circumstances of the German medical community, as attested by the drastic overall salary increase. In 1926 the average physician earned 6,000 RM less than the average lawyer, but by end of war, German doctors were making 2,000 RM more.

A series of racial hygienist measures were passed which had boosted the total number of remunerated positions for physicians, as these were required to administer all such related facilities. For example, the Sterilization Law of July 1933, requiring the sterilization of 'diseased' offspring, led to the formation of 181 Genetic Health Courts where 10-15% of the entire German population became its victim. All were run by doctors. After the creation of the journal *The Genetic Doctor*, 180 PhD theses on sterilization were written.

When the Nuremberg laws of 1935 were passed, making it illegal to marry or have intercourse with a person of Jewish descent, only physicians were endowed with the power to make such determinations. Germany's infamous mass genocide actually began in local German hospitals.

German euthanasia was stimulated by its US counterpart. The Nobel Prize winner Alexis Carrel argued for the 'humane' gassing of invalids. In 1941, the *Journal of the American Psychiatric Association* called for killing of retarded children. The assertive eugenics movement in the United States had a backlash, as the Germans feared United States would become "racially superior" by more successfully eliminating all 'inferiors'. Tragically, German parents routinely took their own invalid children to hospitals to be culled.

Warnings about the dangerous noxiousness of this policy were made as early as 1931 by the psychiatrist Oswald Burnke. Burnke argued that it was wrong to take only 'scientific' and 'economic' factors into consideration by the German medical community. Would the state then kill all widows or injured soldiers, to whom so much was owed? He warned that these views would be generalized into broader definitions of 'Aryan' and 'non Aryan,' as in fact happened.

This is a problem which concerns…the entire future of humanity. One cannot approach this problem either from the point of view of our present scientific opinion nor from the point of view of the still more ephemeral economic crises. [Such a policy would lead to] a

258

quite monstrous logical conclusion: we would then have to put to death not only the mentally sick and the psychopathic personalities but all the crippled, including the disabled veterans, all old maids who do not work, all widows whose children have completed their educations….then pretty soon we will no longer hear about the mentally sick but, instead, about Aryans and non-Aryans, about the blonde Germanic race and about inferior people with round skulls.

Burnke became one of the very few men who warned of the impending tide and its implications for his culture and nation. His warning obviously went unheeded.

The lack of regard for human life reached a point that 70,000 medical patients were killed in Germany so as to make room for wounded German soldiers after the invasion of Poland. The first formal gassings began in January 1940 at the Brandenburg Hospital under the charge of Dr. Victor Barck. The death of psychiatric patients by these means occurred at such a large scale, that it led to concerns within the very psychiatric community as to its future. If all psychiatric patients were killed, psychiatrists would be left without any clients or income; furthermore, the reduced demand also meant that few students would enter the field. In 1941 the Hadamar Psychiatric Institution celebrated the gassing of its 10,000th patient.

In the fall of 1941, once the bulk of its medical euthanasia had been completed, the killing machine apparatus spread throughout hospitals and psychiatric institutions in Germany were transferred to three main industrial parks: Auschwitz, Treblinka, and Majdaneck.

Physicians and psychiatrists provided the means and rationale for Nazi genocide.

The close coordination between the medical community and Nazi public policy occurred even within the concentration camps. The genocides per se begin in 'worker' encampments (Auschwitz), whose killings were initially justified as a means to control the spread of infectious diseases which would otherwise eliminate the involuntary labor force. The spread of typhus could wipe out entire blocks of workers, hurting productivity. While genocide thus began as a 'logical' choice, its rational and method became more widely generalized throughout Germany, as Burnke had warned.

The killings initially occurred on a person to person basis, with the injection of phenol (gasoline). However, the physicians, who pretended to be undertaking routine medical examinations, noted that when the phenol was injected into the blood, the patient's death would occur in such a slow and painful manner, that their shouts of agony and cries for help would warn and alert the rest of the block. Phenol injections had been particularly common with children, who in spite of the pretenses, were well aware of their impending demise. One child, before dying, turns to the doctor and cries out, "Why are you killing me?!," puzzled by the actions of a man who was supposed to be curing him.

A more efficient manner of murder was then discovered and implemented. By injecting the phenol directly into the heart, death would result only within 15 seconds. The bodies, even if still partially alive, were thrown into a pit with the rest of the other corpses. In this grotesque manner, German physicists raised the rate at which they were able to 'treat' patients, to 60-200 assassinations per day. Furthermore, the abundant use of phenol could be 'hidden' among regular solicitations for other medications, and hence evade any unwarranted scrutiny.

Not only did the actions of the German medical community constitute a gross violation of the Hippocratic oath, concentration camp doctors also began using humans as experimental guinea pigs in some of the most horrific cases of torture. Patients were submitted to drastic changes in air pressure chambers, have their genitalia treated with x-rays, placed for hours on end in pools of water at freezing temperatures, shot with infectious bacteria, or have mustard gas inserted into inflicted wounds. Although claiming such research was conducted for medical purposes, few patients ever survived the torturous ordeals. One of the most notorious physicians of all was Joseph Mengele.

Mengele is infamously known for his twin study experiments. Receiving the statistically rare samples of twins from all over German concentration camps, Mengele would inject one twin with a poison or bacteria, using the other twin as a control. When one inevitably died, the other would also be killed so as to perform comparative anatomical analysis at the Kaiser Wilhelm Institute. The prestigious scientific organization, formerly led by Planck prior to the rise Hitler, had been transformed into the instrument of genocide.

Discovery of the horrific human experimentation led to the formation of the Nuremberg Code after the war, and the history of medical ethics can be sharply divided by it. Twenty-five doctors were prosecuted at the Nuremberg's "Doctor's trial," of whom seven were sentenced to death—including Hitler's own personal physician Karl Brandt who so closely coordinated the wartime medical community. Yet, these were only the tip of the iceberg. Many physicians of the Third Reich fled to Latin America, and even became influential within Germany. The unethical attitude underlying such treatments were pervasive yet could not be so easily eradicated.

Yet the most important aim of the trial was to simply expose the rationalized savagery that had occurred, as killing its medical perpetrators would never alleviate or remedy their unspeakable atrocities. Hundreds of thousands of nameless victims had been killed by physicians and psychiatrists, all of whom had sworn to the Hippocratic Oath at the beginning of their medical careers.

The German atomic bomb project

Germany also set about on a path in the atomic arms race, which began as soon as Hahn and Meitner showed that 200 Mev of energy could be obtained from a miniscule uranium sample in December 1938. If Germany had succeeded in first creating an atomic bomb, the implications for twentieth century history were enormous. A victory by Germany of World War II would obviously have led to a radically different world from what we live in today, eliminating all notions of civil rights, establishing a vast number of concentration camps, and so forth. Although this horrific alternative scenario seems be too far-fetched to seriously consider, as it has been portrayed in Quentin Tarantino movies, it was an outcome closer to reality than one might care to imagine.

Quantum mechanics had been a German discovery, and there was no reason to suppose that an atomic bomb was beyond their capacity to achieve. However, due to a long list of contingent and accidental reasons, Germany actually never got past the early stages in its development of the atomic bomb. When compared to the United States effort, one can easily observe in retrospect how backwards the rival effort had been. Hitler never understood the significance of atomic bomb, and likely still associated it with Einstein's "Jewish science." Control of the atomic bomb effort was given to Albert Speer, who was far more sensitive to it's impact than his party's leader. Speer tried to get the party interested, and in 1941 holds a series of conferences by distinguished German physicists who stayed in Germany. The most important of these had been Werner Heisenberg, who despite his young age was one of the founding fathers of quantum mechanics.

Heisenberg's participation in that earlier revolution led to accusations of being a "white Jew" for supporting Jewish physicists. He is even denied a university position in Munich, and is thus forced to turn to Heinrich Himmler, the infamous leader of the German secret police SS, for help. Otto Hahn also helps explain to the German Nazi leadership the bomb's possibilities. Max von Laue, founder of crystallography, and others were also involved, even if their role had been of lesser importance. Yet Hitler did not understand that modern wars were 'industrial wars,' and foolishly believed that Germany could win the war by "blitzkrieg": fast rounds of attacks that would, in theory, quickly overwhelm his opponents. Although initially successful by conquering substantial parts of Europe, the limitations of his blitzkrieg policy were soon exposed when confronted by the Soviets. As Napoleon, Hitler ignored the Russian winter and sent his enormous army to their doom.

By the end of 1941, the attitude towards military alternatives change given the failure of Hitler's strategies. Nazi leaders become more receptive to the atomic effort, and Speer actually asks Heisenberg how much money would be required for its construction. When Heisenberg recommends a sum of some 100,000 Reichmarks, Speer offers two million RM instead. However, as the war drags on, the increasingly precarious financial circumstances of German state places a strain on its projects, leading Speer to withdraw the offer, returning to the originally suggested 100,000 RM figure.

We can obtain a clear impression of the limitations in the German atomic bomb effort when we consider that the United States effort by 1944 had taken up some $ 1.8 B, or roughly twice (4.5 RM) Speer's high offer (2 RM). In other words, the German atomic bomb investment had only been 0.0022% of what was actually required to create an atomic bomb: the making of U235 and U239.

Yet the Allies were fortunate as the range of factors inhibiting the German atomic bomb effort were not in short supply: intercene institutional quarrels, limitations in the expertise required for its construction, and the loss of important component materials required to begin preparing the fissile material in the first place. Control of the effort was also removed from Heisenberg early in the war,

Germany's most important physicists, and handed over to a second rate bureaucrat. Yet, in spite of his abilities, Heisenberg did not fully understand all of the bomb's complexities—specifically the notion of 'critical mass.' While under arrest by the British at Farm Hall, their secretly recorded conversations revealed Heisenberg constantly wavered as to the minimal amount of fissile material required. The successful sale of the entire deuterium stock produced at the Vermok hydroelectric damn to the French certainly helped delay the German effort.

The Allies, however, did gravely underestimate one critical aspect of German military strength. More so than the atomic bomb effort, chemical weapons existed that, if used, would have radically altered the outcome of the war.

But, due to the most fortunate of circumstances, these remained under lock and key. Massive wartime atrocities in the United States were only averted by the most circumspect of reasons.

IG Farben and ZyklonB

To understand role of chemistry during World War II one needs to first turn to the story of Carl Bosch and IG Farben.

The origins of Germany's chemical industry, the principal source of her economic might and political power, actually begin in England. The chemist Karl Perkins accidentally discovers that one could synthetically create mauve, previously derived from organic material as plants and insects as all natural dies had been. The color purple became all the rage in the UK, and even the royal wedding's party were clothed in its color. Yet, in spite of the enormous success of the new organic chemistry, British scientific arrogance neglected to offer systematic support of this science. Its overly individualistic Newtonian legacy presumed that all other sciences also had to operate in such a manner. Hence this arrogance allowed the Germans to usher a new chemical empire, on the basis of a cloth's colors, while Britain lingered behind.

Becoming aware of the importance of systematic research and development, the German government passes a patent law in 1876 that greatly stimulates an already burgeoning industry. Patents would now be issued with regard to chemical process rather than its resulting color, allowing for the rapid growth of a multiplicity of companies producing the same colors, greatly increasing its scope and size. Chemical companies that still exist today were born under those circumstance: Bayer, BASF, Afga, Hoeschst, Cassells and Kalle.

The chemist Fritz Haber invents ammonia, what might be referred to as 'synthetic nitrogen,' ushering new industries around the production of fertilizers and explosives. The required organizational nature of research is aptly demonstrated in Haber's work: thousands of catalysts were tested before the right one, osmium, was identified. His procedure is then scaled up to industrial levels by Carl Bosh, resulting in the now famous Haber-Bosch process.

However, the required investment to discover the process had been very costly, and Bosch at BASF was routinely attacked for his efforts. His eventual success is greatly due to Heinrich von Brunck, the very founder of the company, who sympathized with Bosch's dilemma. The original die of indigo was believed to have been impossible to reproduce synthetically given the huge amount of resources required. When von Brunck succeed, synthetic indigo becomes a huge cash cow for the company and he ascends to its leadership.

The importance organic chemistry by World War II can also be noticed when Haber invents the highly toxic mustard gas, the slightest whiff of which could result in an extended hospital stay. It marks a tragic turning point, both socially and individually. When Haber's wife finds out about the invention, she issues an ultimatum to her husband, and eventually commits suicide in protest. As the Allies had blocked naval trade between Germany and the rest of the world, Haber's inventions became essential to its survival and success; World War I is referred to as the 'chemists war' for its role.

During the interim war period, the six most important German chemical companies combine to form the gargantuan I.G. Farben. Although its existence had been preceded by a similarly named cartel, the company came onto its own in 1925, headed by Carl Bosch.

When the Germans suffered another similar blockade in World War II, Bosch was placed in a predicament similar to the one he had confronted as a younger scientist. Whereas Germany had abundant resources of coal, she lacked petroleum, thus suffered a 'hit to the gut' by the blockade. Bosch, however, realized he could synthetically produce petroleum from coal on a one to one ration using the Bergius hydrogenization process; previous methods used two quantities of coal for every barrel of oil produced, and were half as efficient. It was a problem 'made for' Bosch, but, again, the financial burden made its research a logistical political nightmare.

Bosch would not necessarily receive assistance from the Nazi state, particularly so as Bosch and Hitler detested one another. At a dinner Bosch, as Max Planck had done, warned Hitler of the enormous damage he was making to German science by his dismissal of Jewish scientist, to which Hitler again leaves in a rage. German bureaucrats were told to make sure never to have the two men share a table. The way in which Bosch solves his financial dilemma is, to put it bluntly, by scheming Standard Oil. A deal was reached wherein Standard issued Farben 5% of its stock, with the promise of the Buna rubber and synthetic oil patents. Of the 2,000 patents Standard received, not one of them held it, and all were written in such a way as to be scientifically undecipherable. Upon hearing about the sleight of hand, Standard's board of directors immediately fires its CEO.

During the war, IG Farben developed various manufacturing facilities for synthetics, one of which is Auschwitz where Buna rubber and synthetic oil are produced. However, Leuna becomes the principal manufacturing site of synthetic oil, which Hitler correctly perceives as a key to Nazi survival and success. The end of petroleum had been predicted, globally projected to last only 6 more years upon the dwindling global reserves. The consequent discovery of Texas oil fields made the issue a moot one, but in light of the Allied blockade, Germany was inevitably forced to develop its synthetic alternative. Aware of its strategic

263

importance, the Allies repeatedly bomb the Leuna manufacturing site. In spite of the fact that 75% of its facilities are destroyed, its importance leads the Nazi to move 35,000 men to restore the site in only ten days.

Ultimately, I. G. Farben's link to the Nazi regime was of enormous importance to both German parties. Even when it can be pointed out that Bosch dies early in the war, their mutually beneficial relationship could not be so casually abandoned. A study issued by the Eisenhower Commission after the war revealed that Hitler's continued 'success' would have been impossible without the complicit assistance of the company. Worst yet, the company had in its stores the chemical ZyklonB.

The chemical had been used primarily as a pesticide prior to the war. As its deadly effects were hidden in an odorless and colorless gas, chemical additives were added for immediate detection. However, upon its adoption by the Nazi military, these important sensorial signals were removed, and ZyklonB is turned into a critical ingredient widely used in the gassing of Jews. German labor prisoners were also given what was referred to as the 'Buna diet,' treated as mineral ores whose life was squeezed out of them before being discarded.

Eisenhower insisted on breaking up IG Farben into 47 different companies, a proposal which gets lost during the post-World War II political wrangling.

The most surprising finding of the commission, however, had been the Nazi plan to use German toxic chemicals to widely bomb United States cities. Aside from ZyklonB, the Germans had also developed powerful nerve agents as tabun and sarin, which could kill a person in a few minutes. Martin Boermann, Joseph Goebbels and Robert Ley all propose to Hitler the large scale bombing of major United States cities with these agents: Washington DC, New York City, Philadelphia and so forth. If used, it would have resulted on a scale of military mortality never before seen in US history.

The plan, however, was strongly opposed by Ambrose and Speer, who make four important observations. The patents to these agents were widely known, and in contrast to Germany, the United States had ample petroleum reserves from which to retaliate on a far larger scale should it choose to do so. Essentially, Ambrose and Speer warn Hitler that the United States had the capability to kill every German; knocking out American cities would be a very bad idea.

Hitler was infuriated and, as usual stormed, out of the room—but not for learning of the United States retaliation but rather because I. G. Farben's monopoly of German oil ultimately limited the total quantity being producing. Hitler had come across the textbook case of monopoly power in that I. G. Farben simply had little incentive to maximize production as it so totally dominated the product and charged a fix profit margin in spite of dire wartime necessities.

If the scenario were not bad enough, Hitler then decides to place the SS in change of the German atomic bomb program. Speer, who was not a stupid man, immediately saw the horrific implications such a policy held. The SS were essentially high school brutes with limited cultural background and foresight; their control would have most certainly resulted in the feared worst case scenario. Speer talks to Hitler and convinces him to place the SS only in charge of testing the bomb rather than its production. Speer at one point even begins to plan Hitler's assassination by placing ZyklonB in his bunker. However, the war's end was nearing, and he desists in his plan; its logistics were too complicated and risky. The war would end three months later.

What is surprising about the case is that Ambrose had overestimated the United States ability to produce such agents. Again, the United States patent system had been too ambiguously established, and at the time the United States had not yet successfully deciphered their chemistry.

It was only by most fortunate of incidents, that the United States did not suffer far greater casualties on its own territory than it did during World War II.

Hanford

The failure of the German atomic bomb effort can be better understood by placing it in contrast to that of the United States. By the beginning of World War II, the conception of an atomic bomb by 1939 was not an original idea, as we have seen. The principal difficulty rested mainly in the practical issues, the first of which was in obtaining enough fissionable material, U235 or Plutonium (U239). Another key problem was the trigger mechanism. How does one initiate a bomb with the power of a thousand suns, without first being extinguished in the process or exploding prematurely? Inversely, if the bomb did not have enough fissile material exploding in the correct manner, it would be a dud. Thus, the atomic bomb was not a scientific problem strictly speaking but rather an engineering one; all the parts needed to be put together perfectly in order to work.

Yet these were not trivial difficulties. By 1944, the United States had invested millions of dollars with nothing to show for; $100 M per month were being poured into an effort that seemed to be a never ending barrel. Should the government scrape the project or continue pouring money with the hope that someday it would be completed? The imagined fear of a German menace continued to have its sustained effect on the participants. When the United States Congress began asking questions as to how money was being spent, no answers were provided. Only a few were aware of the tightly kept secret under Franklin Delano Roosevelt's executive control. Even Vice President Harry Truman had not been informed.

Its scale was certainly unprecedented, orders of magnitude greater than that of Germany's effort. The Manhattan Project constituted an enormous coordinated enterprise, involving city-wide factory installations and revolutionary new technologies. By contrast, each German group had fought with each other, and divided resources which if combined would have had a greater effect.

The United States effort can thus be divided into two principal groups: those responsible for creating enough fissionable material and those in charge of preparing a functioning atomic trigger mechanism. The vast industrial site of Hanford in the state of Washington was critical, but had not been the only one in existence. The hydroelectric plants of the Tennessee Valley River Authority (TVA) fed into Oakridge's facilities on a massive scale.

Major General Leslie Groves was placed in charge of the entire project, as overall organizer and administrator. Groves faithfully supported it, obtaining funding even when it all looked hopeless. Many brilliant men would come to collaborate in Los Alamos, New Mexico, headed by the young Robert Oppenheimer who had previously studied in Germany. In the halls of Washington DC, Vannevar Bush and J. D. Conant served as key advisors to the president, keeping him informed of issues and needs.

Various techniques were used to produce the fissionable material at the same time, each with their respective difficulties and benefits: electromagnetic separation, gaseous and thermal diffusion. The second group was located principally at Los Alamos.

Both groups had enormous facilities at their disposal. Los Alamos in New Mexico became a city onto itself, even though built from scratch in the desert. Some 5,000 scientists and engineers worked and lived in facilities that had not been designed to last more than a year. Their average age of 29 meant that most were young, and would easily cohere as a group, forming friendships and romances that would last a lifetime. It was an honor to had been selected, and the experience as a whole was a unique one; constantly debating, interacting, and resolving problems even during hiking excursions into the nearby desert hills. Richard Feynman provides many good anecdotes of his experiences in *Surely You're Joking Mr. Feynman*, appearing some fort years after the event. Most of the issues to be tackled, however, were engineering nightmares.

One could create plutonium in nuclear reactors, but upon neutron bombardment, U235 would not only turn to U239, but would continue to 'mutate' into other elements not suitable for the bomb. EO Lawrence's cyclotron was a powerful machine which, through isotope separation, was able to slowly but surely obtain the much needed material. Its inner walls, however, quickly became filthy with residue, so the cyclotrons had to be routinely turned off, cleaned, and restored; only 0.7% of the material would be separated out by it. The process of gaseous diffusion was much more efficient, but it was beset by the hexachlorine, which kept corroding away its membranes. To properly function, these had to be built from nickel, a costly material. By January 1945, only 200 grams of weapons grade material had been produced, and the war was already in its last days.

The design of the detonation device for the bomb had been equally quirky. Seth Neddermeyer had proposed the use of implosion as a trigger mechanism, which was an original idea but could not personally get it to work. How does one detonate a perfect sphere, perfectly? When explosives were placed around cylindrical pipes, these would only get horribly mangled. The situation seemed hopeless until the Russian émigré George Kistakowsky developed explosive lensing using two different detonators which burned at different rates of contraction.

"Fat man" and "Little boy" were born out pride, fear, and desperation. It would have been difficult for any scientist under those circumstances to have

266

simply walked away from such a momentous project—even if the underlying presumptions of German scientific capabilities were wholly mistaken.

When the captured German scientist at Farm Hall were first informed about the explosions at Nagasaki and Hiroshima, they believed the news to be merely a propaganda trick. When they were eventually made aware of its veracity, some had nervous breakdowns and others took to the bottle. Heisenberg, who had betrayed his closest scientific companions, claimed that to have subconsciously foiled the German effort.

Big Science during the Cold War

Change of Scale, Change in Culture

CONTEMPORARY SCIENCE IS OFTEN REFERRED TO AS 'BIG SCIENCE' given the enormity of its scale, funding, technology, and organization. An example of this might be the recent discovery of Higgs boson or "God particle" giving mass to all objects in the universe, a remarkable finding done at CERN in Europe. The United States was going to build an even larger supercollider in the Texas during the 1990s but its cost was such that Texan legislators, of a seasoned conservative nature, found it hard to support. In spite of the backward attitude commonly seen in the state, which recently opposed the teaching of evolution, there was a legitimate financial concern. How much money should we put into science?

The Texas Supercollider aptly showed that there were limits even to the scientific benefits resulting from such a gargantuan project. Wealth is not an bottomless pit' to be wantonly used, even by developed nations. The scientific benefits were not clear, and some suggested that it would ultimately be more productive simply to astronomically study stars which made for natural experiments at a much reasonable, and reduced, price tag.

Were the conservative Texan representatives correct in their financial concerns over investments in science?

268

CERN and Hubble

The *Conseil Européen pour la Recherche Nucléaire* (CERN) is a $9 B a year facility with a permanent staff of 2,513 staff serving 12,313 fellows, visitors and engineers. Funded by twelve European states, the sole cost of the accelerator itself was half its annual funding ($4.4 B). To place its finances in perspective, its annual budget equals the entire annual budget of the government of Puerto Rico (2015). The Large Haydron Collider (LHC) itself consists of a 27 km tunnel circumference, using 8 teraelectronvolts (8 TeV) which must be kept at subzero temperatures using 120 tons of liquid helium. It is the world's most precise watch.

The LHC generates an enormous amount of data. Its 600 million collisions per second produce more bits than its computers can actually handle; only 30 petabytes per year is actually stored. So much data is generated that 90% of it is discarded—in spite of its enormous computing facilities. It is here where the internet (the world wide web) was born, first created to share data by Tim Berners Lee in 1989 on the earliest 'Mac OSX computer' (NeXT). Even then, CERN's sole purpose is only to store data, and contains only 20% of the project's total computing capacity. It is a computer geek's paradise, with 11,000 servers using 120,000 cores processing one thousand terabytes per day. Its fiber optic connections distribute the information between 170 computing facilities in 36 nations, also making it the world's largest distributed computing grid.

Big science is intimately tied to specific fields as physics, and one cannot help but observe the paradox of how the search for the ever smaller results in ever larger organizations and technologies. A key factor to its expansion is the role of technology, which in turn allows for the expansion of 'human senses' far beyond their natural reach. The increasing technological sophistication, in turn, as other consequences, such as the greater need for funding and ever increasing institutional complexity.

A whole host of specialists are required to construct and maintain its apparatus working. We no longer have the case of a man in his office performing experiments, as Newton with light or Young's two split experiment. Huge teams are required merely to turn the button 'on' or to identify problems and errors which inevitably arise. CERN is an enormous cyclotron that cannot fit on the palm of one's hand, as E. O. Lawrence's first model could.

Yet it is not just the field of physics where the dynamic of big science can be observed; astronomy is also afflicted by it as well. When the Hubble telescope was finally completed, it had been one of the costliest scientific instruments ever constructed. NASA had spent more than $1.17 B before launch, and its final bill in 2010 would amount to some ten times that sum ($10 B).

The sums involved in a single scientific instrument are so large, that we quickly lose a sense of their scope and scale. Again, we have to point out that the same amount of money that was spent by Puerto Rico on its entire annual budget for 2015 is smaller than the total amount of money that was required to build a single scientific instrument. The scale boggles the mind when seen from Puerto Rico's perspective.

Scientometrics

The first to formally analyze the scale of modern science was the physicist Derek de Sola Price in his book *Little Science, Big Science* (1963). What was required to understand this new social phenomenon was in itself a quantitative study of science referred to as 'scientometrics,' applying science back onto itself. There are more practicing scientists today than have ever existed before (87%). When the growth of the number of scientists was plotted, it revealed an exponential curve, doubling itself every 15 years for average science, while taking 20 years for high value science. As a whole, science has grown five orders of magnitude in three centuries. It was not an insignificant growth rate, outpacing ordinary population growth rates.

What is unique about the historical growth of science, at least where compared to other parallel phenomena, is its persistent and continued character. High growth rates are usually short lived. Such continued growth over such a long 300-year time period is simply unusual, and reveals something about the character of modern science: nearly all of science is the result of the most recent generation of scientists. While most great men in history are dead, most of all great scientists are now alive. Whereas Newton had nearly co-existed with many participants of the Scientific Revolution, 'standing on the shoulder of giants,' progress in Greek philosophy had occurred by commenting on authors centuries long ago deceased. The character of science has drastically changed over the ages.

Its rapid growth rate helps account for the nostalgia of early 20th century scientist, as Robert Millikan (1868-1953). Studying in Germany during the quantum revolution, Millikan was aware first hand of the debates in his field. His ingenious oil drop experiment revealed for the first time the mass of an electron, for which he receives the Noble Prize. Although offered an enormous sum of Rockeller money to begin a large institute, Millikan rejects it believing it was better for many small centers to coexist than a single monumental one.

The natural implication of de Sola Price's work is naturally that much of scientific history can be characterized as 'little science,' as seen in the mythical figures of science as Newton or Einstein. The most important feature in this context has been to ask the right questions. Certainly, there is more 'big science' in the past than typically recognized: Napoleon's invasion of Egypt or Davy's use of large batteries in chemical studies. However, what primarily differentiates the two is not so much their scale, as the changes in character which such increases in size implied. Would scientists strike Faustian bargains?

Since the middle of the twentieth century, concerns regarding the impact of scale on scientific activity have been raised. To what degree did the values and principles of science change under the new circumstances? The cases so far mentioned have been positive ones, as CERN's LHC or the Hubble telescope. There are some exceptions to the rule, however, and one of the first to raise such concerns was Alvin Wienberg (1961), one of earliest scholars to use the term 'big science.'

It is recognized that the largest monuments reveal values of their respective societies, as these place resources where their deepest aspirations lie: The Pyramids of Ancient Egypt (gods), the medieval Cathedral of Notre Dame (Catholicism) or the modern skyscrapers of New York City (capitalism). Weinberg as a physicist pushed for salt reactors at Oak Ridge National Laboratory, given these were safer to use but not as powerful. Their institutional rejection led him to a reassessment of his field.

One of his concerns was the issue of 'moneytitis,' where the focus of scientific activity is diverted to spending money on the latest technology rather than asking the best questions to answer. Rather than assisting scientific research, technology was becoming an end in of itself, a distraction from what should be its secondary role.

Weinberg also observed that there was a change in the culture of the scientist, who now became public figures. This change itself required a different set of personality traits from the practitioners of 'little science.' No longer was the scientist just to focus with the topic at hand, but he had to become a public relations specialist concerned with the public perception of science.

The large scale of the scientific enterprise also had a number of problematic institutional implications. Who controls science? Is it genius or technology that dictates the guidelines and paths which science should follow? The need for ever greater funding also implied greater participation in the political process, which in turn led the scientist to become a political figure of sorts. As science became politicized, the entry of nonscientific criteria in project evaluations also emerged.

Large scale scientific projects also tended to be centralized, whose requisite hierarchy implied a great deal of power centered on its administration. A historical example of this dynamic might be Tycho Brahe who as its lord established the goals and methods of the organization for all others to follow. Centralization by nature implied a lesser degree of autonomy for each participating individual scientists, who merely became 'cogs in the wheel,' unable to determine the exact nature and purpose of their work. The relation was typical in the German chemical industry of the nineteenth century, which became the first true large scale research and development enterprises.

It is certainly the case that large scale of modern scientific activity need not necessarily imply centralization, as demonstrated by the large team of collaborators in the atmospheric sciences. These instead reflect 'network' dynamics, where each group and institution operate on their own while at the same time contributing to the overall research agenda. Examples of this phenomena abound in climate change studies of climatology and oceanography (NOAA).

Perhaps the greatest area of concern, however, was over the impact of the 'industrial mentality' on the culture and practice of science. The rise of the industrial 'research and development' laboratory has been a substantive change in science of the twentieth century. In contrast to traditional scientific activity, the promotion of R&D by corporations is designated for particular economic

271

interests rather than for the production of common knowledge broadly shared. The capitalist dynamic naturally is inhibitory of public diffusion of new knowledge, given the corporate concern of competition—in contrast to the philosophy behind Diderot's *Encyclopedie* which sought to 'universalize' the particular knowledge of private craftsmen.

A corporation's implicit privatization of knowledge thus inherently clashes directly with the fundamental open character and ethics of science, referred to as the 'open republic of science' and aptly described by Richard Rhodes. For example, a key trait of science are decisions made by mutual consensus. That scientists abide by the free flow of information as keys to solving scientific problems, also requires abundant public discourse and criticism. Academic training imbues practitioners with these values.

By contrast, all important corporate information is patented, and restricted from broader use and availability. Profit is held as a value in the social hierarchy above the quest for 'truth,' as demonstrated in the tendency of fraud in corporate research. False clinical trials routinely appear in the pharmaceutical industry, whose purpose is to push drugs on which billions of dollars will be made. The FDA rather than critically scrutinizing these procedures often falls prey to them as their officials revolve in and out of industry and government.

The rise of big science is a twentieth century phenomenon. Global RD spending is now at the $1.4 trillion level, of which the United States comprises 32.2% of the total. In 2011, $425 billion dollars (2011) were spent on science, where only $50 B had been spent in 1980; at an adjusted rate, it grew threefold during the period. Most RD funding in the United States actually goes to the life sciences. As a percent of GDP, United States figures are not as favorable, but the continued size of the United States economy retains its scientific investment above the rest—for now. While the United States placed 2.8% of its GDP (2011), Europe spent 2% and Japan 3.5%. Its overall scientific labor force increased from 1 million in 1960 to 5.8 billion in 2011. While we see a positive increase in the rate of Hispanic science BA degrees, from 7% to 9.5% (2000-2011), it is a figure that could be improved.

One of the most remarkable recent trends in RD statistics is the rise of China. During the same period, its scientific GDP% grew from 0.6 to 1.7%, BA science degrees from 60,000 to 310,000, and it witness the largest growth in article publication, from 5,000 to 100,000 per year. Tragically, it has also been characterized by a great deal of fraud, suggesting that its community has not fully internalized science's core value system.

The same could be said for the science in the Soviet Union during the Cold War.

Trofim Lysenko

The case of Trofim Lysenko (1898-1976) is one of the most bizarre and unusual cases ever to have occurred in the history of science. It consists of a hardly educated 'scientist' who obtains control over the whole of Soviet biology during a period of 30 years (1935-1964), tragically resulting in the demise of one of the most promising forefronts of Russians science: genetics. Lysenko accuses genetics of being a "bourgeoisie science," not unlike the Nazi claim that theoretical physics was a "Jewish science. With the support of Joseph Stalin, Lysenko ruthlessly crushes its practitioners with the soviet gulags and the Soviet secret police. It is a case hard to believe, so exceptional that it should be a requirement of every history of science course.

It constitutes a worst case scenario of the noxious results that occur when politics becomes involved in the internal dynamics of scientific practice.

A key to understanding the rise of Lysenkoism is to view its context. The post Bolshevik October Revolution of 1917 had establish collective farming over the vast USSR territory, a large scale reallocation of land taken from elite farmers (called kulaks) who had actually been implementing scientific innovations into their agricultural practices. The kulaks were arrested and exiled to Siberia, and their lands were given to poor and ignorant peasant farmers. Whatever the intention behind the policy, it had horrific consequences for the wellbeing of Soviet life and community.

Gross agricultural output drastically declined, leading to the large scale deaths of domesticated animals. Some 40% of cattle and horses, 65% sheep and goats (100 million) were lost, resulting in a period of unprecedented famine in Soviet history. It is estimated that some 10 million persons died of hunger or its side effects in what is now referred to as the "Great Famine of 1932." While the total number of deaths during the period of Stalin's rule is large, some 40 million, deaths by starvation constituted unnecessary tragedies that could have been avoided.

Soviet genetics at the time operated at a very high level. Theodosius Dobshanzky, who would play a leading part in the merger of Darwinian evolution and Mendelian genetics, originated from such a community of distinguished scientists. If Soviet geneticists had been allowed to obtain some measure of influence and control over Soviet agriculture, the food situation would have drastically improved. In actuality, however, the egregious ignorance of the Soviet peasantry created an apparently unbridgeable gulf between it and the advanced geneticist community. The Soviet peasantry simply did not understand or trust the science of genetics or its practitioners, viewing these with suspicion and disdain. The inability of the two communities to communicate hampered the assistance which geneticists could have offered soviet collectivized farms. In spite of the promise inherent in genetics, the poor outcome also reduced its value for the political leadership of the Soviet Union.

The rise of Lysenko coincided with this state of affairs. In light of the existing food crisis, the Soviet leadership drastically needed a 'savior.' The USSR Commissar of Agriculture, Yakov A. Yakovlev, was naturally under a great deal of pressure to find a solution, and issues a mandate stating that the creation of new plant varieties could take no less than three to five years. From point of view of genetics, Yakovlev was asking for the impossible, as it would take at least 12-16 years for new plant varieties to be developed. Yet Lysenko publicly promises that he could not only do it in 3 years, but could even further shorten the time span to a single year. Yakovlev began publicly supporting Lysenko in spite of criticism from honest scientists. Lysenko, in turn, would come to view all who criticized him as 'enemies' who used covert tactics to undermine his work, which later contributed to the violence of period upon his rise to power.

But Lysenko was a crackpot. His scientific 'experiments' contained many ambiguities and inconsistencies, lacked true scientific analysis, and routinely copied information from more reputable work. One of the first proofs of Lysenko's 'vernalization theory' was undertaken by his own father, a local uneducated farmer. It was often the case that Lysenko's 'data' was in fact merely survey polling figures that had been taken from poor peasant in order to track production. Under the implicit menace of state authority, the peasants were more than happy to oblige, providing Lysenko with the answers he sought to obtain. All of Lysenko's articles appeared in newspapers rather than in academic peer reviewed journals, and would have not withstood genuine scientific scrutiny. His articles reflect the language and appearance of science without actually complying with its spirit.

From a political point of view, however, Lysenko was publicly delivering what the Soviet leadership so desperately wanted to hear. Lysenko's public claims sharply contrasted with those of actual scientists who kept pointing out the harsh biological realities of genetics. Their realist stance hence made Soviet geneticist targets of defamation, subject to the accusations of 'stalling' or of failing to 'work hard enough' to solve the existing crisis. The fact that Lysenko publishes in newspapers rather than academic journals also meant that his work would be directly accessible to the Soviet leadership, in particular Stalin, who begins to publicly praise his work at various important Soviet congresses.

Lysenko's own background also helps explain his contemporary appeal within the context of Stalinist communism. Lysenko was genuinely a 'proletariat' in that his origins resided with the uneducated peasantry of the nation; it would not be until thirteen years of age that he began his formal education. His humble beginnings thus coincide with the ideology of the Bolshevik revolution, and one which Lysenko astutely utilizes for his own self-promotion. He begins to publicly present himself as a humble peasant, and in newspaper photos of his experiments, he could not be immediately differentiated from other workers or peasants. In spite of his uneducated background, Lysenko was a keen political player who could closely 'read' the political climate surrounding him.

It is important to note as well the scientific context in which Lysenkoism occurred. In the 1930s genetics was still a young science that was not well established nor globally recognized. It was still afflicted by many cognitive uncertainties and, as of yet, had been unable to make important contributions to human welfare. Its 'adolescent' status, in turn, placed it in a state of political vulnerability wherein it could be easily attacked and be unable to readily defend itself.

During the early period of Lysenko repression, nearly all geneticists were removed from their positions, many of whom were sent to the soviet gulags; others were persecuted by the Soviet secret police, being either arrested or shot on site. The Soviet repression of genetics could be said to have been far worse

274

than that of Nazi physics, in that the initial period of Nazi repression was characterized by its passive/aggressive character rather than outright violent assaults that were only seen towards the end of the Nazi period.

Incidentally, many Soviet physicists began defending their geneticist counterparts, as the well-known scientist Andrei Sakarov. Yet the Soviet state turns a blind eye to this group given it importance to the Soviet nuclear program. At the height of the Cold War, the Soviet state could not afford to lose its underling nuclear scientific manpower. Thus, whereas Soviet physicists were ignored, Soviet geneticist were brutally repressed.

How did Lysenko get to power?

Lysenko attended the Kiev Agricultural Institute. In spite of having studied biology, his level of ignorance was such that he became aware of Charles Darwin only until after he had become a university professor. He ends up at the agricultural research station in Azerbaijan, where he beings to work on a theory called 'vernalization,' based on Lamarck's notion of inheritance of acquired characteristics.

Traditionally, crops grown in the early winter were sown in spring, but subject to the Russian winter, these tended to freeze. As a solution, peasants would pour water on these, so as to freeze and provide some level protection against the cold, but were now afflicted with the resulting problem of rot. Lysenko comes up with notion of "vernalization": if sown but quickly removed when sprouting, these could be stored and replanted in spring, theoretically leading to a much more successful final harvest.

While appealing, the policy had multiple problems, as the uprooting often destroyed the fragile plants, resulting in a negligible benefit or increase in output. Lysenko mistakenly believed that he had created new species instantaneously and that he could turn 'winter' crops into 'spring' crops by subjecting these to the environment. Vernalization would turn out to be a grotesque charade.

Lysenko obtains his first opportunity when the Soviet journalist Vitaly Fyodorvich, looking for an article topic for *Pravda*, ends up interviewing Lysenko, portraying him as a Soviet hero: a poor peasant scientist working under difficult conditions yet making important contributions to the Soviet nation. The article establishes an important precedent in that it gives Lysenko broad recognition and some popular appeal. Ironically, 'pravda' is the Russian word for 'truth.' Lysenko is thereafter appointed to the Institute of Applied Botanics and New Crops, or the Institute of Genetics, headed by Nikolia Vavilov. It had been the principal home for the development of soviet genetics, with a staff of 700, various research stations, and a budget of 14 M rubles per year.

Lysenko would destroy institute and science of genetics from 'within,' with the naïve assistance of its practitioners.

In sharp contrast to Lysenko, Vavilov was scientists who advances his career through genuine results and discoveries. In 1920 he comes up with the law of homology in variation or, formally, the "Law of Homologous Series in Hereditary Variability." If all known variations of species were placed in table, one could identify traits missing in analogous species—not unlike the manner in which Mendelev's periodic table of elements worked. It was truly groundbreaking in that one could predict varieties of species that not yet empirically identified. Vavilov also travels throughout world to collect seeds, building up one of the world's the largest collections in Leningrad.

Vavilov's personality could be characterized as that of being "a good guy." He is friendly, open to conversation, remembering everyone's names in the institute, and awarding unexpected small birthday gifts. Scientific conversations were often continued in large dinners at his home, which routinely hosted guests. Vavilov creates an open atmosphere of discourse, which drastically contrasted to that of Lysenko's terror and repression wherein anybody could be mercilessly accused and killed at any moment for the slightest critique.

Incidentally, Vavilov had been a true 'soviet' in sense that had supported the Bolshevik Revolution and was an important member of USSR Central Committee. The relationship between Vavilov and Lysenko thus constitutes an important key to understanding Lysenko's rise to power, based on the reputation and corpse of Vavilov.

By the time Lysenko entered the Institute , Vavilov was already an important figure in Soviet science, but had stopped doing research. His role was principally that of an administrator, involved in many projects, attending many issues, and generally a busy person. This, in turn, meant that he was not personally inclined to verify the validity of Lysenko's work.

Vavilov begins inviting Lysenko to conferences so as to demonstrate some of the work done at the Institute. He tended to praise Lysenko to the press, due in part to the apparently social benefits of Lysenko's result, and was generally interested in vernalization or any opportunity to improve Soviet agricultural production. He was certainly keenly aware of the deficiencies in Lysenko's learning, but had faith in man's rationality and the scientific sprit. He was possibly too naïve in believing he could convince Lysenko of the error of his ways by merely providing his colleague with 'data.'

At the beginning of an important conference of 1935, where Lysenko politically stabs Vavilov in the back, Vavilov had placed a display at the entrance specifically for Lysenko to see. He invites Lysenko to look through the microscope slides, to observe 'genetics in action,' hoping Lysenko would be convinced into accepting the validity of the science of genetics. Lysenko looks at the slides for only 5 minutes, perhaps so as to not offend his boss, and casually walks away without a comment.

Vavilov's methods and assumptions sharply contrast to those of Lysenko who apparently believed that any means to power could be used as long as it obtained 'results.' A student of his contacts the United States Department of Agriculture to ask a few questions, for whom the USDA prepares a lengthy memo and even took the time to translate it into Russian. The student ends up including the report verbatim as a doctoral thesis chapter, without any change whatsoever in the text. Upon finding out about the act, Lysenko praises the student for his resourceful plagiarism.

Vavilov perhaps saw himself as a father figure and patiently supports Lysenko in spite of his many failings, continually hoping that his 'son' would somehow reform. Vavilov himself also came from a humble peasant family, and likely had empathy for his younger colleague. Oddly, Lysenko's incorrect allegations at scientific conferences or political congresses are never publicly attacked or criticized by Vavilov, who at the time could have held his colleague in check. However, when Lysenko obtains Stalin's support, he begins to undermine Vavilov's institutional power. By the time leading researchers begin to denounce Lysenko, it was too late; he had obtained too much power so as to be so casually dismissed.

Stalin's rise to power had been curiously similar to that of Lysenko's. Stalin, an ignorant peasant who believed any means to power was valid, joins the secret police. Stalin, as Lysenko, use institutions as a means of acquiring and enhancing their political power to the detriment of the institutions they allegedly serve. As early as 1931, Vavilov begins to be followed by the secret police and is ultimately sent to a gulag where he dies at the age of 54 from malnutrition.

Why did Vavilov take so long to criticize Lysenko?

Vavilov was obviously not aware that a noose was being placed around his neck. One has to distinguish Vavilov's 'early' and 'late' support for Lysenko, however. The early phase is perhaps triggered by fatherly attachments, and in the later phase concerned with the charged atmosphere, making a mistaken political calculation.

With Lysenko's increasing political power, various scientists at the Institute of Genetics are removed, creating an environment of fear as that of the Spanish Inquisition. Living in uncertainty and under the threat of being accused, scientists begin to denounce each other so as to preempt their own accusation. As a leader in these circumstance, Vavilov felt he needed to keep an aura of control, in that most by now were well aware of Stalin's public support for Lysenko. While initially Vavilov might have again held hope of convincing his colleague as to the validity of genetics, his later support is somewhat striking.

When Vavilov finally goes to evaluate Lysenko's experiments at Odessa, he is shocked by the egregious manner in which they were being conducted. Yet, soon after openly criticizing Lysenko in the public media, he then backtracks on his comments—possibly out of fear of Lysenko's newfound political power. Yet these actions lamentably seal his fate. Vavilov's backtracking is used as evidence for the unsound nature of genetics. It is denounced that Vavilov and his colleagues changed opinion because they did not know what they were talking about, taking advantage of the Soviet state.

How was Lysenko able to avoid public scrutiny for so long?

It is clear that his promised returns on agricultural development would never pan out, and evidence inevitably accumulated as to the basic unsoundness of this theories. Commonly known facts should have alerted anyone with a modicum of education as to the foolishness of Lysenko's scheme. Lysenko believed he could easily combine the best traits of each variety, but instead obtained the worst traits of all. In a particularly odd experiment, varieties were lined up aside a long string, which was then pulled to loosen their pollen and mutually cross fertilize. Yet any child knows that the mixing of varieties is a bad idea in biology: horses and donkeys only breed infertile mules. One scientist who witnesses Lysenko's experiments is arrested upon his public denunciation of Lysenko.

277

Yet Lysenko simply could not continually use repression to hide the decline of agricultural productivity, and at one point is forced to confront Stalin over his failures. At the meeting, Lysenko pulls out six potatoes, three potatoes from his right pocket and three potatoes from his left pocked. Lysenko tells Stalin that the smaller left pocket potatoes originated from his colleague's facilities, whereas the larger right pocket potatoes were from his own. Stalin, who lacked any scientific education, accepts this presentation as proof enough of the validity of Lysenko's work.

Stalin's gullibility reflects one of the fundamental problems of autocratic societies: absolute rulers by definition will lack appropriate knowledge of evaluate every issue before them, and hence will be liable to gross errors of judgment.

By 1937, Lysenko had obtained full political power in the USSR, as a member of its Central Committee. Both AI Muralov, who had supported Lysenko at the 1935 congress against his conscience and the opposition of his scientific colleagues, and Yakovlev, the former Agriculture Secretary, are arrested by the Soviet secret police and quietly assassinated.

Anthropology at SLAC

Anthropology has traditionally been used to study 'Other' non-Western societies; both are to some degree the product of colonialism. However, the independence period following the liberation of British colonies after World War II led to an important shift in the social science. Anthropologists began to study local groups within their own communities, analyzing social dynamics, mores, and hidden assumptions as they would a small Swahili tribe in the coast of Africa. During the 1970s Sharon Traweek, a historian of science who switches to anthropology, decides to study the institute she had worked in to pay for her graduate education: the Stanford Linear Accelerator's (SLAC) world of high energy particle physics.

The SLAC had multiple detectors, the oldest of which was the 82-inch bubble chamber, conceived of during a Friday night beer session. Its water was contained at a high temperature and pressure that were sub 'critical,' to the point that where the slightest increase could lead to the chamber's explosion. Its delicate state, however, meant that when a particle passed through it, it would leave a bubble track, captured by the multiple cameras placed at various angles. The chamber contained strong magnets, whose changes of a particle's path could be used to calculate its momentum and mass. Over the years, the capabilities of the chamber improved, increasing the number of events from 6 to 60 per second. A total of some 24 million events were captured over a six-year period.

New detectors during the period were created, however, which drastically improved the SLAC's capture rate, specifically the LARR, a linear accelerator. Although particle paths in the LARR could not be tracked with the same degree of accuracy, their collision against a wire mesh mean that particle detection could be tied to computers, and in turn an increase its collision rate to 100 million events per year. More noteworthy still, a substantial improvement again occurs with the creation of SPEAR, the first particle collision detector. When first proposed in 1963, its reviewers did not believe the machine would work, seeing it as pure fantasy. Its funding agency, the Atomic Energy Agency, was forced to remove the proposal from the PAC program evaluator and assign $ 5 M in funding from its own internal budget.

SPEAR became the grandfather of all particle accelerators thereafter.

As with all technologies, SPEAR had advantages and disadvantages. It took much longer to generate an event, as it would move particles in one loop, then onto to other loops greater in size. Whereas prior instrument had used electrons, SPEAR turned to the proton, 1,800 times heavier but whose lesser levels of radiation made it safer to operate. However, there was a tradeoff. It was not shooting particles at a fixed target but rather at another moving particle, hence requiring an unprecedented greater degree of accuracy. The benefits of this approach, however, are perhaps obvious in retrospect: higher energies of collision. A 9 Gev SPEAR could have more than five times the total collision energies of its predecessors, thus reducing the required amount of electricity to run it. The SPEAR was first used to study hadron collisions, for which the team received a 1966 Nobel Prize.

Traweek had been warned by her dissertation director, the famous Gregory Bateson, of hidden ethnocentric biases, and recommended she visit other particle physics facilities in Japan so as to better place into context the SLAC's practices. Our anthropologist thus compares it to the *Enerugi Kasokuki Kenkyu Kiko*, or KEK particle accelerator. It would turn out to be a very wise decision.

The social context in which a scientist work and live is not a blank slate, but has a powerful influence on their practice.

She discovers monumental differences in their respective cultures. At the SLAC, all technologies, instruments and detectors were built 'in house.' Its resident physicists, be it experimental or theoretical, were in direct control of all aspects of their machines. The facilities housed a local foundry, where molten metals were shaped and forged to the demanding and exact requisites of its designers; parts could be created on the fly, and modified as were required. Its overall aspect, dirty and gritty, sharply contrasted to the stereotypical image of scientists working in white coats, as well as to scientists in the world of biology where all technologies are 'black boxes' that were mass produced and purchased in bulk by catalog. The tight control over the facilities, however, led to clashes with the local student body as well as with visitors from other facilities. Local physicists complained that visitors were not aware of the intricacies of their machine, and thus unable to maximize its use. These clashes would serve as an example for other facilities, who establish radically different policies, as the case of Fermilab in Chicago.

By contrast, all of the technologies used at KEK were never produced in-house but rather obtained from external sources. The KEK, in effect, was being utilized by the Japanese government to help fund and develop industrial concerns—subsidizing these ultimately to the detriment of the facility's own scientific activities. As all parts were built by external corporation, the very timetable dictated by the interests of the manufacturing company rather than the

279

needs of the institution and its scientists. Worst still, only a small amount of funding was provided for maintenance of the machines, so any broken or misconfigured parts would necessitate a repair request to the company, which further slowed down its internal activities.

As a whole, this odd funding structure influenced the KEK's own institutional dynamics. In contrast to SLAC, there was a much greater emphasis on the perfection of the machine, given the uncertainty of the maintenance schedule and timetable. Requests would be placed for items that would last various 'lifetimes.' There was also a consequent lack of urgency in the KEK physics community due to its corporate dependence. Traweek is surprised to see that one working physicists at the facility all alone, as all other important parts were missing and hence lacked a functioning device. The institution was also influenced by a particular funding pattern at major universities. As funds were awarded in block directly to the department, to then be distributed by the director, a distinct culture of technological specialization by department had emerged. In other words, each physics department would focus on one particular type of detector, rather than work on their physics independently of these. One department might focus exclusively on bubble chamber, others on linear accelerators, and so forth—a feature which was particularly striking and bizarre to their United States counterparts.

There were other important cultural differences, perhaps more positive. Higher education in United States physics endow the discipline's values to its students, as with any other field. Undergraduate often saw originality and scientific creation as being beyond their abilities. Experimental anomalies, under this context, were usually feared and detested. However, the ascension to graduate school saw a distinctive behavioral shift in the students, where personal 'myth building' began. While some students excelled in this new environment, others who retained behavioral traits rewarded in their prior context, do not succeed as they were too loyal to their professors. Leading professors tended to pick those who had shown ability for original work, and hence those who had 'stood out' from the crowd, to the detriment of 'loyal' students. A professor's social network also could not be readily 'transferred' onto their students, as each would have to build their own informal peer relations. Piggybacking on former glory is a no-no in the field.

When seen in a cultural perspective, however, United States institutions were characterized by a great deal of competition, which in turn greatly affected its internal social dynamics. The world of United States physics was an environment based on 'team leaders' wherein 'charisma' played a much larger role and leaders were unwilling to accept personal errors. In this context, decisions in United States facilities are made 'high on up' and were typically not discussed with the rest of the group. Graduate students may or may not know the plans or objectives of a given experiment, for example. Because of its competitiveness, team leader also reflected the odd traits of being unwilling to recognizes a subaltern's assistance. Yet, in more cases than one might imagine, graduate students and post-doctoral fellows routinely provided key solutions to critical problems.

280

The Japanese context could not have been more different. The more traditional and conventional social context was placed in the 'kota' context, where a group of strangers defines itself akin to a family. Leadership in this context is based on age and seniority. However, issues are discussed between all members of the group, so that everyone, including graduate students and postdoctoral fellows, were always aware of the exact nature and purpose of any given experiment. Junior members were also recognized for their contributions, in sharp contrast to their United States counterparts. Whereas competition resided at the individual level in the United States, competition in Japan occurred between 'clans' or groups.

This latter trait, however, was one which official at the KEK sought specifically to eliminate and 'Westernize' by forcing individuals from different groups to meet and collaborate. This change of cultural behavior was a very strange experience for many Japanese physicists.

Incidentally, a contrast to European facilities showed that team leaders held so much power in European institutions, directly conflicts could have enormous negative consequences to student's career—effectively being banned form the world of academia.

Conclusion

The Fragility and Resilience of Science

HOW FRAGILE IS SCIENCE? Is it like that of early Rome as expressed by Marcus Aurelius in the movie "Gladiator" (2000): so vulnerable that any more than a whisper could have destroyed it? Typically, science is portrayed as dominant and pervasive, almost hegemonic. Yet we have seen many cases showing this presumption to be historically inaccurate. Nearly every major figure of the Scientific Revolution experienced some potentially decisive obstacle or counterpart. Which of the two versions is the correct one? The answer likely lies somewhere in the middle: not as strong as we presume it to be, but neither as fragile. Perhaps a more appropriate trait to describe its historical character is that of resilience.

Many cases could be used to describe the fragility and vulnerability of the scientific enterprise, which somehow continues to advance in spite of its many obstacles. Galileo is the most prominent example. His clash with Catholic Church could have led to his death, directly or indirectly. The decision to force a sick old man to take a 100-mile journey through plague infested lands could have easily ended far worse than it did. Had he then perished before even confronting the Inquisition of Rome, he certainly would not have published his revolutionary treatise on physics, actually discovered when he was a young man without formally publishing these during his academic lifetime.

Newton is another good example, plagued as he was by both 'internal' and external enemies. The psychological suffering he underwent early in life created a tormented soul. Isaac suffers a nervous breakdown in 1692, likely as a result of his abandonment during infancy. We do not know what might have happened to him had it not been for the intervention of his friends, as John Locke. As in the case of Galileo, had Newton died prior to 1687, the Scientific Revolution would have never occurred. Newton also suffers imprudent and continual attacks by Robert Hooke, whose deficient mathematical abilities raises questions as to the source of his claims and how they had been established. Did Hooke break into

Newton's office and steal his papers? We will never know, although the surrounding circumstances do raise some suspicion. The animosity between the two, alongside the elder's lesser intellectual abilities, clearly led to Hooke's abuse of institutional power. Hooke's actions are not unlike Lysenko's, but obviously of a different degree.

When we turn to the history of kinetic theory, many of its key theorists, Herapath, Waterston, and Boltzmann, were rejected by their respective communities during the nineteenth century. What is most shocking about their cases is the notion that a scientific community would utilize its own institutions to formally ignore or reject revolutionary advancements—a historical case that seems to contradict the very image and definition of science. Although confronting similar abuses, each scientist dealt with it in their own particular manner. Herapath changes career path and enters the railroad industry, allowing for him to personally publish his own work. It is unclear whether Waterston committed suicide after decades of rejection; but it is a possibly that, in light of his circumstances, should not necessarily be discarded.

Boltzmann is certainly the most tragic of the three. The father of kinetic theory and grandfather to the quantum revolution, imitated by both Planckand Einstein, Boltzmann is continually attacked in his day by Ernst Mach's infamous "Vienna Circle." Logical positivism was merely an excuse to undermine genuine originality in physics by a second rate scientist. Beginnings as such are imperfect, and any and all vulnerabilities of kinetic theory were merely used as pretexts for its denunciation. As horrifically witnessed by his youngest daughter, Boltzmann ultimately commits suicide, a genuine tragedy in that only a few months previously Einstein had proven the fundamental soundness of his presumptions. The atom existed. Secondary circumstances certainly played a role in his demise: obesity, diabetes, near blindness, and the loss of manual dexterity must have taken their toll.

As in biological reproduction, death prior to publication would have killed the lines of inheritance in every case reviewed. The observation might even be applied as well to turn of the century physics, relativity and quantum mechanics. Theoretical physics was defined as a Jewish science in Germany. Had the Nazi risen to power years earlier, there can be no doubt that the quantum revolution would have also been exterminated in its infancy. Science in these circumstances teeters too close to the edge of the abyss, vulnerable to finality and intellectual discontinuity.

Yet all were able to publish, before and even after repression, as noted by the case of Jean van Helmont whose son took it as a task to publish all of his father's works posthumously. The same had been the case for Paracelsus, whose non-biological followers also took to heart their role in his intellectual immortality. As in the passing on of genes between generations, it is publication that counts, by contributing to the total sum of human knowledge of consequent generations. The absence of publication cuts short the lines of hereditary intellectual transmission.

The perseverance of our scientific exemplars is thus a testament to the strong will of the scientist, an iron fortitude determined to prevail in light of all apparent insurmountable odds—perhaps the most important trait of the successful scientist.

Does this, however, meant that Toynbee is ultimately correct in that science, as other cultural endeavors, thrives only when confronted with obstacles as his Challenge-Response model suggested? Perhaps, but we should not exaggerate. For however much our imagination might want to assume godly powers, human faculties and limitation are just that, limitations; we are not immortal gods. Ultimately, obstacles took their toll. Galileo loses his beloved daughter, and his life is completely overturned after the encounter. The Royal Society loses years of articles by it most historically important member. The German physics community is exiled from its homeland, altering the course of both European and North American physics.

We should not be so wanton in praising life's personal tragedies.

We also cannot ignore the persistent role disease in our story. James Clerk Maxwell was felled due to cancer at the age of 48, and his 'disciple' Heinrich Hertz died at the age of 36 from granulomatosis. Jean Fresnel at 42, with so much promise, succumbed to cholera, as had the founder of thermodynamics, Sadi Carnot, at the age of 36. Disease has certainly played its notorious role in the history of science. Syphilis, a post-American phenomenon, ruined the life of Copernicus's brother, and could have easily afflicted both brothers. Tycho Brahe's unusual death, by holding his urine for too long, was not unlike that of Francis Bacon's strange "chicken incident" leading to his untimely pneumonia. Poisonings also played a role, as grotesquely shown by Nazi medicine. As the Nuremberg Code noted, any human experimentation required the full consent of the test subject, as well as their capacity to end the experiment should the tested deem necessary. Death should never be a physician's operating presumption.

More importantly is perhaps the question of whether there is a general historical pattern to such vulnerability. How are the obstacles at the beginning of the history of science similar or different to contemporary ones?

However, it is clear that the most obvious change in the historical nature of scientific obstacles occurs in the abuse of institutional or state power.

Up to the Scientific Revolution, oppression tended to originate by external gents, the most common of which were religious institutions which did not share the cultural values of science. In this sense, there is little difference between Arab or Catholic Church persecution of natural philosophers as Giordano Bruno, burned at the stake, or Ibn Sina (Avicenna) whose threat of potential beheading sends him into a miraculous escape. Science had not yet become 'science,' nor acquired influence, social power or public reputation; whenever confronted with some powerful ideological opponents, it would naturally lose given the small population of its practitioners and broader social support.

Yet beginning with its critical formation during the Scientific Revolution, there is a detectable shift from 'external' to 'internal' scientific oppressors. It is the very scientists themselves who become their worst opponents: Hooke to Newton, Kelvin to Darwin, Mach to Boltzmann, Lysenko to Vavilov. Upon the creation of science, scientific institutions are used to both preserve and destroy innovative scientific activity.

284

The cause of the competition was often of a philosophical nature, of a resistance to 'turn the page' away from Kuhn's 'normal science' to anomalous but revolutionary and groundbreaking practices—perhaps most well represented by Philippe Lenard's notions of "Jewish' and "German" science. We should not be fooled by the alleged ethnic face of the conflict, used as a thin veneer hiding the more serious clash of values between scientific generations, of classical Newtonianism versus the quantum revolution or relativity. As in al-Ghazzali's persecution of Averroes, the core values of each respective group were mutually exclusive, and were impossible to be rationally reconciled. One was forced to choose one group or the other; it was impossible, with any sanity of mind, to belong to both.

This change from religious to institutional oppression throughout the *longue duree* of scientific history presents somewhat of a historical irony: a return to the dark ages of the Medieval period. At that time, the most frequent revolts tended to be against the very institutions which were supposed to be guarding over the well-being of its agricultural citizens: monasteries.

Similarly, by creating organizational superstructures designed to 'protect' and 'promote' scientific activity—the institutionalization of science—their very existence implied, by definition, a superstructure which could be co-opted by unscrupulous or unethical individuals towards ends opposite of their originally intended purposes. It is the same dilemma as that of the communist state, pledging popular representation by an exclusive elite. The very organization created to protect a given social activity, in this case science, becomes the very institution that represses its members violently or subtly.

We certainly should not necessarily accept the anarchist's fatalist conclusion, and argue that such institutions should thus be completely discarded. In the majority of instances, scientific institutions have played their respective roles in helping to advance scientific activity. We should not be so foolish as to suggest two thousand years of slow and continued growth of science be wiped out only because of a few bad apples. Everything has a cost and benefit; institutionalism's ultimate costs reside in the never-ending threat of co-optation.

However, the history of Western Science does suggest that 'eternal vigilance' over its organizations is a requisite for its continued long-term existence and success.

Bibliography

IT IS IN AN INTERESTING EXERCISE to read the older histories of science, as Butterfield's *Origins of Modern Science.*(1958), perhaps one of the most cited classics, and look at how consequent authors developed these important works. As is well known, the first histories of science tended to be 'intellectual', and more recent versions have emphasized social aspects, greatly affecting the character and tone of these works. Thus, whereas older books were more 'intellectual,' as Bury's *the Idea of Progress* .(1960), more recent authors enter more into the minutia of the lives of scientists to explore social factors influencing their work, Ede's textbook *from Philosophy to Utility.* (2012) being a good example.

The suggestion that newer works supersede the older ones is not necessarily the case, however, as Derek de Sola aptly notes. This assumption is overly pervasive in the sciences, as they have different purposes in mind. The intellectual experience of reading Toulmin and Goodfield's works (1960s) is extremely rewarding—*The Discovery of Time.* (1965) or the *Architecture of Matter* .(1962), opening up new vistas, whose worth is equally immeasurable.

Inversely, to complain that newer histories have somehow "lost IQ" as a result of their social focus is an equally mistaken claim. It has to be recognized that the social analysis of science, particularly during times of acute political or religious oppression—Galileo's Inquisition, Lysenko's biology, or Nazi German medical practices—are critical ingredients to their genuine understanding. To suggest otherwise is foolish, when seriously considered.

Although we hope our general history of the field appropriately combines both aspects, intellectual and social history, we will here provide a brief list of some of the works used in its preparation. For students first entering the field, we place an asterisk on books which should be considered starting points of their respective topics. We would like to note that this is not a comprehensive bibliography, nor a full list of all the resources that were used to prepare this book, and should be seen simply as a launching point to this fascinating topic.

Finally, a note on resources in the internet. There are many sites where one may obtain original materials and manuscripts. The most comprehensive is perhaps the *Internet Archive* (archive.org). Unfortunately, although the *Gutenberg Project* is promising, its limited collection is of a relatively small scope. Other valuable general 'history of science' online collections include: *University of Oklahoma Libraries, History of Science Collections* (hos.ou.edu) or To*m Settle's Institute and Museum of the History of Science at Firenze* (www.imss.fi.it). Specific topical sites have been created whose rich repositories far supersede those of general sites. Some of these are: *Darwin Online* (darwin-online.org.uk) and the *American Institute of Physics history of physics.* (www.aip.org).

The world of the internet is vast, and a simple search via 'DuckDuckGo' is always the key to finding original primary source material on a specific topic.

If a website on your topic of interest does not exist, create one.

ANCIENT

Crowe, Michael J. *Theories of the World from Antiquity to the Copernican Revolution.* New York: Dover Publications, 1990.

Diamond, Jared. *Guns, Germs, & Steel: The Fates of Human Societies.* New York: W.W. Norton, 1997.*

Goody, Jack *The Domestication of the Savage Mind.* Cambridge, Cambridge University Press, 1977.

McNeill, William H. *The Rise of the West: A History of the Human Community.* Chicago, ILL: University of Chicago Press, 1991.

Neugebauer, O. *The Exact Sciences in Antiquity.* New York: Dover Publications, 1969.

Selfie, Charles *Zero: The Biography of a Dangerous Idea.* New York: Viking, 2000.

Tainter, Joseph A. *The Collapse of Complex Societies.* Cambridge: Cambridge University Press, 1988.

Toynbee, Arnold. *A Study of History.* ed. Jane Caplan., New York: Portland House, 1988.

GREEKS

Boorstin, Daniel J. *The Discoverers.* New York: Vintage Books, 1985.

Bury, J. B. *The Idea of Progress: An inquiry into its origin and growth.* New York: Dover Publications, 1960.

-----, *Aristotle: The Growth & Structure of his Thought.* New York: Cambridge University Press 1990.

Kitto, H. D. F. *The Greeks.* New York: Penguin Books, 1991.*

Lloyd, G.E.R. *Early Greek Science: Thales to Aristotle.* New York: W.W. Norton, 1970.

-----, *Greek Science after Aristotle.* New York: W. W. Norton, 1973.

Russell, Bertrand *A History of Western Philosophy.* New York: Simon & Schuster, 1972.

MEDIEVAL

Barcelo, Miguel. *Arqueologia medieval en las afueras del medievalismo.* Barcelona: Critica, 1988.

Bloch, Marc. *Feudal Society, The Growth of Ties of Dependence.* vol. 1, transl. L.A. Manyon. Chicago: University of Chicago Press, 1961.

Braudel, Fernand. *Civilization and Capitalism, 15th-18th Century,* vol. 1 *The Structures of Everyday Life- The Limits of the Possible.* transl. Sian Reynolds Berkeley, CA: University of California Press, 1992.

Buruma, Ian and Avishai Margalit, *Occidentalism: The West in the Eyes of Its Enemies.* New York: The Penguin Press, 2004.

Dallal, Ahmad. Chpt 4: "Science, Medicine and Technology: The Making of a Scientific Culture" , in *The Oxford History of Islam.* ed. John L. Exposito. New Yok: Oxford University Press, 1999, pp. 155-214.

Gonzalez, Caleb. *Mi Escuela de Medicina.* New Haven, CT: Yale Printing & Publishing Services, 2010.

-----, *UPR School of Medicine: It's Creation.* New Haven, CT: Yale Printing & Publishing Services, 2011.

Hoodbhoy, Pervez. *Islam and Science: Religious Orthodoxy and The Battle for Rationality.* London: Zed Books, 1991.*

Huntington, Samuel P. "The Clash of Civilizations?" *Foreign Affairs.* 7,3 Summer 1993, 22-49.

-----, *The Clash of Civilizations and the Remaking of World Order.* New York: Touchstone, 1997.

Lindberg, David C. "Alhazen's Theory of Vision and Its Reception in the West" *ISIS* 58,3 Autumn 1967, 321-341.*

Lopez, Robert S. *The Commercial Revolution of the Middle Ages, 950-1350.* Cambridge: Cambridge University Press, 1976.

Rashed, Roshdi *The Development of Arabic Mathematics: Between Arithmetic and Algebra.* transl. AFW Armstrong Dordrecth, Kluwer Academic Publishers, 1994.

Romero, José Luis. *Crisis y Orden en el Mundo Feudoburgues.* Buenos Aires: Siglo Veintuno Editores Argentina S.A., 2003.

Sabra, A. I. "Situating Arabic Science: Locality versus Essence," *ISIS* 87, 4 Dec. 1996, 654-670.

Said, Edward W. *Orientalism.* New York: Pantheon Books, 1978.

Smith, A. Mark. "What is the History of Medieval Optics Really about?" *Proceedings of the American Philosophical Societ.* 148, 2 June 2004, 180-194.

Sombart, Werner *El burgués: Contribución a la historia espiritual del hombre económico moderno.* trad. María Pilar Lorenzo Barcelona, España: Alianza Editorial, 1982.

UNESCO, *Islam, Philosophy and Science.* New York: UNESCO Press, 1981.

Valdeón, Julio. *El Feudalismo.* Madrid: Biblioteca Historia, 1992.

RENAISSANCE

Bacon, Francis. *The essays*. New York: Penguin Books, 1985.

Boas Hall, Marie *The Scientific Renaissance, 1450-1630*. London: Dover, 1994.

Butterfield, Herbert. *The Origins of Modern Science, 1300-1800*. New York: MacMillan Co., 1958.*

Burtt, Edwin Arthur. *The Metaphysical Foundations of Modern Science*. Atlantic Highlands, NJ: Humanities Press, 1952.

Debus, Allen G. *Man and Nature in the Renaissance* .New York: Cambridge University Press, 1978.

"Glogal Histories of Science" *ISIS* 101, 1 March 2010.

Koyre, Alexandre. *From the Closed World to the Infinite Universe*. NY: Harper & Brothers, 1957.

Montaigne, Michel de, *The Complete Essays*. New York: Penguin Books, 1993.

Parascandola, John "From Mercury to Miracle Drugs: Syphilis Therapy over the Centuries" *Pharmacy in History*. 51, 1 (2009), 14-23.

Stillman, J. M. "Paracelsus As A Chemist and Reformer Of Chemistry" *The Monist*. 29, 1 (Jan, 1919), 106-124.

-----."The Contributions of Paracelsus to Medical Science and Practice" , *The Monist*. 27, 3 (July 1917), 390-402.

-----."Paracelsus As A Reformer In Medicine" *The Monist*. 29, 4 (Oct., 1919), 526-546.

Tepper, Lloyd B. "Industrial Mercurialism: Agricola to the Danbury Shakes" *The Journal of the Society for Industrial Archeology*. 36, 1 (2010), 47-63.

SCIENTIFIC REVOLUTION

Bowler, Peter J. and Iwan Rhys Morus, *Making Modern Science: A Historical Survey*. Chicago Ill.: University of Chicago Press, 2005.

Ede, Andrew and Leslie b. Cormack, *A History of Science in Society: From Philosophy to Utility*. Onatrio: University of Toronto Press, 2012.

Finocchiaro, Maurice A. ed. *The Essential Galieo*. New York: Hacket Publishing, 2000.

Gleick, James. *Isaac Newton*. New York, Vintage Books, 2003.

Keynes, Milo "The Personality of Isaac Newton," *Notes and Records of the Royal Society of London*. 49,1 1995, 1-56.

-----, "Balancing Newton's Mind: His Singular Behaviour and His Madness of 1692-3," *Notes and Records of the Royal Society of London*. 64, 3 2008, 289-300.

Koestler, Arthur *The Sleepwalkers: A History of Man's Changing Vision of the Universe*. New York: Grosset & Dunlap, 1963.

-----, *The Watershed: A Biography of Johannes Kepler*. Lanham, MD: Univeristy Press of America, 1960.

Sobel, Dava. *Galileo's Daughter: A historical Memoir of Science, Faith and Love.* New York: Bloomsbury, 1999.*

Stillman Drake, ed., *Discoveries and Opinion of Galileo.* New York: Anchor Books, 1958.

-----, *Galileo, Pioneer Scientist.* Toronot, Canada: University of Toronto Press, 1990.

Westfall, Richard S. *The Construction of Modern Science: Mechanisms and Mechanics.* New Yor: Cambridge Unviersity Press, 1971.

-----, *Never at Rest: A Biography of Isaac Newton.* New York: Cambridge University Press, 2010, original: 1980.

-----, *The Life of Isaac Newton.* Cambridge: Cambridge University Press, 1993.*

ENLIGHTENMENT

Bedini, Silvio A. *Thomas Jefferson, Statesman of Science.* New York: MacMillan Publishing Co., 1990.

Fernós, Rodrigo, "Vico's Failed Revolution: Chance, Variability, and Design in Eighteenth Century Historiography." Unpublished esssay. Academia.edu, 2005.

Gay, Peter *The Enlightenment, An Interpretation: The Rise of Modern Paganism.* New York: W. W. Norton & Co., 1966.*

Gillispie, Charles Coulston. *Science and Polity in France at the end of the old regime.* Princeton, NJ: Princeton University Press, 1980.

McDonald, Forrest *Novus Ordo Seclorum: The Intellectual Origins of the Constitution.* Lawrence, KS: University Press of Kansas, 1985.

Segre, Emilio *From Falling Bodies to Radio Waves: Classical Physics and Their Discoverie.s* New York: W. H. Freeman & Co., 1984.

Toulmin, Stephen and June Goodfield, *The Architecture of Matter: The physics, chemistry and physiology of matter, both animate and inanimate as it has evolved since the beginning of science.* New York: Harper & Row, 1962.

Voltaire, *The Age of Louis. XIV*, transl. Martyn P. Pollack, ed. Ernest Rhys. New York: E. P. Dutton & Co., 1935.

Woolf Harry. *The Transits of Venus: A Study of Eighteenth Century Science.* Princeton: Princeton U Press, 1959.

Wright, Robert *The Moral Animal: Evolutionary Psychology and Everyday Life.* NY: Vintage Books, 1994.

19TH CENTURY

Adas, Michael. *Machines as the Measure of Men: Science, Technology, and Ideologies of Western Dominance.* Ithaca: Cornell University Press 1989.*

Bowler, Peter J. *Evolution: The History of an Idea.* Berkeley: University of California Press, 1989.

-----, *The Eclipse of Darwinism: Anti-Darwinian Evolution Theories in the Decades around 1900.* Baltimore: Johns Hopkins University Press, 1983.

Bock, W. H. "Humboldt and the British: A Note on the Character of British Science" *Annals of Science.* 50,4 1993, 365-372.

Browne, Janet. *Charles Darwin Voyaging: A Biography.* Princeton, NJ: Princeton University Press, 1995.*

Brush, Stephen G. "Foundations of Statistical mechanics, 1845-1915" *Archive for History of Exact Sciences.* 4,3 March 1967, 145-183.

-----. *The kind of motion we call heat: A History of the Kinetic Theory of Gases in the 19th Century.* 2 vols. New York: North Holland Publishing Company, 1976

Darwin, Charles. *The Origin of Species by Means of Natural Selection or The Preservation of Favoured Races in the Struggle for Life.* New York: Penguin Books, 1985.

-----, *The Voyage of the Beagle* New York:Bantam Books, 1972.

Dennett, Daniel C., *Darwin's Dangerous Idea: Evolution and the Meanings of Life.* NY: Simon & Schuster, 1995.

Hopkins, Robert S. *Darwin's South America* NY: John Day Co., 1969.

Gould, Steven Jay. *The panda's thumb: more reflections in natural history.* New York: Norton, 1980.

-----. *Wonderful lfie: the Burgess Shale and the nature of history.* New York: W.W. Norton, 1989.

von Humoldt, Alexander. *Personal Narrative of a Journey to the Equinoctial Regions of the New Continent.* abridged and transl by Jason Wilson. New York: Penguin Books, 1995.

Lindley,David *Boltzmann's Atom: The Great Debate that Launched a Revolution in Physics.* New York: The Free Press, 2001.*

Mayer, Ernst *The Growth of Biological Thought: Diversity, Evolution and Inheritance.* Cambridge, MA: Belknap Press of Harvard Unviersity, 1982.

Misa, Thomas J., Philip Brey, and Andrew Feenberg, eds., *Modernity and Technology.* Cambridge, MA: MIT Press, 2003.

Ruse, M. "Were Owen and Darwin Naturphilosophen?" . *Annals of Science.* 50,4 1993, 383-388.

20TH CENTURY

Annas, George J. & Micheal A. Drodin, eds., *The Nazi Doctors and the Nuermberg Code: Human Rights in Human Experimentation.* New York: Oxford University Press, 1992.

Banesh Hoffman, *Albert Einstein: Creator and Rebel.* New York: Penguin, 1972.*

Borkin, Joseph. *The Crime and Punishment. of I.G. Farben.* New York: The Free Press, 1978.*

Capshew James H. and Karen A. Rader, "Big Science: Price to the Present" *OSIRIS* Science after '40, Second Series, vol. 7 1992, 3-25.

Dupree, A. Hunter *Science in the Federal Government.* Baltimore, MD: Johns Hopkins, 1980.

Feynman, Richard *QED: The Strange Theory of Light and Matter*. Princeton: Princeton University Press, 1985.

Fernos, Rodrigo. "Nuestra Telefónica": La Nacionalización de la Puerto Rico Telephone Company (PRTC), 1974." PhD Thesis, University of Puerto Rico, 2011.

Forman, Paul John Heilbron, and Spencer Weart, "Physics ca 1900. Personnel, Funding, and Productivity of the Academic Establishments." *Historical Studies in the Physical Sciences*. ed Russell McCormach Princeton: Princeton University Press, 1975.

Gilman, Nils *Mandarins of the Future: Modernization Theory in Cold War America*. Baltimore: Johns Hopkins University Press, 2003.

Goldberg, Stanley Roger H Stuewer, eds., *The Michelsonian Era in Aemrican Science, 1870-1930*. NY: American Institute of Physics, 1988.

Jaffe, Bernard. *Michelson and the Speed of Light*. Westport, Connecticut: Greenwood Press, 1960.

Kevles, Daniel J. *The Physicists: The History of a Scientific Community in Modern America*. Cambridge, MA: Harvard University Press, 1987.

Kraugh, Helge. *Quantum Generations: A History of Physics in the Twentieth Century*. Princeton, NJ: Princeton University Press, 1999.

Lederer, Susan E. *Subjected to Science: Human Experimentation in America Before the Second World War*. Baltimore: Johns Hopkins University Press, 1995.

Lifton, Robert Jay *The Nazi Doctors: Medical Killing and the Psychology of Genocide*. New York: Basic Books Inc., 1986.

Macrakis, Kristie. *Surviving the Swastika: Scientific Research in Nazi Germany*. New York: Oxford University Press, 1993.

National Science Board. 2014. *Science and Engineering Indicators 2014*. Arlington VA: National Science Foundation (NSB 14-01).

Oleson, Alexandra and John Voss, eds., *The Organization of knoweldge in Modern America, 1896-1920*. Batlimore: Johns Hopkins University Press, 1979.

Rhodes, Richard *The Making of the Atomic Bomb*. New York: Touchstone Book, 1986.

Schrodinger, Erwin. *What is Life?* Cambridge: Cambridge University Press, 1967.

Singh, Simon. *Big Bang: The Origin of the Universe*. New York: 4th Estate, 2004.

Simonton, Dean Keith. *Scientific genius: A psychology of science*. Cambridge: Cambridge University Press, 1988.

Skloot, Rebecca *The Immortal Life of Henrietta Lacks*. New York: Broadway Books, 2011.*

Soyfer, Valery N. *Lysenko and the Tragedy of Soviet Science*. transl. Leo Gruliow and Rebecca Gruliow New Brunsuick, Rutgers University Press, 1994.

de Solla Price, *Little Science, Big Science…and Beyond*. New York: Columbia University Press, 1986.

Traweek, Sharon *Beamtimes and Lifetimes: The World of High Energy Physics*. Cambridge: Harvard University Press, 1992.

Classics : Scientific Revolution

Pierre Duhem, *Etudes sur Leonard de Vinci* (1913)
Edwin Arthur Burrt, *Metaphysical Foundations of Modern Science* (1925)
Martha Ornstein, *The Role of Scientific Societies in the 17th Century* (1928)
Robert K. Merton, *Science, Technology and Society in Seventeenth Century England* (1935)
Edgar Zilsel, *Sociological Roots of Science* (1942)
Herbert Butterfield, *The Origins of Modern Science, 1300-1800* (1949)
E. J. Dijksterhuis, *Mechanization of the World Picture* (1950)
Alexandre Koyre, *Galilean Studies* (1952)
Maurice Dumas, *Scientific Instruments of the Seventeenth and Eighteenth Century* (1953)
AC Crombie, *Robert Grosseteste and the Origins of Experimental Science, 1100-1700* (1953)
Joseph Needham, *Science and Civilization in China* (1954) [15 vols]
Alexandre Koyre, *From the Closed World to the Infinite Universe* (1957)
Lynn Thorndike, *The History of Magic and Experimental Science* (1958)
Richard Westfall, *Science and Religion in Seventeenth Century England* (1958)
John Greene, *The Death of Adam: Evolution and Its Impact on Western Thought* (1959)
Richard Popkin, *The History of Skepticism from Erasmus to Spinoza* (1960)
Thomas Kuhn, *Structure of Scientific Revolutions* (1962)
Marie Boas, *The Scientific Renaissance, 1450-1630* (1962)
Leslie White, *Medieval Technology and Social Change* (1962)
Francis Yates, *Giordano Bruno and the Hermetic Tradition* (1964)
Paul Hazard, *The European Mind* (1963)
R. Hooykaas, *Portuguese discoveries and the Rise of Modern Science* (1966)
S. H. Nasr, *Science and Civilization in Islam* (1968)
P. Rossi, *Francis Bacon: From Magic to Science* (1968)
Joseph Needham, *The Grand Titration: Science and Society in East and West* (1969)
Joseph Ben-David, *The Scientist's Role in Society* (1971)
R. Hooykaas, *Religion and the Rise of Modern Science* (1972)
BJTeeter Dobbs, *The Foundation of Newton's Alchemy* (1975)
Elizabeth Einsenstein, *The Printing Press as an Agent of Change* (1979)
Richard Westfall, *Never at Rest: A Biography of Isaac Newton* (1980)
Carolyn Merchant, *The Death of Nature* (1980).
Anneliese Maier, *On the Threshold of Exact Science* (1982)
David Landes, *Revolution in Time* (1983)
Shapin - Shaffer, *Leviathan and the Air Pump* (1985)
Amos Funkenstein, *Theology and the Scientific Imagination from the Middle Ages to the Seventeenth Century* (1986)
William Barrett, *Death of the Soul, From Descartes to the Computer* (1987)
Margaret C. Jacob, *The Cultural Meaning of the Scientific Revolution* (1988)

INDEX

295

297

298

303

307

308

316

317

www.ingramcontent.com/pod-product-compliance
Lightning Source LLC
Chambersburg PA
CBHW071534200326
41519CB00021BB/6478